Studies in Fuzziness and Soft Computing

Volume 400

Series Editor

Janusz Kacprzyk, Systems Research Institute, Polish Academy of Sciences, Warsaw, Poland

The series "Studies in Fuzziness and Soft Computing" contains publications on various topics in the area of soft computing, which include fuzzy sets, rough sets, neural networks, evolutionary computation, probabilistic and evidential reasoning, multi-valued logic, and related fields. The publications within "Studies in Fuzziness and Soft Computing" are primarily monographs and edited volumes. They cover significant recent developments in the field, both of a foundational and applicable character. An important feature of the series is its short publication time and world-wide distribution. This permits a rapid and broad dissemination of research results.

Indexed by ISI, DBLP and Ulrichs, SCOPUS, Zentralblatt Math, GeoRef, Current Mathematical Publications, IngentaConnect, MetaPress and Springerlink. The books of the series are submitted for indexing to Web of Science.

More information about this series at http://www.springer.com/series/2941

Sunil Jacob John

Soft Sets

Theory and Applications

Sunil Jacob John
Department of Mathematics
National Institute of Technology Calicut
Calicut, Kerala, India

ISSN 1434-9922 ISSN 1860-0808 (electronic)
Studies in Fuzziness and Soft Computing
ISBN 978-3-030-57656-1 ISBN 978-3-030-57654-7 (eBook)
https://doi.org/10.1007/978-3-030-57654-7

This Springer imprint is published by the registered company Springer Nature Switzerland AG
The registered company address is: Gewerbestrasse 11, 6330 Cham, Switzerland

Dedicated to

My beloved father Late Mr. Jacob John

&

My respected Guru Prof. T. Thrivikraman.

Foreword

G. J. Klir has stated that among the various paradigmatic changes in science and mathematics in the twentieth century, one such change concerned the concept of uncertainty. In science, this change has been manifested by a gradual transition from the traditional view, which states that uncertainty is undesirable in science and should be avoided by all possible means, to an alternative which is tolerant of uncertainty and insists that science cannot avoid it. Uncertainty is essential to science and has great utility. An important point in the evolution of the modern concept of uncertainty was a publication of a seminal paper by Lotfi Zadeh.

Soft set theory was proposed by Molodtsov in 1999 to deal with uncertainty in a parametric manner. A soft set is a parameterized family of sets, intuitively soft because the boundary of the set depends on the parameters. One notion of a set is the concept of vagueness. This vagueness or the representation of imperfect knowledge has been a problem for a long time for philosophers, logicians, and mathematicians. Recently, it became a crucial issue for computer scientists particularly in the area of artificial intelligence. To handle situations like this, many tools have been suggested. They include fuzzy sets, multisets, rough sets, soft sets, and many more.

Molodtsov proposed soft set as a completely generic mathematical tool for modeling uncertainties. There is no limited condition to the description of objects. Thus researchers can choose the form of parameters they need. This simplifies the decision-making process and makes the process more efficient in the absence of partial information.

A soft set can be considered as an approximate description of an object precisely consisting of two parts, namely, predicate and approximate value set. Exact solutions to the mathematical models are needed in classical mathematics. If the model is so complicated that we cannot get an exact solution, we can derive an approximate solution and there are many methods for this. On the other hand, in soft set theory as the initial description of object itself is of an approximate nature, we need not have to introduce the concept of an exact solution.

Soft set theory has rich potential for application in many directions, some of which are reported by Molodtsov in his work. He successfully applied soft set theory in areas such as the smoothness of functions, game theory, operation research, Riemann integration, and elsewhere. Later he presented some definitions on soft sets as a subset, the complement of a soft set and discussed in detail the application of soft theory in decision-making problems. Applications have been made to decision-making, business competitive capacity information systems, classification of natural textures, optimization problems, data analysis, similarity measures, algebraic structures of soft sets, soft matrix theory, parameter reduction in soft set theory, classification of natural textures, and soft sets and their relation to rough and fuzzy sets.

The book, Soft Sets: Theory and Applications, by Prof. S. J. John is a strong contribution to the development of soft set theory. It examines the algebraic and topological structure of soft sets. It also considers some hybrid structures of soft sets. The book contains interesting applications to decision-making, medical and financial diagnosis problems. It is my hope that researchers will apply the concepts of soft set theory to the existential problem of climate change and related problems such as world hunger, coronavirus, modern slavery, and human trafficking.

Creighton University, USA John N. Mordeson
June 2020

Preface

A challenging problem faced by researchers while working towards efficient computational systems is the need to model inherent uncertainty, imprecision and partial information in the computational problem itself. The methods developed in this context, broadly known as reasoning under uncertainty or approximate reasoning or imprecise reasoning, bring about a significant paradigm shift, which reflected the remarkable reasoning ability of the human mind for information processing and analysis which is often better than present day computers. In this context only the idea of Soft computing emerged whose guiding principle is exploiting tolerance for imprecision, uncertainty, partial truth, and approximation to achieve tractability, robustness and low solution cost. Traditionally, the major ingredients of Soft Computing include fuzzy logic, neuro-computing and probabilistic reasoning. More recently we have genetic algorithms, belief networks, chaotic systems and many more added up.

As the role model for this type of reasoning is the human mind and the fact that human reasoning is not crisp and admits degrees. Obviously, fuzzy logic and fuzzy set theory introduced by Zadeh play a major role in soft computing. Owing to the fact that there are limitations of fuzzy set theory such as non-clarity in assigning membership values, lack of parameterization techniques, soft techniques should have the freedom to draw ideas from other generalizations of sets like rough sets, multisets, soft sets, etc., and potential hybridization among these. Among these, rough set theory is well developed and applications of rough sets in soft computing are well explored. But other structures like multisets, soft sets, fuzzy multisets, fuzzy soft sets, etc., are still in developing stages and their applications in various soft computing scenarios need to be explored.

The development of soft set theory by Molodtsov in 1999 is significant in this context. Taking the advantage of parameterization technique, soft set theory evolved as a powerful tool for decision making in information systems, data mining and reasoning from data, especially when uncertainty is involved. Before the effective utilization of any mathematical technique for real life applications, it is desirable to have a strong theoretical support of the developed concepts. Apart from the basic operations and notions that are relevant in the context, structure studies involving algebraic, topological and lattice theoretic concepts together with possible hybridization of the novel concept with already existing well established techniques are also most relevant. This book is a humble attempt towards consolidating all these in the context of soft sets. For this, this monograph relies heavily on many published works of the author, doctoral thesis of author's students and works of many other colleagues and researchers in this newly emerging area.

As such, this book contains 6 chapters covering various aspects of soft sets from theoretical to application problems. Apart from that, a brief historic development of soft sets and related structures together with some future directions in the development of soft set theory and applications is also provided.

Chapter 1 introduces the basic definitions and notions of soft structure. Tabular representation, operations and many results including analogue of DeMorgan laws and results involving Cartesian product, relations and functions are provided. The notions of distance, similarity and entropy also form a part of this chapter. Chapter concludes with the representation of fuzzy sets, rough sets and topological spaces as particular types of soft set, justifying the fact that soft set is a generalized tool.

With the intention of enriching the theoretical studies, the algebraic structures of soft sets are studied in Chap. 2. They include soft groups, normalistic soft groups, soft BCK/BCI algebras, soft rings and modules and soft lattices.

Topology is a major branch of mathematics with many applications in the fields of physical and computer sciences. Topological structures on soft sets are more generalized methods and they can be useful for measuring the similarities and dissimilarities between the objects in a universe which are soft sets. Chapter 3 discusses two different approaches to soft topology. The basic difference in these approaches is that one of them considers a subcollection of a set of all soft sets in an initial universe with a fixed set of parameters and the other one considers a subcollection from the set of all soft subsets of a given soft set in a universe. In this chapter, both approaches are considered with respect to some standard typical topological notions.

Category theory brings together various branches of mathematics into a united whole and paves the way to describe and compare objects with similar and different properties. Chapter 4 is an attempt to accommodate categorical concepts in the context of soft sets and soft graphs. Further, the relationship between soft sets and classical information systems is also explored.

A usual practice in applications of uncertainty modelling problems is the hybridization of existing structures with the intention that the evolving hybrid structure will have advantages of the constituent ones. Soft sets are also not an exception and there are many hybrid structures involving soft sets which yielded better results. Chapter 5 gives a panoramic view of these structures. They include hybridization including fuzzy sets, intuitionistic fuzzy sets, hesitant fuzzy sets, rough sets, etc.

In order to justify the relevance and importance of the introduced concepts in various application scenarios and to show the relationships of soft sets with other related fields, Chap. 6 discusses various applications of soft sets in many real problems like decision making, parameter reduction, game theory and studies involving incomplete data.

The book is primarily designed for scientists, researchers and students working in the field of soft sets and other related areas like rough sets, fuzzy sets, graph structures and hybrid models involving them. I sincerely hope that this book will certainly be an important source for graduate and postgraduate students, teachers and researchers in colleges/universities in various fields of engineering as well as mathematics/physics. I believe that with the help of the global reputed nature of the publisher, the cutting edge ideas consolidated in this book will find ways to create a stimulating atmosphere for further active development of soft computing techniques round the globe.

Calicut, India
July 2020

Sunil Jacob John

Acknowledgements

I would like to express my deep sense of gratitude to all my co-authors, and collaborators for their help, suggestion and active participation in developing a book on Soft sets. Special thanks are due to Prof. Janusz Kacprzyk, the Editor-in-chief of the book series for the encouragement and the support provided. I am indebted to Prof. Sivaji Chakravorti, Director, National Institute of Technology Calicut, India for providing the facilities and kind understanding. My colleagues at the Department of Mathematics were always a source of inspiration while the preparation of this title. I sincerely hope that the foreword to the title provided by Prof. John N. Mordeson will add up to the readability and the reachability of the book and I am greatly indebted to him.

The timely help and support provided by my postgraduate and doctoral students were really remarkable. The patience and love of my family members, in particular my wife Jinta and son Sujin is gratefully appreciated. I am most appreciative of Dr. Leontina Di Cecco, Ms. Jayarani Premkumar and other personnel of Springer Nature for their help during the preparation of this book.

Contents

Acronyms

ar(R)	Anti-reflexive kernel of R
Cov(U)	Set of all coverings of the universe U
CS(U)	Set of all covering type soft sets over U
F(X×Y)	Family of all fuzzy relationships from X to Y
F(U)	Set of all fuzzy subsets of U
FS(U)	Set of all fuzzy soft sets over U
Fset	Category of fuzzy sets
Grp	Category of groups
HFE	Hesitant fuzzy element
HFS	Hesitant fuzzy set
$\widetilde{H}(U)$	Set of all hesitant fuzzy sets in U
IFS(U)	Intuitionistic fuzzy power set of U
PAS(U)	Set of all Pawlak approximation spaces
Par(U)	Set of all partitions of the universe U
P(U)	Power set of U
PS(U)	Set of all partition type soft sets over U
$pr(f_A)$	Set of all parameter reductions of soft set (F, A)
RED(P)	Family of all reductions of P
R^+	Set of all non-negative real numbers
radM	Jacobson radical of module
Set	Category of sets
SGr(U)	Category of soft graphs over U
S(P)	Strength of path P
$SM_{m\times n}$	Set of all $m \times n$ soft matrices over U
SMMDM	Soft max-max decision making
SMmDM	Soft max-min decision making
SmMDM	Soft min-max decision making
SmmDM	Soft min-min decision making
SSR(F,A)	Collection of all soft set relations defined on (F,A)
Sset(U)	Category of soft sets over U

$S_L(E)$	Family of all soft lattices over a lattice L with parameter set E
s(R)	Symmetric kernel of R
$S(U)_E$	Family of all soft sets over U with parameter set E
soc(M)	Socle of module
Vec	Category of vector spaces

Part I
Historical Perspective of Soft Sets

Introduction of fuzzy set theory by Zadeh [161] in 1965 by giving room for partial membership for better handling practical situations made a paradigm shift in mathematics. The rationale behind this idea was the need of modelling imprecise human knowledge. The diffusion of this concept to various applied sciences and industry was rapid and Zadeh himself contributed much for this. This was followed by many successful generalizations of fuzzy sets and literature is abundant on these. These generalizations include *L*-fuzzy sets by Goguen [47], intuitionistic fuzzy sets by Atanassov [11], type two fuzzy sets [163], bipolar fuzzy sets [166], hesitant fuzzy sets by Torra [144], pythagorean fuzzy sets [159], picture fuzzy sets [35], spherical fuzzy sets [71], fermatean fuzzy sets [123] and many more. Apart from these another set of generalized structures useful for approximate reasoning was also developed parallel mostly in a complementary manner to fuzzy set theory. They include rough sets by Pawlak [98], multisets by Yager [158], genuine sets by Demirci et al. [37], neutrosophic and plithogenic logic by Smarandache [121], multiple sets by Shijina et al. [128, 129] etc.

Molodtsov developed soft set theory in a fundamentally different perspective. The application of this theory can be used for meaningfully interpreting real life problems in pure and applied sciences involving imprecise data. Current studies shows that ambiguities in data mining problems can also be solved using soft set theory techniques. The soft set theory could be used to interrogate and extend the idea of probability, fuzzy set, rough set and intuitionistic fuzzy set further. The disadvantage of lack of parameterization tool related to the concepts mentioned above gave a higher realm to soft set theory. In short, unlimited nature of approximate description is the greatest advantage of soft set theory.

While pondering over difficulties related to modelling uncertainties, eighteenth century mathematicians identified probability theory as a solution, which addressed uncertainty via randomness. The prominence of this was unchallenged till mid-twentieth century. In 1965, Zadeh [161] introduced fuzzy sets for addressing imprecision comprehensively. He expressed fuzziness via partial membership of an element in a set. Basically a fuzzy set can be identified with a class fitted with an ordering for elements which expresses the more or less belongingness of them in to that class under consideration.

In contrast to Aristotelian classical bi-valued logic, Polish mathematician Jan Lukasiewicz (1878–1956) introduced three-valued logic. Lukasiewicz is regarded as the main founder and contributor of multi-valued logic, which was later extended by Zadeh to fuzzy or infinite valued logic. In fuzzy logic, reasoning of false and truth are considered in a graded fashion but in classical logic absolutely true or false statements only are considered. Fuzzy logic can be considered as a branch of multi-valued logic based on the paradigm of inference under vagueness. Further, the introduction of fuzzy sets led to the development of many hybrid mathematical structures also.

By employing the notion of an equivalence relation called indiscernibility relation, Pawlak [98] in 1982 brought in theory of rough sets by means of lower and upper approximations and boundary region of a set. Lower approximation consists of all elements which surely belongs to the concepts and upper approximation consists of all elements which possibly belongs to the concepts. The difference of lower and upper approximations is the boundary region. If boundary is empty then set is crisp, otherwise it is rough. One benefit of the rough set theory is that it does not require any additional parameter or details regarding the data to extract information.

In 1994 Pawlak [101] published a paper titled "Hard and soft sets", in which he used a unified approach by taking ideas from classical set theory, rough sets and fuzzy sets for representing soft sets. Motivated by this work, D. Molodtsov [89] in 1999 published the paper titled "Soft set theory: first results", which is considered as the origin of theory of soft sets. Apart from the basic notions of the theory, some of its possible applications and some problems of the future research directions are also discussed in this paper. This theory was further solidified by P. K. Maji et al. [81] in 2003 by defining some fundamentals of the theory such as equality of two soft sets, subset and super set of a soft set etc. As continuation of these ideas, many extensions, hybridizations and extensions were put forward by many authors, some of them are the following: Maji, Biswas and Roy [79] introduced fuzzy soft sets, Wang, Li and Chen [148] introduced hesitant fuzzy soft sets and Pei and Miao [103] explore the relationship between fuzzy soft sets and classical information systems. The theory of soft sets is still developing very rapidly both in theoretical as well as application perspectives.

The never ending probe of researchers for better and better modeling of uncertainty, ambiguity and vagueness may add more and more structures similar to soft sets which will more specifically and accurately solve many problems of real world. The relevance of soft sets in this context is always worth mentioning.

Chapter 1
Soft Sets

The aim of introducing a soft structure over a set is to make a certain discretization of such fundamental mathematical concepts with effectively continuous nature and thus providing new tools for the use of the technology of mathematical analysis in real applications involving uncertainty or imperfect data. This is achieved through a certain parameterization of a given set. As usual, this new perspective of ideas draw attention of both pure and applied mathematicians and researchers in many related areas as well. Specifically, the specialists found the concept of a soft set well coordinated with many other modern mathematical concepts such as fuzzy sets, rough sets and many more. Further, this resulted in a series of works where soft versions of mathematical concepts were realized.

1.1 Basic Definitions and Examples

A soft set gives an approximate description of an object under consideration in two precise parts, namely predicate and approximate value set. Classical Mathematics always need exact solutions to mathematical models. Increasing level of complexity or complications in model makes it difficult to get exact solutions and one may go for approximate solutions and there are many methods for this. On the other hand, in soft set theory as the initial description of object itself is of approximate nature, we need not have to introduce the concept of exact solution.

Soft set theory, which was introduced by Russian researcher Molodtsov [88] in 1999 is a completely generic mathematical tool for modeling uncertainties. There is no condition imposed on the description of objects; so researchers can choose any form of parameters they needed, which greatly simplifies the decision-making process and make the process more efficient and reliable in the presence of partial

S. J. John, *Soft Sets*, Studies in Fuzziness and Soft Computing 400,
https://doi.org/10.1007/978-3-030-57654-7_1

information. There are many techniques available for modeling real world complex systems, such as the classical probability theory, fuzzy set theory introduced by Zadeh [160], interval mathematics [61, 96, 135] etc. Major drawback of all these techniques is the lack of parameterization of the tools and hence they could not be applied successfully in tackling problems especially in areas like economics, environmental and social sciences. Soft set theory is relatively free from the difficulties associated with above mentioned techniques and has a wider scope for many applications in a multidimensional way.

In this section basic definitions, an example and a tabular representation as introduced by Molodtsov [88], Maji et al. [80], and Babitha and Sunil [15] are mentioned.

Definition 1.1 Let U be an initial universe set and E be a set of parameters. Let $P(U)$ denotes the power set of U and $A \subset E$. A pair (F, A) is called a soft set over U, where F is a mapping given by $F : A \to P(U)$.

In other words, a soft set over U is a parameterized family of subsets of the universe U. For $\epsilon \in A$, $F(\epsilon)$ may be considered as the set of ϵ-approximate elements of the soft set (F, A).

Example 1 Let U be a set of all students under consideration. E is a set of parameters. Each parameter can be a word or sentence. $E = \{brilliant, average, healthy\}$. In this case, we can define a soft set (F, A) to point out the Nature of students as follows: Suppose that there are six students in the universe U given by $U = \{x_1, x_2, , x_3, x_4, x_5, x_6\}$ and $E = \{e_1, e_2, e_3\}$ where e_1 stands for brilliant, e_2 stands for average and e_3 stands for healthy. The soft set (F, A) where $A = E$ defined as $F(e_1) = \{x_1, x_2, , x_5\}$, $F(e_2) = \{x_3, x_4, x_6\}$, $F(e_3) = \{x_1, x_4, x_5, x_6\}$ gives the soft set representing the nature of students. The soft set (F, E) is a parametrized family $\{F(e_i) : i = 1, 2, 3\}$ of subsets of the set U and gives us a collection of approximate descriptions of an object.

Here note that for each $e \in E$, $F(e)$ is a crisp set. So the soft set (F, A) is called a standard soft set.

1.1.1 Tabular Representation of a Soft Set

For the purpose of storing a soft sets in computers, one may need the representation of a soft set in the form of a matrix or a table. The (i, j)th entry in table

$$t_{i,j} = \begin{cases} 1 \text{ if } x_i \in F(e_j) \\ 0 \text{ otherwise} \end{cases}$$

With reference to Example 1 given above, the tabular representation of the soft set is given Table 1.1.

Table 1.1 Tabular Representation of the soft set in Example 1

$U \downarrow$	e_1 (brilliant)	e_2 (average)	e_3 (healthy)
x_1	1	0	1
x_2	1	0	0
x_3	0	1	0
x_4	0	1	1
x_5	1	0	1
x_6	0	1	1

1.2 Operations of Soft Sets

Various operations analogous to union, intersection, complement, difference etc. in set theory will be discussed in the context of soft sets. Definitions and results given in this section are due to [7, 15, 80, 123].

Definition 1.2 For two soft sets (F, A) and (G, B) over a common universe U, we say that (F, A) is a soft subset of (G, B) if

(i) $A \subseteq B$, and
(ii) $\forall \epsilon \in A$, $F(\epsilon)$ and $G(\epsilon)$ are identical approximations.

We write $(F, A)\tilde{\subseteq}(G, B)$.

(F, A) is said to be a soft super set of (G, B), if (G, B) is a soft subset of (F, A). We denote it by $(F, A)\tilde{\supseteq}(G, B)$).

Two soft sets (F, A) and (G, B) over a common universe U are said to be soft equal if (F, A) is a soft subset of (G, B) and (G, B) is a soft subset of (F, A).

Definition of soft subset can be modified with replacing condition (ii) in Definition 1.2 "$F(\epsilon)$ and $G(\epsilon)$ are identical approximations" by "$F(\epsilon) \subseteq G(\epsilon)$".

Definition 1.3 Let $E = \{e_1, e_2, e_3, .., e_n\}$ be a set of parameters. The NOT set of E denoted by $\neg E$ is defined by $\neg E = \{\neg e_1, \neg e_2, \neg e_3, .., \neg e_n\}$ where $\neg e_i =$ not $e_i \forall i$.

Definition 1.4 The complement of a soft set (F, A) is denoted by $(F, A)^c$ and is defined by $(F, A)^c = (F^c, \neg A)$ where $F^c : \neg A \to P(U)$ is a mapping given by $F^c(\neg\alpha) = U - F(\alpha), \forall \neg\alpha \in \neg A$.

We call F^c to be the soft complement function of F. Clearly $(F^c)^c$ is the same as F and $((F, A)^c)^c = (F, A)$. It is also known as neg-complement as F^c is defined on the NOT set of the parameter set.

Definition 1.5 Let U be an initial universe set, E be the set of parameters, and $A \subset E$.

(i) (F, A) is said to be a relative null soft set (with respect to the parameter set A), denoted by $\tilde{\Phi}_A$, if $\forall \epsilon \in A$, $F(\epsilon) = \phi$, (null-set).

(ii) (F, A) is said to be a relative whole soft set (with respect to the parameter set A), denoted by $\widetilde{U}_A$, if $\forall \epsilon \in A, F(\epsilon) = U$.

The relative null soft set with respect to E denoted by $\widetilde{\Phi}_E$ is called the null soft set over U.

The relative whole soft set with respect to E denoted by $\widetilde{U}_E$ is called the absolute soft set over U.

Definition 1.6 The relative complement of a soft set (F, A) is denoted by $(F, A)^r$ or $(F, A)'$ and is defined by $(F, A)^r = (F^r, A)$ where $F^r : A \to P(U)$ is a mapping given by $F^r(\alpha) = U - F(\alpha), \forall \alpha \in A$.

Clearly, we have the following propositions.

Proposition 1.1 *If A and B are two sets of parameters then we have the following:*

(i) $\neg(\neg A) = A$
(ii) $\neg(A \cup B) = (\neg A) \cup (\neg B)$
(iii) $\neg(A \cap B) = (\neg A) \cap (\neg B)$

Proposition 1.2 *Let U be a universe, E a set of parameters,* $A, B, C \subset E$. *If* (F, A), (G, B) *and* (H, C) *are soft sets over U, Then*

(i) $(F, A)\widetilde{\subseteq}\widetilde{U}_A$.
(ii) $\widetilde{\Phi}_A\widetilde{\subseteq}(F, A)$.
(iii) $(F, A)\widetilde{\subseteq}(F, A)$.
(iv) $(F, A)\widetilde{\subseteq}(G, B), (G, B)\widetilde{\subseteq}(H, C)$ *implies* $(F, A)\widetilde{\subseteq}(H, C)$.
(v) $(F, A) = (G, B)$ *and* $(G, B) = (H, C)$ *implies* $(F, A) = (H, C)$.

Definition 1.7 The union of two soft sets (F, A) and (G, B) over the common universe U is a soft set (H, C), where $C = A \cup B$ and for each $e \in C$,

$$H(e) = \begin{cases} F(e), \text{ if } e \in A - B \\ G(e), \text{ if } e \in B - A \\ F(e) \cup G(e), \text{ if } e \in A \cap B \end{cases}.$$

We write $(F, A)\widetilde{\cup}(G, B) = (H, C)$,

Definition 1.8 The intersection of two soft sets (F, A) and (G, B) over the common universe U is a soft set (H, C), where $C = A \cap B$, and $H(e) = F(e) \cap G(e), \forall e \in C$.

We write $(F, A)\widetilde{\cap}(G, B) = (H, C)$.

Definition 1.9 Let (F, A) and (G, B) be soft sets over a common universe U such that $A \cap B \neq \phi$. Then the restricted union of (F, A) and (G, B) denoted by $(F, A) \cup_R (G, B)$ and is defined as $(F, A) \cup_R (G, B) = (H, C)$ where $C = A \cap B$ and for all $c \in C$, $H(c) = F(c) \cup G(c)$.

Definition 1.10 Extended intersection of two soft sets (F, A) and (G, B) over the common universe U, denoted by $(F, A) \cap_E (G, B)$ and is the soft set (H, C), where

$$C = A \cup B, \text{ and } \forall e \in C, H(e) = \begin{cases} F(e), \text{ if } e \in A - B \\ G(e), \text{ if } e \in B - A \\ F(e) \cap G(e), \text{ if } e \in A \cap B \end{cases}.$$

Definition 1.11 Let (F, A) and (G, B) be soft sets over a common universe U such that $A \cap B \neq \phi$. Then the restricted difference of (F, A) and (G, B) denoted by $(F, A)\tilde{-}_R(G, B)$ and is defined as $(F, A)\tilde{-}_R(G, B) = (H, C)$ where $C = A \cap B$ and $\forall c \in C$, $H(c) = F(c) - G(c)$, the difference of the sets $F(c)$ and $H(c)$.

Definition 1.12 If (F, A) and (G, B) are soft sets over a common universe U, then $(F, A) AND (G, B)$ denoted by $(F, A) \wedge (G, B)$ is defined as $(F, A) \wedge (G, B) = (H, A \times B)$ where $H(a, b) = F(a) \cap G(b)$ for every $(a, b) \in A \times B$.

Definition 1.13 If (F, A) and (G, B) are soft sets over a common universe U, then $(F, A) OR (G, B)$ denoted by $(F, A) \vee (G, B)$ is defined as $(F, A) \vee (G, B) = (K, A \times B)$ where $K(a, b) = F(a) \cup G(b)$ for every $(a, b) \in A \times B$.

For soft sets (F, A), (G, B) and (H, C) over the same universe U with A, B, C subsets of the parameter set E, the following theorems hold:

Theorem 1.1 *Properties of union operation*

(a) $(F, A)\tilde{\cup}((G, B)\tilde{\cup}(H, C)) = ((F, A)\tilde{\cup}(G, B))\tilde{\cup}(H, C)$
(b) $(F, A)\tilde{\cup}\tilde{U}_A = \tilde{U}_A$, $(F, A)\tilde{\cup}\tilde{U}_E = \tilde{U}_E$, $(F, A)\tilde{\cup}\tilde{\Phi}_A = (F, A)$
(c) (F, A) *need not be a soft subset of* $(F, A)\tilde{\cup}(G, B)$. *But if* $(F, A)\tilde{\subset}(G, B)$, *then* $(F, A)\tilde{\subset}(F, A)\tilde{\cup}(G, B)$, *moreover* $(F, A) = (F, A)\tilde{\cup}(G, B)$
(d) $(F, A)\tilde{\cup}(G, A) = \tilde{\Phi}_A$ *if and only if* $(F, A) = \tilde{\Phi}_A$ *and* $(G, A) = \tilde{\Phi}_A$
(e) $(F, A)\tilde{\cup}((G, B)\tilde{\cap}(H, C)) = ((F, A)\tilde{\cup}(G, B))\tilde{\cap}((F, A)\tilde{\cup}(H, C))$
(f) $((F, A)\tilde{\cap}(G, B))\tilde{\cup}(H, C) = ((F, A)\tilde{\cup}(H, C))\tilde{\cap}((G, B)\tilde{\cup}(H, C))$

Proof Proof of (a), (b), (e) and (f) are straight forward and follows easily from definitions.

(c) Let $(F, A)\tilde{\cup}(G, B) = (H, C)$ where $C = A \cup B$ and

$$H(e) = \begin{cases} F(e) & \text{if } e \in A - B \\ G(e) & \text{if } e \in B - A \\ F(e) \cup G(e) & \text{if } e \in A \cap B \end{cases}.$$

It is obvious that if $e \in A \cap B$, then $H(e) = F(e) \cup G(e)$, thus $F(e)$ and $H(e)$ need not be the same approximations. Thus (F, A) need not be a soft subset of $(F, A)\tilde{\cup}(G, B)$.

Now let $(F, A)\tilde{\subset}(G, B)$. Then, it is clear that $A \subset A \cup B = A$. We need to show that $F(e)$ and $H(e)$ are the same approximations for all $e \in A$. Let $e \in A$, then $e \in A \cap B = A$, since $A \subset B$ implies $A - B = \phi$. Thus, $H(e) = F(e) \cup G(e) = F(e) \cup F(e) = F(e)$, as $G(e)$ and $F(e)$ are the same approximations for all $e \in A$. This follows that H and F are the same set-valued mapping for all $e \in A$, as required.

(d) Suppose that $(F, A)\tilde{\cup}(G, A) = (H, A)$, where $H(x) = F(x) \cup G(x)$ for all $x \in A$. Since $(H, A) = \tilde{\Phi}_A$ from the assumption, $H(x) = F(x) \cup G(x) = \phi \Leftrightarrow F(x) = \phi$ and $G(x) = \phi \Leftrightarrow (F, A) = \tilde{\Phi}_A$ and $(G, A) = \tilde{\Phi}_A$ for all $x \in A$. Now assume that $(F, A) = \tilde{\Phi}_A$ and $(G, A) = \tilde{\Phi}_A$ and $(F, A)\tilde{\cup}(G, A) = (H, A)$. Since $F(x) = \phi$ and $G(x) = \phi$ for all $x \in A$, $H(x) = F(x) \cup G(x) = \phi$ for all $x \in A$. Therefore, $(F, A)\tilde{\cup}(G, A) = \tilde{\Phi}_A$. □

Theorem 1.2 *Properties of restricted union operation*

(a) $(F, A) \cup_R (G, B) \cup_R (H, C) = (F, A) \cup_R (G, B) \cup_R (H, C)$.
(b) $(F, A) \cup_R \widetilde{U}_A = \widetilde{U}_A, (F, A) \cup_R \widetilde{U}_E = \widetilde{U}_A, (F, A) \cup_R \widetilde{\Phi}_A = (F, A).,$
$(F, A) \cup_R \widetilde{\Phi}_E = (F, A)$
(c) $(F, A) \not\subseteq (F, A) \cup_R (G, B)$, *in general. But if* $(F, A)\widetilde{\subset}(G, B)$, *then* $(F, A)\widetilde{\subset}$ $(F, A) \cup_R (G, B)$, *moreover* $(F, A) = (F, A) \cup_R (G, B)$
(d) $(F, A) \cup_R (G, A) = \widetilde{\Phi}_A \Leftrightarrow (F, A) = \widetilde{\Phi}_A$ *and* $(G, A) = \widetilde{\Phi}_A$
(e) $(F, A) \cup_R ((G, B)\widetilde{\cap}(H, C)) = (F, A) \cup_R (G, B)\widetilde{\cap}((F, A) \cup_R (H, C)$
(f) $((F, A)\widetilde{\cap}(G, B)) \cup_{\mathcal{R}} (H, C) = ((F, A) \cup_{\mathcal{R}} (H, C))\widetilde{\cap}((G, B) \cup_{\mathcal{R}} (H, C))$
(g) $(F, A) \cup_{\mathcal{R}} ((G, B) \cap_E (H, C)) = \big((F, A) \cup_{\mathcal{R}} (G, B)\big) \cap_E \big((F, A) \cup_{\mathcal{R}} (H, C)\big)$
(h) $((F, A) \cap_E (G, B)) \cup_{\mathcal{R}} (H, C) = \big((F, A) \cup_{\mathcal{R}} (H, C)\big) \cap_E \big((G, B) \cup_{\mathcal{R}} (H, C)\big)$

Proof (a) First, we investigate the left-hand side of the equality. Suppose that $(G, B) \cup_R (H, C) = (T, B \cap C)$, where $T(x) = G(x) \cup H(x)$ for all $x \in B \cap C \neq \phi$. And assume $(F, A) \cup_{\mathcal{R}} (T, B \cap C) = (W, A \cap (B \cap C))$, where $W(x) = F(x) \cup T(x) = F(x) \cup (G(x) \cup H(x))$ for all $x \in A \cap (B \cap C) \neq \phi$.

Now consider the right-hand side of the equality. Suppose that $(F, A) \cup_{\mathcal{R}} (G, B) = (M, A \cap B)$, where $M(x) = F(x) \cup G(x)$ for all $x \in A \cap B \neq \phi$ · And let $(M, A \cap B) \cup_{\mathcal{R}} (H, C) = (N, (A \cap B) \cap C)$, where $N(x) = M(x) \cup H(x) = (F(x) \cup G(x)) \cup H(x)$ for all $x \in (A \cap B) \cap C \neq \phi$. Since W and N are the same mapping for all $x \in A \cap (B \cap C) = (A \cap B) \cap C$, the proof is completed.

(b) Proof of (b) follows directly from the definitions.

(c) Since $A \not\subseteq A \cap B$ without any extra condition being given, $(F, A) \not\subseteq (F, A) \cup_{\mathcal{R}} (G, B)$ in general. Now assume that (F, A) is a soft subset of (G, B) and $(F, A) \cup_{\mathcal{R}} (G, B) = (H, A \cap B = C)$, where $H(x) = F(x) \cup G(x)$ for all $x \in C$. Then,
$(F, A)\widetilde{\subset}(G, B) \Leftrightarrow A \subset A \cap B = A$ and $F(e)$ and $G(e)$ are the same approximations for all $e \in A \Leftrightarrow H(e) = F(e) \cup G(e) = F(e) \cup F(e) = F(e)$ for all $e \in A$. Thus, F and H are the same set-valued mapping for all $e \in A$, so the proof is completed.

(d) Proof follows from the fact that $(F, A) \cup_{\mathcal{R}} (G, A) = (F, A) \widetilde{\cup} (G, A))$ and Theorem 1.1(d).

(e) First, we handle the left-hand side of the equality. Suppose that $(G, B)\widetilde{\cap}$ $(H, C) = (T, B \cap C)$, where $T(x) = G(x) \cap H(x)$ for all $x \in B \cap C$. Let (F, A) $\cup_{\mathcal{R}} (T, B \cap C) = (W, A \cap (B \cap C))$, where $W(x) = F(x) \cup T(x) = F(x)$ $\cup (G(x) \cap H(x))$ for all $x \in (A \cap B) \cap C$.

Now consider the right-hand side of the equality. Assume that $(F, A) \cup_{\mathcal{R}} (G, B) = (M, A \cap B)$, where $M(x) = F(x) \cup G(x)$ for all $x \in A \cap B \neq \phi$. And let $(F, A) \cup_{\mathcal{R}}$ $(H, C) = (N, A \cap C)$, where $N(x) = F(x) \cup H(x)$ for all $x \in A \cap C \neq \phi$. Suppose that $(M, A \cap B)\widetilde{\cap}(N, B \cap C) = (K, (A \cap B) \cap (A \cap C)) = (K, (A \cap B)$ $\cap C)$, where $K(x) = M(x) \cap N(x) = (F(x) \cup G(x)) \cap (F(x) \cup H(x)) = F(x)$ $\cup(G(x) \cap H(x))$ for all $x \in (A \cap B) \cap C$. Since W and K are the same set-valued mapping, the proof is completed.

(f) By similar techniques used to prove (e), (f) can be illustrated, and is therefore omitted.

(g) Suppose that $(G, B) \cap_E (H, C) = (T, B \cup C)$, where

$$T(e) = \begin{cases} G(e) & \text{if } e \in B - C \\ H(e) & \text{if } e \in C - B \\ G(e) \cap H(e) & \text{if } e \in B \cap C \end{cases}$$

Assume that $(F, A) \cup_R (T, B \cup C) = (M, A \cap (B \cup C))$, where $M(x) = F(x) \cup T(x)$ for all $x \in A \cap (B \cup C)$. By taking into account the properties of operations in set theory and the definitions of M along with T and considering that T is a piecewise function, we can write the below equalities for M:

$$M(e) = \begin{cases} F(e) \cup G(e) & \text{if } e \in A \cap (B - C) = (A \cap B) - (A \cap C) \\ F(e) \cup H(e) & \text{if } e \in A \cap (C - B) = (A \cap C) - (A \cap B) \\ F(e) \cup (G(e) \cap H(e)) & \text{if } e \in A \cap (B \cap C) \end{cases}$$

for all $e \in A \cap (B \cup C)$.

Now consider the right-hand side of the equality. Suppose that $(F, A) \cup_{\mathcal{R}} (G, B) = (Q, A \cap B)$, where $Q(x) = F(x) \cup G(x)$ for all $x \in A \cap B \neq \phi$. Assume $(F, A) \cup_{\mathcal{R}} (H, C) = (W, A \cap C)$, where $W(x) = F(x) \cup H(x)$ for all $x \in A \cap C \neq \phi$. Let $(Q, A \cap B) \cap_E (W, A \cap C) = (N, (A \cap B) \cup (A \cap C))$, where

$$N(e) = \begin{cases} Q(e) & \text{if } e \in (A \cap B) - (A \cap C) \\ W(e) & \text{if } e \in (A \cap C) - (A \cap B) \\ Q(e) \cap W(e) & \text{if } e \in (A \cap B) \cap (A \cap C) = A \cap (B \cap C) \end{cases}$$

for all $x \in (A \cap B) \cup (A \cap C)$. By taking into account the definitions of Q and W, we can rewrite N as below:

$$N(e) = \begin{cases} F(e) \cup G(e) & \text{if } e \in (A \cap B) - (A \cap C) \\ F(e) \cup H(e) & \text{if } e \in (A \cap C) - (A \cap B) \\ (F(e) \cup G(e)) \cap (F(e) \cup H(e)) & \text{if } e \in A \cap (B \cap C) \end{cases}$$

This follows that N and M are the same set-valued mapping when considering the properties of operations on set theory, which completes the proof.

(h) By similar techniques used to prove (g), (h) can be illustrated, and is therefore omitted. □

Similar theorems follow for extended intersection and intersection also. Proofs are in similar lines and hence omitted.

Theorem 1.3 *Properties of extended intersection operation*

(a) $(F, A) \cap_E ((G, B) \cap_E (H, C)) = ((F, A) \cap_E (G, B)) \cap_E (H, C)$
(b) $(F, A) \cap_E \widetilde{U}_A = (F, A)$, $(F, A) \cap_E \widetilde{\Phi}_A = \widetilde{\Phi}_A$
(c) $(F, A) \cap_E (G, B) \not\subseteq (G, B)$, *in general. But if* $(F, A) \widetilde{\subseteq} (G, B)$, *then* $(F, A) \cap_E (G, B) \widetilde{\subseteq} (G, B)$, *moreover* $(F, A) \cap_E (G, B) = (G, B)$
(d) $(F, A) \cap_E ((G, B) \cup_{\mathcal{R}} (H, C)) = ((F, A) \cap_E (G, B)) \cup_{\mathcal{R}} ((F, A) \cap_E (H, C))$
(e) $((F, A) \cup_{\mathcal{R}} (G, B)) \cap_E (H, C) = ((F, A) \cap_E (H, C)) \cup_{\mathcal{R}} ((G, B) \cap_E (H, C))$

Theorem 1.4 *Properties of intersection operation*

(a) $(F, A) \widetilde{\cap} ((G, B) \widetilde{\cap} (H, C)) = ((F, A) \widetilde{\cap} (G, B)) \widetilde{\cap} (H, C)$
(b) $(F, A) \widetilde{\cap} \widetilde{U}_A = (F, A)$, $(F, A) \widetilde{\cap} \widetilde{U}_E = (F, A)$, $(F, A) \widetilde{\cap} \widetilde{\Phi}_A = \widetilde{\Phi}_A$, $(F, A) \cap \widetilde{\Phi}_E = \widetilde{\Phi}_A$
(c) $(F, A) \widetilde{\cap} (G, B) \not\subseteq (F, A)$, *in general. But if* $(F, A) \widetilde{\subseteq} (G, B)$, *then* $(F, A) \widetilde{\cap} (G, B) \widetilde{\subseteq} (F, A)$ *moreover* $(F, A) \widetilde{\cap} (G, B) = (F, A)$.

(d) $(F, A)\widetilde{\cap}((G, B) \cup_{\mathcal{R}} (H, C)) = ((F, A)\widetilde{\cap}(G, B)) \cup_{\mathcal{R}} ((F, A)\widetilde{\cap}(H, C))$
(e) $((F, A) \cup_{\mathcal{R}} (G, B))\widetilde{\cap}(H, C) = ((F, A)\widetilde{\cap}(H, C)) \cup_{\mathcal{R}} ((G, B)\widetilde{\cap}(H, C))$
(f) $(F, A)\widetilde{\cap}((G, B)\widetilde{\cup}(H, C)) = ((F, A)\widetilde{\cap}(G, B))\widetilde{\cup}((F, A)\widetilde{\cap}(H, C))$
(g) $((F, A)\widetilde{\cup}(G, B))\widetilde{\cap}(H, C) = ((F, A)\widetilde{\cap}(H, C))\widetilde{\cup}((G, B)\widetilde{\cap}(H, C))$
(h) $(F, A)\widetilde{\cap}((G, B) \sim_{\mathcal{R}} (H, C)) = ((F, A)\widetilde{\cap}(G, B)) \sim_{\mathcal{R}} ((F, A)\widetilde{\cap}(H, C))$
(i) $((F, A) \sim_{\mathcal{R}} (G, B))\widetilde{\cap}(H, C) = ((F, A)\widetilde{\cap}(H, C)) \sim_{\mathcal{R}} ((G, B)\widetilde{\cap}(H, C))$

Proposition 1.3 *Let (F, A) be a soft set over U. Then we have the following;*

(i) $(F, A)\widetilde{\cup}(F, A)^r = (F, A) \cup_R (F, A)^r = \widetilde{U}_A$
(ii) $(F, A) \cap_E (F, A)^r = (F, A)\widetilde{\cap}(F, A)^r = \widetilde{\Phi}_A$
(iii) $(\widetilde{U}_E)^r = \widetilde{\Phi}_E$, $(\widetilde{U}_A)^r = \widetilde{\Phi}_A$

Proof Obvious. □

1.2.1 De Morgan Laws

In this sub section, we show that the following De Morgan's type of results hold in soft set theory for different types of union, intersection, complements, AND and OR operations. Results given in this section are taken from Ali et al. [7] , Maji et al. [80] and Sezgin et al. [123].

Let (F, A) and (G, B) be two soft sets over a common universe U. Then we have the following:

Theorem 1.5 (De Morgan laws with respect to relative complement, restricted union and intersection)

(a) $[(F, A) \cup_R (G, B)]^r = (F, A)^r\widetilde{\cap}(G, B)^r$
(b) $[(F, A)\widetilde{\cap}(G, B)]^r = (F, A)^r \cup_R (G, B)^r$

Proof (a) Let $(F, A)\widetilde{\cup}_R(G, B) = (H, C)$ where $H(c) = F(c) \cup G(c)$ for all $c \in C = A \cap B \neq \emptyset$. Since $((F, A) \cup_R (G, B))^r = (H, C)^r$, by definition $H^r(c) = U - [F(c) \cup G(c)] = [U - F(c)] \cap [U - G(c)]$ for all $c \in C$.

Now $(F, A)^r\widetilde{\cap}(G, B)^r = (F^r, A)\widetilde{\cap}(G^r, B) = (K, C)$ where $C = A \cap B$. So by definition, we have,

$$\begin{aligned} K(c) &= F^r(c) \cap G^r(c) \\ &= (U - F(c)) \cap (U - G(c)) \\ &= H^r(c) \forall c \in C \end{aligned}$$

Hence $[(F, A)\widetilde{\cup}_R(G, B)]^r = (F, A)^r\widetilde{\cap}(G, B)^r$.

(b) Let $(F, A)\widetilde{\cap}(G, B) = (H, C)$ where $H(c) = F(c) \cap G(c)$ for all $c \in C = A \cap B \neq \emptyset$. Since $((F, A)\widetilde{\cap}(G, B))^r = (H, C)^r$, by definition $H^r(c) = U - (F(c) \cap G(c)) = [U - F(c)] \cup [U - G(c)]$ for all $c \in C$. Now $(F, A)^r \cup_R (G, B)^r = (F^r, A) \cup_R (G^r, B) = (K, C)$ where $C = A \cap B$. So by definition, we have $K(c) = F^r(c) \cup G^r(c) = (U - F(c)) \cup (U - G(c)) = H^r(c)$ for all $c \in C$. Hence $[(F, A)\widetilde{\cap}(G, B)]^r = (F, A)^r\widetilde{\cup}_R(G, B)^r$. □

Theorem 1.6 (De Morgan laws with respect to extended intersection, union and neg-complement)

(a) $[(F, A) \cap_E (G, B)]^C = (F, A)^C \tilde{\cup} (G, B)^C$
(b) $[(F, A) \tilde{\cup} (G, B)]^C = (F, A)^C \cap_E (G, B)^C$

Proof (a) Suppose that $(F, A) \cap_E (G, B) = (H, A \cup B)$. Then $((F, A) \cap_E (G, B))^C = (H, A \cup B)^c = (H^c, \rceil(A \cup B)) = (H^c, \rceil A \cup \rceil B)$ where $H^c(\rceil e) = U - H(e)$ for all $\rceil e \in \rceil A \cup \rceil B$.

By definition, $H(e) = \begin{cases} F(e), & \text{if } e \in A - B \\ G(e), & \text{if } e \in B - A \\ F(e) \cap G(e), & \text{if } e \in A \cap B \end{cases}$

Thus we have $H^c(\rceil e) = \begin{cases} U \backslash F(e) = F^c(\rceil e), & if \rceil e \in \rceil A - \rceil B \\ U \backslash G(e) = G^c(\rceil e), & if \rceil e \in \rceil B - \rceil A \\ U \backslash (F(e) \cap G(e)) = F^c(\rceil e) \cup G^c(\rceil e), & if \rceil e \in \rceil A \cap \rceil B \end{cases}$

Moreover, let $(F, A)^c \tilde{\cup} (G, B)^c = (F^c, \rceil A) \tilde{\cup} (G^c, \rceil B) = (K, \rceil A \cup \rceil B)$.

Then $K(\rceil e) = \begin{cases} F^c(\rceil e), & if \rceil e \in \rceil A - \rceil B \\ G^c(\rceil e), & if \rceil e \in \rceil B - \rceil A \\ F^c(\rceil e) \cup G^c(\rceil e), & if \rceil e \in \rceil A \cap \rceil B \end{cases}$

Since H^c and K are indeed the same set-valued mapping, we conclude that $((F, A) \cap_E (G, B))^c = (F, A)^c \tilde{\cup} (G, B)^c$ as required.
(b) By using a similar technique, part (b) can be proved. □

Theorem 1.7 (De Morgan laws with respect to AND, OR, and neg complement)

(a) $[(F, A) \vee (G, B)]^C = (F, A)^C \wedge (G, B)^C$
(b) $[(F, A) \wedge (G, B)]^C = (F, A)^C \vee (G, B)^C$

Proof (a) Suppose that $(F, A) \vee (G, B) = (O, A \times B)$. Therefore, $((F, A) \vee (G, B))^c = (O, A \times B)^c = (O^c, \rceil(A \times B))$. Now

$$\begin{aligned}(F, A)^c \wedge (G, B)^c &= \left(F^c, \rceil A\right) \wedge \left(G^c, \rceil B\right) \\ &= (J, \rceil A \times \rceil B), \text{ where } J(x, y) = F^c(x) \cap G^c(y) \\ &= (J, \rceil(A \times B)\end{aligned}$$

Now, take $(\rceil\alpha, \rceil\beta) \in \rceil(A \times B)$. Therefore,

$$\begin{aligned}O^c(\rceil\alpha, \rceil\beta) &= U - O(\alpha, \beta) \\ &= U - [F(\alpha \cup G(\beta)] \\ &= [U - F(\alpha)] \cap [U - G(\beta)] \\ &= F^c(\rceil\alpha) \cap G^c(\rceil\beta) \\ &= J(\rceil\alpha, \rceil\beta)\end{aligned}$$

Thus O^c and J are same. Hence, proved.
(b) By using a similar technique, part (b) can be proved. □

Theorem 1.8 (De Morgan laws with respect to AND, OR, and relative complement)

(a) $[(F, A) \vee (G, B)]^r = (F, A)^r \wedge (G, B)^r$
(b) $[(F, A) \wedge (G, B)]^r = (F, A)^r \vee (G, B)^r$

Proof (a) Suppose that $(F, A) \vee (G, B) = (O, A \times B)$. Therefore, $((F, A) \vee (G, B))^r = (O, A \times B)^r = (O^r, A \times B)$. Now, $(F, A)^r \wedge (G, B)^r = (F^r, A) \wedge (G^r, B), = (J, A \times B)$, where $J(x, y) = F^r(x) \cap G^r(y)$. Let $(\alpha, \beta) \in A \times B$. Then, $O^r(\alpha, \beta) = U \setminus O(\alpha, \beta) = U \setminus [F(\alpha) \cup G(\beta)] = [U \setminus F(\alpha)] \cap [U \setminus G(\beta)] = F^r(\alpha) \cap G^r(\beta) = J(\alpha, \beta)$. Since O^r and J are indeed the same set-valued mapping, $((F, A) \vee (G, B))^r = (F, A)^r \wedge (G, B)^r$.
(b) By using a similar technique, part (b) can be proved. □

In most of the studies related to algebraic and topological structures, we often need to handle indexed family of entities. Feng et al. [41] gives various union, intersection and/or operations for indexed families of soft sets.

Definition 1.14 Let $(F_i, A_i)_{i \in I}$ be a nonempty family of soft sets over a common universe U. The union of these soft sets is defined to be the soft set (G, B) such that $B = \bigcup_{i \in I} A_i$ and, for all $x \in B$, $G(x) = \bigcup_{i \in I(x)} F_i(x)$, where $I(x) = \{i \in I | x \in A_i\}$. In this case, we write $\widetilde{\bigcup}_{i \in I} (F_i, A_i) = (G, B)$.

Definition 1.15 Let $(F_i, A_i)_{i \in I}$ be a nonempty family of soft sets over a common universe set U. The AND- soft set $\widetilde{\bigwedge}_{i \in I} (F_i, A_i)$ of these soft sets is defined to be the soft set (H, B) such that $B = \prod_{i \in I} A_i$ and $H(x) = \bigcap_{i \in I} F_i (x_i)$ for all $x = (x_i)_{i \in I} \in B$.

Definition 1.16 Let $(F_i, A_i)_{i \in I}$ be a nonempty family of soft sets over a common universe set U. The OR- soft set $\widetilde{\bigvee}_{i \in I} (F_i, A_i)$ of these soft sets is defined to be the soft set (H, B) such that $B = \prod_{i \in I} A_i$ and $H(x) = \bigcup_{i \in I} F_i (x_i)$ for all $x = (x_i)_{i \in I} \in B$.

Note that, if $A_i = A$ and $F_i = F$ for all $i \in I$, then $\widetilde{\bigwedge}_{i \in I} (F_i, A_i)$ (respectively, $\widetilde{\bigvee}_{i \in I} (F_i, A_i)$) is denoted by $\widetilde{\bigwedge}_{i \in I}(F, A)$ (respectively, $\widetilde{\bigvee}_{i \in I}(F, A)$). In this case, $\prod_{i \in I} A_i = \prod_{i \in I} A$ means the direct power A^I.

Definition 1.17 The restricted union of a nonempty family of soft sets $(F_i, A_i)_{i \in I}$ over a common universe set U is defined as the soft set $(H, B) = \bigcup_{R_{i \in I}} (F_i, A_i)$ where $B = \bigcap_{i \in I} A_i \neq \emptyset$ and $H(x) = \bigcup_{i \in I} F_i(x)$ for all $x \in B$.

Definition 1.18 The extended intersection of a nonempty family of soft sets $(F_i, A_i)_{i \in I}$ over a common universe set U is defined as the soft set $(H, B) = \bigcap_{E i \in I} (F_i, A_i)$ such that such that $B = \bigcup_{i \in I} A_i$ and $H(x) = \bigcup_{i \in I(x)} F_i(x)$, where $I(x) = \{i \in I | x \in A_i\}$ for all $x \in B$.

Definition 1.19 Let $(F_i, A_i)_{i \in I}$ be a nonempty family of soft sets over a common universe U. The intersection of these soft sets is defined to be the soft set (G, B) such that $B = \widetilde{\bigcap}_{i \in I} A_i \neq \emptyset$ and, for all $x \in B$, $G(x) = \bigcap_{i \in I} F_i(x)$. In this case, we write $\widetilde{\bigcap}_{i \in I} (F_i, A_i) = (G, B)$.

1.3 Cartesian Product, Relations and Partitions

Mathematically, a relation is an association between objects based on certain properties of various objects. In most of naturally occurring phenomena, two variables may be linked by some type of relationship, which can be represented as a set of ordered pairs. A function is a particular kind of relation that associates each input with exactly one output. The theory of relations and functions undoubtedly forms the foundation of mathematics.

This section provides theoretical aspects of soft sets by extending the notions of cartesian products, relations, functions etc. in classical sets to that of soft sets. Transitive closure of a soft set relation is defined and many theorems related to it are obtained. An important related contribution is the analogous of Warshall algorithm to construct the transitive closure of a soft set relation. Further, orderings on soft sets is defined and discussed. For the fundamentals concepts of set theory, we refer Karel Hrbacek [51], Patrick Suppes [94] and Paul R Halmos[95]. Most of the results given in this section is due to Babitha and Sunil [16, 17] and Hai-Long Yang and Zhi-Lian Guo [155].

1.3.1 Soft Set Relations

Definition 1.20 Let (F, A) and (G, B) be two soft sets over U, then the Cartesian product of (F, A) and (G, B) is defined as $(F, A) \times (G, B) = (H, A \times B)$, where $H : A \times B \to P(U \times U)$ and $H(a, b) = F(a) \times G(b)$, where $(a, b) \in A \times B$. i.e., $H(a, b) = \{(h_i, h_j) : h_i \in F(a) \text{ and } h_j \in G(b)\}$

The Cartesian product of three or more nonempty soft sets can be defined by generalizing the definition of the Cartesian product of two soft sets.

The Cartesian product $(F_1, A_1) \times (F_2, A_2) \times, \ldots, \times (F_n, A_n)$ of non empty soft sets $(F_1, A_1), (F_2, A_2), \ldots, (F_n, A_n)$ is denoted by $\tilde{\prod}_i (F_i, A_i)$.

Example 2 Consider the soft set (F, A) which describes the "cost of the houses" and the soft set (G, B) which describes the "attractiveness of the houses". Suppose that $U = \{h_1, h_2, h_3, h_4, h_5, h_6, h_7, h_8, h_9.h_{10}\}$, $A =$ {very costly; costly; cheap} and $B =$ {beautiful; in the green surroundings; cheap}. Let F (very costly) $= \{h_2, h_4, h_7, h_8\}$, F (costly) $= \{h_1, h_3, h_5\}$, F (cheap) $= \{h_6, h_9, h_{10}\}$, and G (beautiful) $= \{h_2, h_3, h_7\}$, G (in the green surroundings) $= \{h_6, h_5, h_8\}$, G (cheap) $= \{h_6, h_9, h_{10}\}$. Now $(F, A) \times (G, B) = (H, A \times B)$ where a typical element will look like H(very costly, beautiful)$= \{h_2, h_4, h_7, h_8\} \times \{h_2, h_3, h_7\} = \{(h_2, h_2), (h_2, h_3), (h_2, h_7), (h_4, h_2), (h_4, h_3), (h_4, h_7), (h_7, h_2), (h_7, h_3), (h_7, h_7), (h_8, h_2), (h_8, h_3), (h_8, h_7)\}$.

Definition 1.21 Let (F, A) and (G, B) be two soft sets over U, then a relation from (F, A) to (G, B) is a soft subset of $(F, A) \times (G, B)$.

In other words, a relation from (F, A) to (G, B) is of the form (H_1, S) where $S \subset A \times B$ and $H_1(a, b) = H(a, b), \forall (a, b) \in S$ where $(H, A \times B) = (F, A) \times (G, B)$ as defined in Definition 1.20. Any subset of $(F, A) \times (F, A)$ is called a relation on (F, A).

In an equivalent way, we can define the relation R on the soft set (F, A) in the parameterized form as follows: If $(F, A) = \{F(a), F(b)\}$ then $F(a)RF(b)$ if and only if $F(a) \times F(b) \in R$. The collection of all soft set relations defined on (F, A) is denoted by $SSR(F, A)$.

Definition 1.22 Let R be a soft set relation from (F, A) to (G, B). Then the Domain of R $(domR)$ is defined as the soft set (D, A_1) where $A_1 = \{a \in A : H(a, b) \in R$ for some $b \in B\}$ and $D(a_1) = F(a_1) \forall a_1 \in A_1$.

The range of R $(ranR)$ is defined as the soft set (RG, B_1), where $B_1 \subset B$ and $B_1 = \{b \in B : H(a, b) \in R$ for some $a \in A\}$ and $RG(b_1) = G(b_1) \forall b_1 \in B_1$.

Consider the following to illustrate a relation on soft sets.

Example 3 Let U denotes set of people in a social gathering. Take $U = \{p_1, p_2, p_3, p_4, p_5, p_6, p_7, p_8, p_9, p_{10}\}$. Let A denotes different job categories. Take $A =$ {chartered accountant, doctors, engineers, teachers} i.e., $A = \{c, d, e, t\}$ Let B denotes qualification of people. Take $B = \{B.Sc, B.Tech, MBBS, M.Sc\}$ i.e., $B = \{b_1, b_2, m_1, m_2\}$. Then the soft set (F, A) which gives the nature of jobs of people is given by $\{F(c) = \{p_1, p_2\}; F(d) = \{p_4, p_5\}; F(e) = \{p_7, p_9\}; F(t) = \{p_3, p_4, p_7\}$ and the soft set (G, B) which gives the qualification of people is $\{G(b_1) = \{p_1, p_6, p_8\}; G(b_2) = \{p_3, p_6, p_7, p_9\}; G(m_1) = \{p_3, p_4, p_5, p_8\}, G(m_2) = \{p_3, p_8\}\}$. Define a relation R from (F, A) to (G, B) as follows: $F(a)RG(b)$ iff $F(a) \subseteq G(b)$. Then $R = \{F(d) \times G(m_1), F(e) \times G(b_2)\}$ Then $domR = (D, A_1)$ where $A_1 = \{doctors, engineers\}$ and $D(a) = F(a)$ for every $a \in A_1$. Similarly $ranR = (RG, B_1)$ where $B_1 = \{B.Tech, MBBS\}$ and $RG(b) = B_1(b)$ for every $b \in B_1$.

Definition 1.23 1. The complement of a soft set relation R on (F, A) denoted as R^c is defined by $R^c = \{F(a) \times F(b) : F(a) \times F(b) \notin R, a, b \in A\}$.
2. The union of two soft set relations R and Q on (F, A) denoted as $R \cup Q$ is defined by $R \cup Q = \{F(a) \times F(b) : F(a) \times F(b) \in R$ or $F(a) \times F(b) \in Q\}$.
3. The intersection of two soft set relations R and Q on (F, A) denoted as $R \cap Q$ is defined by $R \cap Q = \{F(a) \times F(b) : F(a) \times F(b) \in R$ and $F(a) \times F(b) \in Q\}$.
4. Let R, Q be two soft set relations on (F, A). $\forall a, b \in A$, if $F(a) \times F(b) \in R \Longrightarrow F(a) \times F(b) \in Q$, then we call $R \subset Q$ or $R \leq Q$.

It is easy to verify that the union and the intersection of soft set relations satisfy commutative, associative and distributive laws.

1.3.2 *Induced Relations from Universal Set and the Attribute Set*

In soft sets, one is dealing with two kinds of ordinary sets, universal set and the attribute set. Hence we may think about relations defined on universal set as well as the attribute set. Corresponding to these, we can induce some relation on the soft set also.

Definition 1.24 Let (F, A) be a soft set defined on the universal set and R be a relation defined on U. (i.e., $R \subset U \times U$). Then the induced soft set relation R_U on (F, A) is defined as follows: $F(a) R_U F(b) \Longleftrightarrow uRv$ for every $u \in F(a)$ and $v \in F(b)$.

In a similar manner we can define the induced soft set relation corresponding to a relation in the attribute set also.

Definition 1.25 Let (F, A) be a soft set defined on the universal set and R be a relation defined on A. (i.e., $R \subset A \times A$). Then the induced soft set relation R_A on (F, A) is defined as follows: $F(a) R_A F(b) \Longleftrightarrow aRb$ for every $a, b \in A$.

1.3.3 *Equivalence Relations and Partitions on Soft Sets*

Definition 1.26 Let R be a relation on (F, A), then

1. R is reflexive if $F(a) \times F(a) \in R$ for every $a \in A$
2. R is symmetric if $F(a) \times F(b) \in R \implies F(b) \times F(a) \in R$.
3. R is transitive if $F(a) \times F(b) \in R$ and $F(b) \times F(c) \in R \implies F(a) \times F(c) \in R$ for every $a, b, c \in A$.

Definition 1.27 A soft set relation R on a soft set (F, A) is called an equivalence relation if it is reflexive, symmetric and transitive.

Example 4 Consider a soft set (F, A) over U where $U = \{c_1, c_2, c_3, c_4, c_5, c_6, c_7, c_8, c_9\}$, $A = \{m_1, m_2\}$ and $F(m_1) = \{c_1, c_2, c_5, c_6\}$, $F(m_2) = \{c_3, c_4, c_7, c_8, c_9\}$. Consider a relation R defined on (F, A) as $\{F(m_1) \times F(m_2), F(m_2) \times F(m_1), F(m_1) \times F(m_1), F(m_2) \times F(m_2)\}$. This relation is a soft set equivalence relation.

Definition 1.28 Let (F, A) be a soft set, then equivalence class of $F(a)$ denoted by $[F(a)]$ is defined as $[F(a)] = \{F(b) : F(b) R F(a)\}$.

In Example 4, we have $[F(m_1)] = \{F(m_1), F(m_2)\} = [F(m_2)]$.

Lemma 1.1 *Let R be an equivalence relation on a soft set (F, A) , then for any* $F(a), F(b) \in (F, A)$, $F(a) R F(b)$ *iff* $[F(a)] = [F(b)]$.

Proof Suppose $[F(a)] = [F(b)]$. Since R is reflexive $F(b)RF(b)$ Hence $F(b) \in [F(b)] = [F(a)]$ which gives $F(a)RF(b)$.

Conversely suppose $F(a)RF(b)$. Let $F(a') \in [F(a)]$. Then$F(a')RF(a)$. Using the transitive property of R this gives $F(a') \in [F(b)]$. Hence $[F(a)] \subseteq [F(b)]$. Using a similar argument, $[F(b)] \subseteq [F(a)]$. Hence $[F(a)] = [F(b)]$. □

Definition 1.29 A collection of nonempty soft sub sets $P = \{(F_i, A_i) : i \in I\}$ of a soft set (F, A) is called a partition of (F, A) if (i) $(F, A) = \widetilde{\cup}_i(F_i, A_i)$ and (ii) $A_i \cap A_j = \phi$ whenever $i \neq j$.

Example 5 Let $U = \{h_l, h_2, h_3, h_4, h_5\}$, $A = \{a_1, a_2, a_3, a_4\}$. $(F, A) = \{F(a_1), F(a_2), F(a_3), F(a_4)\}$, $F(a_1) = \{h_1, h_2\}$, $F(a_2) = \{h_3\}$, $F(a_3) = \{h_3, h_4\}$, $F(a_4) = \{h_4, h_5\}$, $A_1 = \{a_1, a_2\}$, $A_2 = \{a_3, a_4\}$, $(F_1, A_1) = \{F_1(a_1), F_1(a_2)\}$, $(F_2, A_2) = \{F_2(a_3), F_2(a_4)\}$, so that $F_i = F \forall i = 1, 2$. $(F, A) = (F_1, A_1)\widetilde{\cup}(F_2, A_2)$ and $A_1 \cap A_2 = \phi$. Thus $\{(F_1, A_1), (F_2, A_2)\}$ is a soft set partition.

Remark 1.1 Members of the partition are called blocks of (F, A). Moreover corresponding to a partition $\{(F_i, A_i)\}$ of a soft set (F, A), we can define a soft set relation on (F, A) by $F(a)RF(b)$ iff $F(a)$ and $F(b)$ belongs to the same block.

The relation defined in this manner is an equivalence relation.

Theorem 1.9 *Let $\{(F_i, A_i) : i \in I\}$ be a partition of soft set (F, A). The soft set relation defined on (F, A) as $F(a)RF(b)$ iff $F(a)$ and $F(b)$ are the members of the same block is an equivalence relation.*

Proof Reflexive: Let $F(a)$ be any element of (F, A). It is clear that $F(a)$ is in the same block of itself. Hence $F(a)RF(a)$.

Symmetric: If $F(a)RF(b)$, then $F(a)$ and $F(b)$ are in the same block. Therefore $F(b)RF(a)$.

Transitive: If $F(a)RF(b)$, $F(b)RF(c)$ then $F(a)$, $F(b)$, $F(c)$ must lie in the same block. Therefore $F(a)RF(c)$ □

Remark 1.2 The equivalence soft set relation defined in Theorem 1.9 is called the equivalence soft set relation determined by the partition P.

Theorem 1.10 *Corresponding to every equivalence relation R defined on a soft set (F, A), there exists a partition on (F, A) and this partition precisely consists of the equivalence classes of R.*

Proof Let $[F(a)]$ be an equivalence class with respect to a relation R on (F, A). Let A_a denotes all those elements in A corresponds to $[F(a)]$. i.e., $A_a = \{b \in A : F(b)RF(a)\}$. Thus we can denote $[F(a)]$ as (F, A_a). So we have to show that the collection $\{(F, A_a) : a \in A\}$ of such distinct sets form a partition P of (F, A). In order to prove this we should prove

1. $(F, A) = \widetilde{\cup}_{a \in A}(F, A_a)$
2. If A_a, A_b are not identical then $A_a \cap A_b = \phi$

Since R is reflexive, $F(a)RF(a)\forall a \in A$ so that (1) follows.

Now for (2), Let $x \in A_a \cap A_b$. Then $F(x) \in (F, A_a)$ and $F(x) \in (F, A_b) \Longrightarrow F(x)RF(a)$ and $F(x)RF(b)$. Using the transitivity of R, $F(a)RF(b)$. By Lemma 1.1 we have $[F(a)] = [F(b)]$. This gives $A_a = A_b$. □

Remark 1.3 The partition constructed in Theorem 1.10 consists of all equivalence classes of R and is called quotient soft sets of (F, A) and is denoted by $(F, A)/R$.

Definition 1.30 Let (F, A), (G, B) and (H, C) be three soft sets. Let R be a soft set relation from (F, A) to (G, B) and S be a soft set relation from (G, B) to (H, C). Then a new soft set relation, the composition of R and S expressed as $S \circ R$ from (F, A) to (H, C) is defined as follows : If $F(a)$ is in (F, A) and $H(c)$ is in (H, C) then $F(a)S \circ RH(c)$ iff there is some $G(b)$ in (G, B) such that $F(a)RG(b)$ and $G(b)RH(c)$.

Definition 1.31 The inverse of a soft set relation R denoted as R^{-1} is defined by $R^{-1} = \{(F(b) \times F(a)) : F(a)RF(b)\}$

Theorem 1.11 *Let R, Q be two soft set relations on (F, A). Then*

(a) *R is symmetric iff $R = R^{-1}$.*
(b) $(R^c)^{-1} = \left(R^{-1}\right)^c$.
(c) $\left(R^{-1}\right)^{-1} = R$, $(R^c)^c = R$
(d) $R \cup Q \supset R$, $R \cup Q \supset Q$
(e) $R \cap Q \subset R$, $R \cap Q \subset Q$
(f) $R \subset Q \Longrightarrow R^{-1} \subset Q^{-1}$
(g) *If $P \supset Q$ and $P \supset R$, then $P \supset R \cup Q$*
(h) *If $P \subset Q$ and $P \subset R$, then $P \subset R \cap Q$*
(i) *If $R \subset Q$, then $R \cup Q = Q$ and $R \cap Q = R$*
(j) $(R \cup Q)^{-1} = R^{-1} \cup Q^{-1}$, $(R \cap Q)^{-1} = R^{-1} \cap Q^{-1}$
(k) $(R \cup Q)^c = R^c \cap Q^c$, $(R \cap Q)^c = R^c \cup Q^c$.

Proof Obviously, $(c) - (i)$ hold. We only show (a), (b), (j), and (k)

(a) Now $\forall F(a) \times F(b) \in R$, by the symmetry of R, we have $F(b) \times F(a) \in R$, then $F(a) \times F(b) \in R^{-1}$. So $R \subset R^{-1}$. Conversely $\forall F(a) \times F(b) \in R^{-1}$, then $F(b) \times F(a) \in R$. By the symmetry of R, $F(a) \times F(b) \in R$. So $R^{-1} \subset R$. Thus $R = R^{-1}$.

Conversely assume $R = R^{-1}.\forall a, b \in A$, if $F(a) \times F(b) \in R$, then $F(a) \times F(b) \in R^{-1}$, so $F(b) \times F(a) \in R$. Hence R is symmetric.

(b) $\forall F(a) \times F(b) \in (R^c)^{-1}$, $a, b \in A$, we have $F(b) \times F(a) \in R^c$, which implies that $F(b) \times F(a) \notin R$. Thus $F(a) \times F(b) \notin R^{-1}$, which implies that $F(a) \times F(b) \in \left(R^{-1}\right)^c$. Hence $(R^c)^{-1} \subset \left(R^{-1}\right)^c$. Conversely, $\forall F(a) \times F(b) \in \left(R^{-1}\right)^c$, $a, b \in A$, we have $F(a) \times F(b) \notin R^{-1}$ which implies that $F(b) \times F(a) \notin R$. So $F(b) \times F(a) \in R^c$, which implies that $F(a) \times F(b) \in (R^c)^{-1}$. Hence $\left(R^{-1}\right)^c \subset (R^c)^{-1}$ Therefore $(R^c)^{-1} = \left(R^{-1}\right)^c$.

(j) $\forall a, b \in A$, $F(a) \times F(b) \in (R \cup Q)^{-1} \Longleftrightarrow F(b) \times F(a) \in R \cup Q \Longleftrightarrow F(b) \times F(a) \in R$ or $F(b) \times F(a) \in Q \Longleftrightarrow F(a) \times F(b) \in R^{-1}$ or $F(a) \times F(b)$

$\in Q^{-1} \Longleftrightarrow F(a) \times F(b) \in R^{-1} \cup Q^{-1}$. So $(R \cup Q)^{-1} = R^{-1} \cup Q^{-1}$. The proof of $(R \cap Q)^{-1} = R^{-1} \cap Q^{-1}$ is similar.

(k) $\forall a, b \in A, F(a) \times F(b) \in (R \cup Q)^c \Longleftrightarrow F(a) \times F(b) \notin R \cup Q$ and $\Longleftrightarrow F(a) \times F(b) \notin R$ $F(a) \times F(b) \notin Q \Longleftrightarrow F(a) \times F(b) \in R^c$ and $F(a) \times F(b) \in Q^c \Longleftrightarrow F(a) \times F(b) \in R^c \cap Q^c$. So $(R \cup Q)^c = R^c \cap Q^c$. The proof of $(R \cap Q)^c = R^c \cup Q^c$ is similar. □

From part (a) and (b) of Theorem 1.11, it follows that, the complement of a symmetric soft set relation is again a symmetric soft set relation.

Theorem 1.12 *Let R be soft set relation from (F, A) to (G, B) and S be a soft set relation from (G, B) to (H, C). Then $(SoR)^{-1} = R^{-1}oS^{-1}$.*

Proof Clearly $(SoR)^{-1}$ is a soft set relation from (H, C) to (F, A). Now let $H(c)$ be any element in (H, C) and $F(a)$ be an element in (F, A).Then $H(c)(SR)^{-1}F(a)$ if $F(a)SoRH(c)$. This by definition exists if there is some $G(b)$ in (G, B) such that $F(a)RG(b)$ and $G(b)RH(c)$. This is equivalent to $G(b)R^{-1}F(a)$ and $H(c)S^{-1}G(b)$ Then $H(c)R^{-1}oS^{-1}F(a)$. Hence $(SoR)^{-1} = R^{-1}oS^{-1}$. □

Definition 1.32 The identity relation I_{F_A} on any soft set (F, A) is defined as follows: $F(a)I_{F_A}F(b)$ iff $a = b$.

Definition 1.33 Let (F, A) and (G, B) be two non empty soft sets. Then a soft set relation from (F, A) to (G, B) is called a soft set function if every element in domain has a unique element in range.

A theory analogous to that of set-functions can be developed for soft functions also. Further, owing to the fact that, associated with two soft sets we can think of two ordinary mapping, one on the universe sets and and other on the parameter sets. Combining these two, a more natural analogue of the concept of functions in soft context can be obtained. We reserve the name soft mappings for that and that will be discussed in a later section.

1.3.4 Kernels of Soft Set Relations

Here we introduce the notions of anti-reflexive kernel and symmetric kernel of a soft set relation, and discuss their properties.

Definition 1.34 Let R be a soft set relation on (F, A), then R is anti-reflexive if $F(a) \times F(a) \notin R, \forall a \in A$.

Let us denote the minimal (resp. the maximal) soft set relation on (F, A), by m(resp. M) then clearly $m = \phi$, and $M = (F, A) \times (F, A) = \{F(a) \times F(b) | a, b \in A\}$. It is easy to verify that m and I^c are two anti-reflexive soft set relations, m, M, and I are three symmetric soft set relations, M and I are two reflexive soft set relations.

If R is not an anti- reflexive soft set relation, then there is no anti-reflexive soft set relation containing R. If R is not a reflexive soft set relation, then there is no reflexive soft set relation contained in R. Moreover, if R is a reflexive soft set relation, then $R \supset I$, and if R is an anti-reflexive soft set relation, then $R \subset I^c$.

Definition 1.35 Let R be a soft set relation on (F, A) .

(1) The maximal anti-reflexive soft set relation contained in R is called anti-reflexive kernel of R, denoted by $ar(R)$.
(2) The maximal symmetric soft set relation contained in R is called symmetric kernel of R, denoted by $s(R)$.

Theorem 1.13 *Let R be a soft set relation on* (F, A). *Then*
(1) $ar(R) = R \cap I^c$. *Therefore we obtain a mapping (called anti-reflexive kernel operator)* $ar\colon \mathcal{SSR}(F, A) \longrightarrow \mathcal{SSR}(F, A)$.
(2) $s(R) = R \cap R^{-1}$. *Therefore we obtain a mapping (called symmetric kernel operator)* $s : \mathcal{SSR}(F, A) \longrightarrow \mathcal{SSR}(F, A)$

Proof (1) By Theorem 1.11(e), $R \cap I^c \subset R$ and $R \cap I^c \subset I^c$. Now $\forall a \in A$, by the definition of I, $F(a) \times F(a) \in I$, so $F(a) \times F(a) \notin I^c$. Hence $F(a) \times F(a) \notin R \cap I^c$, i.e., $R \cap I^c$ is an anti-reflexive soft set relation on (F, A). If T is an anti-reflexive soft set relation on (F, A) and $T \subset R$. Then $T \subset I^c$. Hence $T \subset R \cap I^c$. So $ar(R) = R \cap I^c$.
(2) By Theorem 1.11(j) and (c) $\left(R \cap R^{-1}\right)^{-1} = R^{-1} \cap \left(R^{-1}\right)^{-1} = R^{-1} \cap R = R \cap R^{-1}$ i.e., $R \cap R^{-1}$ is a symmetric soft set relation on (F, A) by Theorem 1.11(a). By Theorem 1.11(e), $R \cap R^{-1} \subset R$. On the other hand, if T is a symmetric soft set relation on (F, A) and $T \subset R$. By Theorem 1.11(f), $T^{-1} \subset R^{-1}$. Then by Theorem 1.11(a) and (h), $T = T^{-1} \subset R \cap R^{-1}$. So $s(R) = R \cap R^{-1}$. □

Example 6 Let $U = \{u_1, u_2, u_3, u_4, u_5, u_6, u_7\}$, $A = \{e_1, e_2, e_3\}$. The soft set (F, A) is given by $\{F(e_1) = \{u_1, u_2\}, F(e_2) = \{u_2, u_3, u_4\}, F(e_3) = \{u_1, u_5, u_6, u_7\}\}$. A soft set relation R on (F, A) is given by $R = \{F(e_1) \times F(e_1), F(e_2) \times F(e_3), F(e_3) \times F(e_1)\}$. Then $ar(R) = R \cap I^c = \{F(e_1) \times F(e_1), F(e_2) \times F(e_3), F(e_3) \times F(e_1)\} \cap \{F(e_1) \times F(e_2), F(e_1) \times F(e_3), F(e_2) \times F(e_1), F(e_2) \times F(e_3), F(e_3) \times F(e_1)\}, \quad F(e_3) \times F(e_2) = \{F(e_2) \times F(e_3), F(e_3) \times F(e_1)\}$ $s(R) = R \cap R^{-1} = \{F(e_1) \times F(e_1), F(e_2) \times F(e_3), F(e_3) \times F(e_1)\} \cap \{F(e_1) \times F(e_1), F(e_3) \times F(e_2), F(e_1) \times F(e_3)\} = \{F(e_1) \times F(e_1)\}$.

Theorem 1.14 *The anti-reflexive kernel operator ar has the following properties:*
(1) $ar(m) = m$, $ar(I^c) = I^c$.
(2) $\forall R \in \mathcal{SSR}(F, A)$, $ar(R) \subset R$
(3) $\forall R, Q \in \mathcal{SSR}(F, A)$, $ar(R \cup Q) = ar(R) \cup ar(Q)$, $ar(R \cap Q) = ar(R) \cap ar(Q)$
(4) $\forall R, Q \in \mathcal{SSR}(F, A)$, *if* $R \subset Q$, *then* $ar(R) \subset ar(Q)$
(5) $\forall R \in \mathcal{SSR}(F, A)$, $ar(ar R)) = ar(R)$

Proof (1) By the anti-reflexivity of m and I^c, obviously, $ar(m) = m$, $ar(I^c) = I^c$.
(2) $\forall R \in \mathcal{SSR}(F, A)$, by Theorem 1.11($a$) and ($e$), $ar(R) = R \cap I^c \subset R$.
(3) $\forall R, Q \in \mathcal{SSR}(F, A)$, by Theorem 1.11($a$) $ar(R \cup Q) = (R \cup Q) \cap I^c = (R \cap I^c) \cup (Q \cap I^c) = ar(R) \cup ar(Q)$, $ar(R \cap Q) = (R \cap Q) \cap I^c = (R \cap I^c) \cap (Q \cap I^c) = ar(R) \cap ar(Q)$.
(4) $\forall R, Q \in \mathcal{SSR}(F, A)$, $R \subset Q$, by (3) and Theorem 1.11(d) and (i), $ar(Q) = ar(R \cup Q) = ar(R) \cup ar(Q) \supset \text{ar}(R)$
(5) $\forall R \in \mathcal{SSR}(F, A)$, by Theorem 1.11($a$), $ar(R) = R \cap I^c$. Hence $ar(ar(R)) = ar\,(R \cap I^c) = (R \cap I^c) \cap I^c = R \cap I^c = ar(R)$. □

Theorem 1.15 *The symmetric kernel operator s has the following properties:*
(1) $s(m) = m$, $s(M) = M$, $s(I) = I$.
(2) $\forall R \in \mathcal{SSR}(F, A)$, $s(R) \subset R$.
(3) $\forall R, Q \in \mathcal{SSR}(F, A)$, $s(R \cap Q) = s(R) \cap s(Q)$.
(4) $\forall R, Q \in \mathcal{SSR}(F, A)$, *if* $R \subset Q$, *then* $s(R) \subset s(Q)$.
(5) $\forall R \in \mathcal{SSR}(F, A)$, $s(s(R)) = s(R)$.

Proof (1) By the symmetry of m, M and I, $s(m) = m$, $s(M) = M$, $s(I) = I$.
(2) $\forall R \in \mathcal{SSR}(F, A)$, by Theorems 1.13(2) and 1.11(e), $s(R) = R \cap R^{-1} \subset R$.
(3) $\forall R, Q \in \mathcal{SSR}(F, A)$, by Theorems 1.13(2) and 1.11(j), we have $s(R \cap Q) = (R \cap Q) \cap (R \cap Q)^{-1} = (R \cap Q) \cap (R^{-1} \cap Q^{-1}) = (R \cap R^{-1}) \cap (Q \cap Q^{-1}) = s(R) \cap s(Q)$.
(4) $\forall R, Q \in \mathcal{SSR}(F, A)$, $R \subset Q$, by (3) and Theorem 1.11(e) and (i), $s(Q) = s(R \cap Q) = s(R) \cap s(Q) \subset s(Q)$.
(5) $\forall R \in \mathcal{SSR}(F, A)$, by Theorem 1.13(2), $s(R) = R \cap R^{-1}$. Hence $s(s(R)) = s\left(R \cap R^{-1}\right) = \left(R \cap R^{-1}\right) \cap \left(R \cap R^{-1}\right)^{-1} = \left(R \cap R^{-1}\right) \cap \left(R^{-1} \cap \left(R^{-1}\right)^{-1}\right) = \left(R \cap R^{-1}\right) \cap \left(R^{-1} \cap R\right) = R \cap R^{-1} = s(R)$. □

1.3.5 Closures of Soft Set Relations

In this subsection, we study the concepts of reflexive closure, symmetric closure and transitive closure of a soft set relation and investigate their properties.

Definition 1.36 Let R be a soft set relation on (F, A). The minimal reflexive soft set relation containing R is called reflexive closure of R, denoted by $\overline{r}(R)$.

Definition 1.37 Let R be a soft set relation on (F, A). The minimal symmetric soft set relation containing R is called symmetric closure of R, denoted by $\overline{s}(R)$.

Theorem 1.16 *Let R be a soft set relation on* (F, A). *Then*

(1) $\overline{r}(R) = R \cup I$. *Therefore we obtain a mapping (called reflexive closure operator)* $\overline{r} : \mathcal{SSR}(F, A) \longrightarrow \mathcal{SSR}(F, A)$

(2) $\overline{s}(R) = R \cup R^{-1}$. *Therefore we obtain a mapping (called symmetric closure operator)* $\overline{s} : \mathcal{SSR}(F, A) \longrightarrow \mathcal{SSR}(F, A)$.

Proof (1) By Theorem 1.11(d), $R \cup I \supset R.\forall a \in A$, $F(a) \times F(a) \in I \subset R \cup R$, so $R \cup I$ is reflexive. On the other hand, if T is a reflexive soft set relation on (F, A) and $T \supset R$. By the reflexivity of $T, T \supset I$, thus by Theorem 1.11(g), we have $T \supset R \cup I$. So $\overline{r}(R) = R \cup I$.

(2) By Theorem 1.11(j), $\left(R \cup R^{-1}\right)^{-1} = R^{-1} \cup \left(R^{-1}\right)^{-1} = R^{-1} \cup R = R \cup R^{-1}$, i.e., $R \cup R^{-1}$ is a symmetric soft set relation on (F, A), and $R \cup R^{-1} \supset R$ by Theorem 1.11(d). If T is a symmetric soft set relation on (F, A) and $T \supset R$. By Theorem 1.11(f), $T^{-1} \supset R^{-1}$. According to Theorem 1.11(a) and (g), $T = T^{-1} \supset R \cup R^{-1}$. So $\overline{s}(R) = R \cup R^{-1}$. □

Example 7 Let $U = \{u_1, u_2, u_3, u_4, u_5, u_6, u_7\}$, $A = \{d, e, f, g\}$. The soft set (F, A) is given by $\{F(d) = \{u_1, u_2\}, F(e) = \{u_2, u_3, u_4\}, F(f) = \{u_1, u_5, u_6\}, F(g) = \{u_7\}\}$. Consider a soft set relation R defined on (F, A) as $R = \{F(d) \times F(e), F(e) \times F(g), F(f) \times F(f)\}$. Then $\overline{r}(R) = R \cup I = \{F(d) \times F(e), F(e) \times F(g), F(f) \times F(f)\} \cup \{F(d) \times F(d), F(e) \times F(e), F(f) \times F(f), F(g) \times F(g)\} = \{F(d) \times F(d), F(d) \times F(e), F(e) \times F(e), F(e) \times F(g), F(f) \times F(f), F(g) \times F(g)\}$. $\overline{s}(R) = R \cup R^{-1} = \{F(d) \times F(e), F(e) \times F(g), F(f) \times F(f)\} \cup \{F(e) \times F(d), F(g) \times F(e), F(f) \times F(f)\} = \{F(d) \times F(e), F(e) \times F(g), F(e) \times F(d), F(g) \times F(e), F(f) \times F(f)\}$.

Theorem 1.17 *The reflexive closure operator $\overline{r}$ has the following properties:*
(1) $\overline{r}(M) = M, \overline{r}(I) = I$.
(2) $\forall R \in \mathcal{SSR}(F, A)$, $R \subset \overline{r}(R)$.
(3) $\forall R, Q \in \mathcal{SSR}(F, A)$, $\overline{r}(R \cup Q) = \overline{r}(R) \cup \overline{r}(Q)$, $\overline{r}(R \cap Q) = \overline{r}(R) \cap \overline{r}(Q)$.
(4) $\forall R, Q \in \mathcal{SSR}(F, A)$, if $R \subset Q$, then $\overline{r}(R) \subset \overline{r}(Q)$.
(5) $\forall R \in \mathcal{SSR}(F, A)$, $\overline{r}(\overline{r}(R)) = \overline{r}(R)$.

Proof (1) By the reflexivity of M and $I, \overline{r}(M) = M, \overline{r}(I) = I$.

(2) $\forall R \in \mathcal{SSR}(F, A)$, by Theorems 1.16(1) and 1.11(d) $\overline{r}(R) = R \cup I \supset R$.

(3) $\forall R, Q \in \mathcal{SSR}(F, A)$, by Theorem 1.16(1), $\overline{r}(R \cup Q) = (R \cup Q) \cup I = (R \cup I) \cup (Q \cup I) = \overline{r}(R) \cup \overline{r}(Q)$, $\overline{r}(R \cap Q) = (R \cap Q) \cup I = (R \cup I) \cap (Q \cup I) = \overline{r}(R) \cap \overline{r}(Q)$.

(4) $\forall R, Q \in \mathcal{SSR}(F, A)$, $R \subset Q$, by (3) and Theorem 1.11(i), $\overline{r}(Q) = \overline{r}(R \cup Q) = \overline{r}(R) \cup \overline{r}(Q) \supset \overline{r}(R)$.

(5) $\forall R \in \mathcal{SSR}(F, A)$, by Theorem 1.16(1), $\overline{r}(R) = R \cup I$. Hence $\overline{r}(\overline{r}(R)) = \overline{r}(R \cup I) = (R \cup I) \cup I = R \cup I = \overline{r}(R)$. □

An analogous theorem for symmetric closure is given below, for which the proof is similar and is omitted.

Theorem 1.18 *The symmetric closure operator $\overline{s}$ has the following properties:*
(1) $\overline{s}(m) = m, \overline{s}(M) = M, \overline{s}(I) = I$.
(2) $\forall R \in \mathcal{SSR}(F, A)$, $\overline{s}(R) \supset R$.
(3) $\forall R, Q \in \mathcal{SSR}(F, A)$, $\overline{s}(R \cup Q) = \overline{s}(R) \cup \overline{s}(Q)$.
(4) $\forall R, Q \in \mathcal{SSR}(F, A)$, if $R \subset Q$, then $\overline{s}(R) \subset \overline{s}(Q)$.
(5) $\forall R \in \mathcal{SSR}(F, A)$, $\overline{s}(\overline{s}(R)) = \overline{s}(R)$.

Lemma 1.2 $\forall R \in \mathcal{SSR}(F, A)$, *we have*
(1) $(\overline{r}\,(R^c))^c = ar(R)$.
(2) $\overline{r}(ar(R)) = \overline{r}(R)$.
(3) $ar(\overline{r}(R)) = ar(R)$.

Proof (1) By Theorem 1.16(1), $\overline{r}\,(R^c) = R^c \cup I$. By Theorems 1.11(*k*) and 1.13(1), $(\overline{r}\,(R^c))^c = (R^c \cup I)^c = (R^c)^c \cap I^c = R \cap I^c = ar(R)$,

(2) By Theorems 1.13(1) and 1.16(1) $\overline{r}(ar(R)) = \overline{r}\,(R \cap I^c) = (R \cap I^c) \cup I = (R \cup I) \cap (I^c \cup I) = (R \cup I) \cap M = \overline{r}(R)$.

(3) By Theorems 1.13(1) and 1.16(1) $ar(\overline{r}(R)) = ar(R \cup I) = (R \cup I) \cap I^c = (R \cap I^c) \cup (I \cap I^c) = (R \cap I^c) \cup m = ar(R)$. □

Theorem 1.19 (Six-relations Theorem) $\forall R \in \mathcal{SSR}(F, A)$, *six different soft set relations are the most soft set relations that can be obtained by using anti-reflexive kernel operator, reflexive closure operator, and complement operator.*

Proof $\forall R \in \mathcal{SSR}(F, A)$, by Lemma 1.2(1), $(\overline{r}\,(R^c))^c = ar(R)$, so we can replace the anti-reflexive kernel operator with the complement operator and the reflexive closure operator.

(1) Take the complement operator first, then the reflexive closure operator on R. The following five different soft set relations are the most soft set relations that can be constructed: R^c, $\overline{r}\,(R^c)$, $(\overline{r}\,(R^c))^c$, $\overline{r}\left((\overline{r}\,(R^c))^c\right)$, $\left(\overline{r}\left((\overline{r}\,(R^c))^c\right)\right)^c$ It is because that by Lemma 1.2(1, 2) and Theorem 1.11(*c*), $\overline{r}\left(\left(\overline{r}\left(\overline{r}\,(R^c))^c\right)\right)^c\right) = \overline{r}\,((\overline{r}(ar(R)))^c) = \overline{r}\,((\overline{r}(R))^c) = \overline{r}\,(ar\,(R^c)) = \overline{r}\,(R^c)$, which implies that the sixth is the same as the second. This is repeated emergence.

(2) Take the reflexive closure operator first, then the complement operator on R. By Lemma 1.2(1) and (2), $\overline{r}(R) = \overline{r}(ar(R)) = \overline{r}\left((\overline{r}\,(R^c))^c\right)$, which implies that the first is the same as the fourth in (1). This is repeated emergence.

(3) Take the reflexive closure operator successively or the complement operator successively on R. By Theorems 1.11(*c*) and 1.17 (5), $(R^c)^c = R, \overline{r}(\overline{r}(R)) = \overline{r}(R)$. This is repeated emergence. Thus the proof is complete. □

Results similar to Lemma 1.2 and Theorem 1.19 follows for symmetric kernel , symmetric closure , and complement operators also.

Lemma 1.3 $\forall R \in \mathcal{SSR}(F, A)$, *we have*
(1) $(\overline{s}\,(R^c))^c = s(R)$.
(2) $\overline{s}(s(R)) = s(R)$.
(3) $s(\overline{s}(R)) = \overline{s}(R)$.

Theorem 1.20 $\forall R \in \mathcal{SSR}(F, A)$, *six different soft set relations are the most soft set relations that can be obtained by using symmetric kernel operator, symmetric closure operator, and complement operator.*

Theorem 1.21 *The relation R on a soft set (F, A) is transitive iff* $R^n \subset R$ *for every* $n \in N$.

Proof Suppose that $R^n \subset R$ for every $n \in N$. In particular $R^2 \subset R$. Let $F(a) \times F(b) \in R$ and $F(b) \times F(c) \in R$, then by definition of composition, $F(a) \times F(c) \in R^2$. Since $R^2 \subset R$, this means that $F(a) \times F(c) \in R$, proving R is transitive.

Conversely suppose that R is transitive. We prove $R^n \subset R$ by induction. For n=1 it is true. Assume that $R^n \subset R$ and $F(a) \times F(b) \in R^{n+1}$. Since $R^{n+1} = R^n o R$, there is an element $F(x)$ such that $F(a) \times F(x) \in R$ and $F(x) \times F(b) \in R^n$. Now $R^n \subset R$ gives $F(x) \times F(b) \in R$. Furthermore, since R is transitive and $F(a) \times F(x) \in R$, it follows that $F(a) \times F(b) \in R$ showing that $R^{n+1} \subset R$. □

Theorem 1.22 *If T, U are two soft set relations from (F, A) to (G, B) and R, S are two soft set functions from (G, B) to (H, C) then $R \subset S$ and $T \subset U \implies RoT \subset SoU$.*

Proof Suppose that $F(a) \times H(c) \in RoT$. This implies that there exists $G(b) \in (G, B)$ such that $F(a) \in G(b) \in T$ and $G(b) \in H(c) \in R$. Now $R \subset S \implies G(b) \times H(c) \in S$ and $T \subset U \implies F(a) \times G(b) \in U$. Then $F(a) \times H(c) \in SoU$, showing that $RoT \subset SoU$. □

Corollary 1.1 *If $R \subset S$ then $R^n \subset S^n$.*

Definition 1.38 Let R be a relation on (F, A). The transitive closure of R denoted by R^+ is the smallest soft set relation containing R that is transitive.

Definition 1.39 (*Matrix Representation of a Soft Set Relation*) Every soft set relation R on a soft set (F, A) denoted as $\{F(a_1), F(a_2), \ldots, F(a_n)\}$ can be represented in matrix form as follows: $M = \begin{bmatrix} F_{11} & F_{12} & \ldots & F_{1n} \\ F_{21} & F_{22} & \ldots & F_{2n} \\ . & . & . & \\ . & . & . & \\ . & . & . & \\ F_{n1} & F_{n2} & \ldots & F_{nn} \end{bmatrix}$ where $F_{ij} = \begin{cases} 1 \text{ if } F(a_i) \times F(a_j) \in R \\ 0 \; otherwise \end{cases}$

Above representation is useful for storing soft set relations in computers.

Extension of the Warshall Algorithm to Soft Set Relations

Warshall's algorithm [33] is an efficient method of finding the adjacency matrix of transitive closure of a relation on a finite set from the adjacency matrix of the relation. Now the same algorithm will be extended to construct the transitive closure of a soft set relation.

Notation: Let m_i and m_j denote the ith and jth rows of the matrix M corresponding to the soft set relation R.

Warshall algorithm:

Input: Matrix M corresponding to a relation R on soft set (F, A).

Output: Matrix T corresponding to transitive closure of R.

Algorithm body:

$T := M$ [initialize T to M]

for $j := 1$ to n

for $i := 1$ to n
if $T_{ij} = 1$ then
$a_t = a_i \vee a_j$ [form the boolean OR of row i and row j store it in a_t]
next i
next j
end algorithm Warshall

1.3.6 Orderings on Soft Sets

Definition 1.40 A binary soft set relation R on (F, A) is antisymmetric if $F(a) \times F(b) \in R$ and $F(b) \times F(a) \in R$ for every $F(a), F(b)$ in (F, A) imply $F(a) = F(b)$.

Definition 1.41 A binary soft set relation R on (F, A) which is reflexive, antisymmetric and transitive is called a partial ordering of (F, A). The triple (F, A, R) is called a partially ordered soft set.

Example 8 Consider a soft set (F, A) given by $U = \{2, 4, 6, 9, 10, 16\}$ and $A = \{2, 3, 8\}$ with $F(2) = \{2, 4, 6, 10, 16\}$, $F(3) = \{6, 9\}$, $F(8) = \{16\}$. Define a relation R on (F, A) as $F(a)RF(b)$ iff a divides b. Then $R = \{F(2) \times F(2), F(3) \times F(3), F(8) \times F(8), F(2) \times F(8)\}$ is a partial ordering on (F, A) . We denote a partial ordering on soft set (F, A) as $\subseteq$.

Definition 1.42 A binary soft set relation R on (F, A) is asymmetric if for every $F(a), F(b)$ in (F, A), $F(a)RF(b)$ implies that $F(b)RF(a)$ does not hold. That is $F(a)RF(b)$ and $F(b)RF(a)$ can never both true simultaneously.

Definition 1.43 A binary soft set relation R on (F, A) is called a strict ordering if it is asymmetric and transitivc.

The following theorem gives relationships between orderings and strict orderings.

Theorem 1.23 *(a) Let R be an ordering of (F, A). Then the soft set relation S on (F, A) defined by $F(a)SF(b)$ iff $F(a)RF(b)$ and $F(a) \neq F(b)$ is a strict ordering of (F, A).*
(b) Let S be a strict ordering of (F, A). Then the soft set relation R defined by $F(a)RF(b)$ iff $F(a)SF(b)$ or $F(a) = F(b)$ is an ordering of (F, A).

Proof (a) To show that S is asymmetric, assume that both $F(a)SF(b)$ and $F(b)SF(a)$ for some $F(a), F(b)$ in (F, A). Then we have $F(a)RF(b)$ and $F(b)RF(a)$ by definition and $F(a) = F(b)$ since R is antisymmetric. This contradicts the definition of $F(a)SF(b)$. To prove the transitivity of S, suppose that $F(a)SF(b)$ and $F(b)SF(c)$. Then we have $F(a)RF(b)$ and $F(b)RF(c)$. So we have $F(a)RF(c)$ by transitivity of R. Thus we have $F(a)SF(c)$.

(b) Proof is similar and is omitted. □

Definition 1.44 Let $\subseteq$ be an ordering of (F, A) and $F(a)$ and $F(b)$ be any two elements in (F, A). We say that $F(a)$ and $F(b)$ are comparable in the ordering if $F(a) \subseteq F(b)$ or $F(b) \subseteq F(a)$. We say that $F(a)$ and $F(b)$ incomparable if they are not comparable.

Example 9 Consider a soft set (F, A) given by $U = \{p_1, p_2, p_3, p_4, p_5 p_6, p_7, p_8\}$ and $A = \{\text{young, smart, weak}\} = \{y, s, w\}$. (F, A) denotes people of different types. $F(y) = \{p_1, p_4, p_7, p_8\}$, $F(s) = \{p_1, p_7\}$, $F(w) = \{p_4, p_5, p_7\}$. Define a relation R on (F, A) as $F(a)RF(b)$ iff $F(a) \cap F(b) \neq \phi$. Then $R = \{F(y) \times F(s), F(y) \times F(w), F(s) \times F(w), F(s) \times F(y), F(w) \times F(y), F(w) \times F(s), F(y) \times F(y), F(w) \times F(w), F(s) \times F(s)\}$ Here every element in (F, A) is comparable.

Definition 1.45 Let R be a partial ordering on soft set (F, A). Then R is called a total ordering on (F, A) if every element in (F, A) is comparable in the ordering R.

Definition 1.46 Let (F, A, R) be a partially ordered soft set. Then

(a) $F(a)$ is the least element of (F, A) if $F(a)RF(x)$ for every $F(x)$ in (F, A).
(b) $F(a)$ is a minimal element of (F, A) if there exists no $F(x)$ such that $F(x)RF(a)$ and $F(x) \neq F(a)$.
(c) $F(a)$ is the greatest element of (F, A) if $F(x)RF(a)$ for every $F(x)$ in (F, A).
(d) $F(a)$ is a maximal element of (F, A) if there exists no $F(x)$ such that $F(a)RF(x)$ and $F(x) \neq F(a)$.

Theorem 1.24 *Let R be a reflexive and antisymmetric relation on (F, A). Then the following are equivalent.*

(i) R is a total order on (F, A)
(ii) R and its complimentary soft set relation R^c are both transitive.

Proof $(i) \implies (ii)$ Clearly R is transitive. Let $F(a)$, $F(b)$, $F(c)$ be in (F, A). Let $F(a)R^cF(b)$ and $F(b)R^cF(c)$. Then neither $F(a)RF(b)$ nor $F(b)RF(c)$ would hold. Therefore $F(a)$ is not R related to $F(c)$. Thus $F(a)R^cF(c)$ and R^c is transitive. $(ii) \implies (i)$ Suppose R and its complimentary soft set relation R^c are both transitive. If $F(a)$ and $F(b)$ are distinct elements of (F, A) then either $F(a)RF(b)$ or $F(b)RF(a)$ must hold. Otherwise we would have $F(a)R^cF(b)$ and $F(b)R^cF(a)$ Hence $F(a)R^cF(a)$ (since R^c is transitive). But this contradicts $F(a)RF(a)$. So R is a total order. □

Theorem 1.25 *Let (F, A) be a soft set defined on universal set U and R be an ordering on A. Then the induced relation R_A is an ordering on (F, A). If (A, R) is a lattice then (F, A) is also a lattice with meet $\sqcap$ and join $\sqcup$ defined as $F(a) \sqcap F(b) = F(a \wedge b)$ and $F(a) \sqcup F(b) = F(a \vee b)$ where $\wedge$ and $\vee$ are the corresponding meet and join on (A, R).*

Proof By definition, $F(a)R_AF(b) \iff aRb$. Clearly $F(a)R_AF(a)$ as aRa. So R_A is reflexive. If $F(a)R_AF(b)$ and $F(b)R_AF(a)$ .Then aRb and bRa, since R is anti symmetric, $a = b$ and so $F(a) = F(a)$. Thus R_A is anti symmetric. If

$F(a)R_AF(b)$ and $F(b)R_AF(c)$. Then aRb and bRc, since R is transitive, aRc and so $F(a)R_AF(c)$. So R_A is transitive. Hence R_A is an ordering on (F, A). Suppose (A, R) is a lattice with meet and join represented by $\wedge$ and $\vee$ respectively. Now we define the corresponding meet and join of any two elements $F(a)$ and $F(b)$ of soft set as $F(a) \sqcap F(b) = F(a \wedge b)$ and $F(a) \sqcup F(b) = F(a \vee b)$. Hence (F, A) is a lattice. □

1.4 Soft Mappings

In this section we discuss mappings between families of soft sets with different parameter and universe sets. This will form fundamental in further studies related to algebraic a well as topological structures defined on soft sets. A soft mapping basically consist of two ordinary mappings, one defined on universal sets and other on parameter sets. Results given here are taken from Kharal and Ahmad [65] , Aygunoglu and Aygun [14], Zorlutuna et al. [168] and Thomas [136].

Definition 1.47 Let $S(U)_E$ and $S(V)_K$ be the families of all soft sets over U and V with parameter sets E and K respectively. Let $\psi : U \to V$ and $\chi : E \to K$ be two mappings. Then the soft mapping $\psi_\chi : S(U)_E \to S(V)_K$ is defined as follows:

(i) Let (F, A) be a soft set in $S(U)_E$. The image of (F, A) under the soft mapping ψ_χ is the soft set over V, denoted by $\psi_\chi(F, A)$ and is defined by, for all $k \in K$,

$$\psi_\chi(F, A)(k) = \begin{cases} (\cup_{e \in \chi^{-1}(k) \cap A}) \psi(F(e)), & \chi^{-1}(k) \cap A \neq \phi \\ \phi \text{ otherwise} \end{cases}.$$

(ii) Let (G, B) be a soft set in $S(V)_K$. The inverse image of (G, B) under the soft mapping ψ_χ is the soft set over U, denoted by $(\psi_\chi)^{-1}(G, B)$ and is defined by,

$$\text{for all } e \in E,\ (\psi_\chi)^{-1}(G, B)(e) = \begin{cases} \psi^{-1}(G(\chi(e))), & \chi(e) \in B \\ \phi \text{ otherwise} \end{cases}$$

The soft mapping ψ_χ is called soft injective if ψ and χ are both injective. The soft mapping ψ_χ is called soft surjective if ψ and χ are both surjective. The soft mapping ψ_χ is called soft bijective iff ψ_χ is soft injective and soft surjective. A soft mapping from $S(U)_E$ to itself may be denoted by $\psi : S(U)_E \to S(U)_E$.

Definition 1.48 Let $\psi : U \to V, \tau : V \to W, \chi : E \to K$ and $\sigma : K \to L$ be mappings. Let $\psi_\chi : S(U)_E \to S(V)_K$ and $\tau_\sigma : S(V)_K \to S(W)_L$, then the soft composition of the soft mappings ψ_χ and τ_σ, denoted by $\tau_\sigma \circ \psi_\chi$, is defined by $\tau_\sigma \circ \varphi_\chi = (\tau \circ \psi)_{(\sigma \circ x)}$.

Theorem 1.26 *If $(F, A), (F, B), (F, A_i) \in S(U)_E$ and $(G, A), (G, B), (G, B_i) \in S(V)_K$ Then for soft mappings $\psi_\chi : S(U)_E \to S(V)_K$,$\tau_\sigma : S(V)_K \to S(W)_L$, the following statements are true:*

(i) *If $(F, B) \subseteq (F, A)$, then $\psi_\chi(F, B) \subseteq \psi_\chi(F, A)$.*

(ii) $\psi_\chi(\widetilde{\Phi}_E) = \widetilde{\Phi}_K$.
(iii) $\psi_\chi(\widetilde{U}_E) = \widetilde{U}_K$ *, if* ψ_χ *is soft surjective.*
(iv) $\psi_\chi(F, A)^c = (\psi_\chi(F, A))^c$*, if* ψ_χ *is soft bijective.*
(v) $\psi_\chi((F, A_1)\widetilde{\cup}(F, A_2)) = (\psi_\chi(F, A_1))\widetilde{\cup}(\psi_\chi(F, A_2))$.
In general, $\psi_\chi(\widetilde{\cup}_{i\in J}(F, A_i) = \widetilde{\cup}_{i\in J}(\psi_\chi(F, A_i)$.
(vi) $\psi_\chi((F, A_1)\widetilde{\cap}(F, A_2)) \subseteq (\psi_\chi(F, A_1))\widetilde{\cap}(\psi_\chi(F, A_2))$.
In general $\psi_\chi(\widetilde{\cap}_{i\in J}(F, A_i) \subseteq \widetilde{\cap}_{i\in J}(\psi_\chi(F, A_i))$*,*
equality holds if ψ_χ *is soft injective.*
(vii) $(F, A) \subseteq \psi_\chi^{-1}(\psi_\chi(F, A))$*, equality holds if* ψ_χ *is soft injective.*
(viii) $\psi_\chi(\psi_\chi^{-1}(F, A)) \subseteq (F, A)$*, equality holds if* ψ_χ *is soft surjective.*
(ix) If $(G, B) \subseteq (G, A)$*, then* $\psi_\chi^{-1}(G, B) \subseteq \psi_\chi^{-1}(G, A)$.
(x) $\psi_\chi^{-1}(\widetilde{\Phi}_K) = \widetilde{\Phi}_E$
(xi) $\psi_\chi^{-1}(G, B)^c = (\psi_\chi^{-1}(G, B))^c$
(xii) $\psi_\chi^{-1}((G, B_1)\widetilde{\cup}(G, B_2)) = (\psi_\chi^{-1}(G, B_1))\widetilde{\cup}(\psi_\chi^{-1}(G, B_2))$.
In general, $\psi_\chi^{-1}(\widetilde{\cup}_{i\in J}(G, B_i) = \widetilde{\cup}_{i\in J}\psi_\chi^{-1}(G, B_i)$.
(xiii) $\psi_\chi^{-1}((G, B_1)\widetilde{\cap}(G, B_2)) = (\psi_\chi^{-1}(G, B_1))\widetilde{\cap}(\psi_\chi^{-1}(G, B_2))$.
In general, $\psi_\chi^{-1}(\widetilde{\cap}_{i\in J}(G, B_i) = \widetilde{\cap}_{i\in J}\psi_\chi^{-1}(G, B_i)$.
(xiv) $(\tau_\sigma o\psi_\chi)^{-1} = (\psi_\chi)^{-1}o(\tau_\sigma)^{-1}$.

Proof We prove (v) and (vi) only. Other proofs follow directly from corresponding definitions.

(v) For $\beta \in K$, we will show that $\psi_\chi((F, A)\widetilde{\cup}(G, B))\beta = (\psi_\chi(F, A)\widetilde{\cup}\psi_\chi(G, B))\beta$.

Consider $\psi_\chi((F, A)\widetilde{\cup}(G, B))\beta = \psi_\chi(H, A \cup B)\beta(\text{say}) \;= \psi\left(\bigcup_{\alpha\in\chi^{-1}(\beta)\cap(A\cup B)} H(\alpha)\right)$ where

$$H(\alpha) = \begin{cases} F(\alpha), & \alpha \in A - B \\ G(\alpha), & \alpha \in B - A \\ F(\alpha) \cup G(\alpha), & \alpha \in A \cap B \end{cases}$$

Then $\psi_\chi((F, A)\widetilde{\cup}(G, B))\beta = \psi\left(\bigcup\begin{cases} F(\alpha), & \alpha \in (A - B) \cap \chi^{-1}(\beta) \\ G(\alpha), & \alpha \in (B - A) \cap \chi^{-1}(\beta) \\ F(\alpha) \cup G(\alpha), & \alpha \in (A \cap B) \cap \chi^{-1}(\beta) \end{cases}\right)$

Next, for $\beta \in K$, we have $(\psi_\chi(F, A)\widetilde{\cup}\psi_\chi(G, B))\beta = \psi_\chi(F, A)\beta \cup \psi_\chi(G, B)\beta$ $= \psi\left(\bigcup_{\alpha\in\chi^{-1}(\beta)\cap A} F(\alpha)\right) \cup \psi\left(\bigcup_{\alpha\in\chi^{1}(\beta)\cap B} G(\alpha)\right) = \psi\left(\bigcup_{\alpha\in\chi^{-1}(\beta)\cap A} F(\alpha)\right) \cup \left(\bigcup_{\alpha\in\chi^{-1}(\beta)\cap B} G(\alpha)\right)$

$$= \psi\left(\bigcup\begin{cases} F(\alpha), & \alpha \in (A - B) \cap \chi^{-1}(\beta) \\ G(\alpha), & \alpha \in (B - A) \cap \chi^{-1}(\beta) \\ F(\alpha) \cup G(\alpha), & \alpha \in (A \cap B) \cap \chi^{-1}(\beta) \end{cases}\right)$$

Thus proof is complete.

(vi) For $\beta \in K$, we show that $\psi_\chi((F, A)\widetilde{\cap}(G, B))\beta \subseteq (\psi_\chi(F, A)\widetilde{\cap}\psi_\chi(G, B))\beta$.
Consider $\psi_\chi((F, A)\widetilde{\cup}(G, B))\beta = \psi_\chi(H, A \cap B)\beta = \psi\left(\bigcup_{\alpha\in\chi^{-1}(\beta)\cap(A\cap B)} H(\alpha)\right)$ where $H(\alpha) = F(\alpha) \cap G(\alpha)$.

Thus $\psi_\chi(H, A \cap B)\beta = \psi\left(\bigcup_{\alpha\in\chi^{-1}(\beta)\cap(A\cap B)} H(\alpha)\right) = \psi\left(\bigcup_{\alpha\in\chi^{-1}(\beta)\cap(A\cap B)}(F(\alpha) \cap G(\alpha))\right)$ or $\psi_\chi((F, A)\widetilde{\cap}(G, B))\beta = \psi\left(\bigcup_{\alpha\in\chi^{-1}(\beta)\cap(A\cap B)}(F(\alpha) \cap G(\alpha))\right)$ On the other hand, we have $(\psi_\chi((F, A))\widetilde{\cap}\phi_\chi(G, B))\beta = \psi_\chi(F, A)\beta \cap \psi_\chi(G, B)\beta =$

$\left(\psi\left(\bigcup_{\alpha\in\chi^{-1}(\beta)\cap A}F(\alpha)\right)\right)\cap\psi\left(\bigcup_{\alpha\in\chi^{-1}(\beta)\cap B}G(\alpha)\right)$ we get $(\psi_\chi(F,A)\widetilde{\cap}\psi_\chi(G,B))$ $\beta=\psi\left(\bigcup_{\alpha\in\chi^{-1}(\beta)\cap A}F(\alpha)\right)\cap\psi\left(\bigcup_{\alpha\in\chi^{-1}(\beta)\cap B}G(\alpha)\right)\supseteq\psi\left(\bigcup_{\alpha\in\chi^{-1}(\beta)\cap(A\cap B)}(F(\alpha)\right.$ $\left.\cap G(\alpha))\right)=\psi_\chi((F,A)\widetilde{\cap}(G,B))\beta$. This completes the proof. □

1.5 Distance and Similarity Measures

The extent to which different sets, objects, signals and patterns are alike or similar is quantified by similarity measures. This measure is extensively used in applications of various generalized set structures like fuzzy sets [160], intuitionistic fuzzy sets [11], vague sets [44] etc. The applications involve problems of pattern recognition, signal detection, medical diagnosis and security verification systems. The idea of similarity measures for fuzzy sets, intuitionistic fuzzy sets and vague sets were studied by many authors [30, 31, 71].

Similarity measure of soft sets was introduced by Majumdar and Samanta [81]. They used matrix/tabular representation-based distances of soft sets to introduce similarity measures. Similarity measure between two fuzzy sets has been defined by many authors. Further, there are several techniques for defining similarity measures. Some based on distances and some others are based on matching functions. There are techniques based on set-theoretic approach also. Results, theorems and examples given in this section are taken from [64, 81, 87].

1.5.1 Similarity Measure of Two Soft Sets

Let U be the universe and E be the parameter set, where both are non-empty. Clearly we can express a soft set over U as a matrix using the tabular representation. Each column of the membership matrix will be represented by the vector $\overrightarrow{F}(e_i)$. As the matching function-based similarity measure has some advantages over the distance-based similarity measure, Majumdar and Samantha [81] gives the matching function-based similarity as

Definition 1.49 If $E_1=E_2$, then similarity between (F_1,E_1) and (F_2,E_2) is defined by $S(F_1,F_2)=\frac{\sum_i\overrightarrow{F_1}(e_i)\bullet\overrightarrow{F_2}(e_i)}{\sum_i\left[\overrightarrow{F_1}(e_i)^2\vee\overrightarrow{F_2}(e_i)^2\right]}$, where $\overrightarrow{F_1}(e_i)$ represents the ith column vector corresponding to the tabular/matrix representation of the soft set (F_1,E_1). If $E_1\neq E_2$ and $E=E_1\cap E_2\neq\phi$, then we first define $\overrightarrow{F}_1(e)=\underline{0}$ for $e\in E_2\backslash E$ and $\overrightarrow{F_2}(f)=\underline{0}$ for $f\in E_1\backslash E$. Then $S(F_1,F_2)$ is defined by the formula for $S(F_1,F_2)$ given above.

We observe the following:

(i) If $E_1\cap E_2=\phi$, then $S(F_1,F_2)=0$
(ii) $S\left(F_1,F_1^r\right)=0$ as $\overrightarrow{F_1}(e_i)\cdot\overrightarrow{F}_1^r(e_i)=0\forall i$

Example 10 Consider two soft sets (F, E) and (G, E) over U, where $U = \{x_1, x_2, x_3\}$, $E = \{e_1, e_2, e_3\}$. Let $A = \begin{pmatrix} 1\,1\,0 \\ 0\,1\,0 \\ 0\,0\,1 \end{pmatrix}$ and $B = \begin{pmatrix} 0\,1\,1 \\ 1\,1\,0 \\ 0\,1\,1 \end{pmatrix}$ be their representing matrices. $S(F, G) = \frac{\sum_{i=1}^{3} \mathbf{F}(e_i)\cdot\mathbf{G}(e_i)}{\sum_{i=1}^{3} [\mathbf{F}(e_i)^2 \vee \mathbf{G}(e_i)^2]} = \frac{3}{6} = 0.5$

Lemma 1.4 *Let (F_1, E_1) and (F_2, E_2) be two soft sets over the same finite universe U. Then the following hold: (i) $S(F_1, F_2) = S(F_2, F_1)$; (ii) $0 \le S(F_1, F_2) \le 1$; and (iii) $S(F_1, F_1) = 1$ provided both soft sets are non null.*

Proof Trivially follows from the definition. □

Theorem 1.27 *Let $(F, E), (G, E), (H, E)$ be three soft sets such that (F, E) is a soft subset of (G, E) and (G, E) is a soft subset of (H, E) then , $S(F, H) \le S(G, H)$*

Proof The proof is straightforward. □

Definition 1.50 Two soft sets (F_1, E_1) and (F_2, E_2) in $S(U)_E$ are said to be α-similar, denoted as $(F_1, E_1) \overset{\alpha}{\approx} (F_2, E_2)$, iff $S(F_1, F_2) \ge \alpha$ for $\alpha \in (0, 1)$.

clearly we have

Lemma 1.5 $\overset{\alpha}{\approx}$ *is reflexive and symmetric, but not transitive.*

To show that $\overset{\alpha}{\approx}$ is not transitive we give an example.

Example 11 Let $U = \{x_1, x_2, x_3\}$ be the universe and $E = \{e_1, e_2, e_3\}$ be the set of parameters. Let $\alpha = \frac{1}{2}$. We define three soft sets (F_1, E_1), (F_2, E_2) and (F_3, E_3) such that $E_1 = E_2 = E_3 = E$ with A, B, C as membership matrices as follows:

$$A = \begin{pmatrix} 1\,0\,0 \\ 0\,1\,0 \\ 0\,0\,1 \end{pmatrix}, \quad B = \begin{pmatrix} 1\,0\,0 \\ 0\,1\,0 \\ 0\,0\,0 \end{pmatrix} \quad \text{and} \quad C = \begin{pmatrix} 1\,0\,0 \\ 0\,0\,0 \\ 0\,0\,0 \end{pmatrix}.$$

Then $S(F_1, F_2) = \frac{2}{3} > \frac{1}{2}$, $S(F_2, F_3) = \frac{1}{2}$ but $S(F_1, F_3) = \frac{1}{3} < \frac{1}{2}$.

1.5.2 Distances Between Soft Sets

Here we take our universe U to be finite, namely $U = \{x_1, x_2, \ldots, x_n\}$ and parameter set $E = \{e_1, e_2, \ldots, e_m\}$. Now for any soft set $(F, A) \in S(U)_E$, A is a subset of E. Next we extend the soft set (F, A) to the soft set $(\hat{F}, E)$ where $\hat{F}(e) = \phi, \forall e \notin A$, i.e., $\hat{F}(e)\left(x_j\right) = 0 \forall j = 1, 2, \ldots, n \forall e \notin A$.Thus without loss of generality we can take the parameter subset of any soft set over $S(U)_E$ to be the same as the parameter set E. Now we define the following distances between two soft sets $\left(\hat{F}_1, E\right)$ and $\left(\hat{F}_2, E\right)$ as follows:

Definition 1.51 For two soft sets $\left(\hat{F}_1, E\right)$ and $\left(\hat{F}_2, E\right)$ we define:

(a) the mean Hamming distance $D^S\left(\hat{F}_1, \hat{F}_2\right)$ between two soft sets as: $D^s\left(\hat{F}_1, \hat{F}_2\right) = \frac{1}{m}\left\{\sum_{i=1}^m \sum_{j=1}^n \left|\hat{F}_1(e_i)(x_j) - \hat{F}_2(e_i)(x_j)\right|\right\}$

(b) the normalized Hamming distance $L^s(F_1, F_2)$ as: $L^S\left(\hat{F}_1, \hat{F}_2\right) = \frac{1}{m.n}\left\{\sum_{i=1}^m \sum_{j=1}^n \left|\hat{F}_1(e_i)(x_j) - \hat{F}_2(e_i)(x_j)\right|\right\}$

(c) the Euclidean distance $E^s\left(\hat{F}_1, \hat{F}_2\right)$ as:

$$E^s\left(\hat{F}_1, \hat{F}_2\right) = \sqrt{\frac{1}{m}\sum_{i=1}^m \sum_{j=1}^n \left(\hat{F}_1(e_i)(x_j) - \hat{F}_2(e_i)(x_j)\right)^2}$$

(d) the normalized Euclidean distance $Q^s\left(\hat{F}_1, \hat{F}_2\right)$ as:

$$Q^s\left(\hat{F}_1, \hat{F}_2\right) = \sqrt{\frac{1}{m \cdot n}\sum_{i=1}^m \sum_{j=1}^n \left(\hat{F}_1(e_i)(x_j) - \hat{F}_2(e_i)(x_j)\right)^2}.$$

Example 12 Let $U = \{x_1, x_2, x_3\}$ and $E = \{e_1, e_2, e_3\}$. Let (F_1, E) and (F_2, E) be two soft sets defined below by their membership matrices

(F_1, E)	$F_1(e_1)$	$F_1(e_2)$	$F_1(e_3)$
x_1	1	1	0
x_2	0	1	0
x_3	0	0	1

and

(F_2, E)	$F_2(e_1)$	$F_2(e_2)$	$F_2(e_3)$
x_1	0	1	1
x_2	1	1	0
x_3	0	1	1

$$D^S(F_1, F_2) = \frac{4}{3}.$$

$$L^s(F_1, F_2) = \frac{4}{9}.$$

Now,

$$E^s(F_1, F_2) = \sqrt{\frac{4}{3}}.$$

$$Q^s(F_1, F_2) = \sqrt{\frac{4}{9}} = \frac{2}{3}.$$

Theorem 1.28 *Let $m = |E|$, $n = |U|$. Then for any two soft sets (F_1, E) and (F_2, E) we have*

(1) $D^s(F_1, F_2) \in \left\{\frac{k}{m} | k = 0, 1, 2, \ldots, mn\right\}$
(2) $L^s(F_1, F_2) \in \left\{\frac{k}{mn} | k = 0, 1, 2, \ldots, mn\right\}$
(3) $E^s(F_1, F_2) \in \left\{\sqrt{\frac{k}{m}} | k = 0, 1, 2, \ldots, mn\right\}$
(4) $Q^s(F_1, F_2) \in \left\{\sqrt{\frac{k}{mn}} | k = 0, 1, 2, \ldots, mn\right\}$

Proof (1) The smallest and the largest distances are given as

$$D^s(\widetilde{\Phi}_E, \widetilde{\Phi}_E) = \frac{1}{m}\left\{\sum_{i=1}^m \sum_{j=1}^n \left|\widetilde{\Phi}_E(e_i)(x_j) - \widetilde{\Phi}_E(e_i)(x_j)\right|\right\} = 0$$

$$D^s(\widetilde{\Phi}_E, \widetilde{U}_E) = \frac{1}{m}\left\{\sum_{i=1}^m \sum_{j=1}^n \left|\widetilde{\Phi}_E(e_i)(x_j) - \widetilde{U}_E(e_i)(x_j)\right|\right\}$$

$$= \frac{1}{m}\left\{\sum_{i=1}^{m}\sum_{j=1}^{n}|0-1|\right\}, \quad \text{since } \widetilde{\Phi}_E(e_i)(x_j) = 0 \text{ and } \widetilde{U}_E(e_i)(x_j) = 1 \forall i, j$$

$$= \frac{1}{m}\underbrace{(|0-1| + |0-1| + \cdots + |0-1|)}_{mn \text{ times}} = \frac{mn}{m} = n$$

Furthermore, suppose the arrangement of entries in matrix representation of two arbitrary soft sets (F_1, E) and (F_2, E) is such that the term $\left|F_1(e_i)(x_j) - F_2(e_i)(x_j)\right|$ evaluates to 1, k times. Then, we can re-arrange the terms in expansion of $D^s(F_1, F_2)$ to get $D^s(F_1, F_2) = \frac{1}{m}((\underbrace{|0-1| + |1-0| + |0-1| + \cdots + |1-0|}_{k \text{ times}})$

$$+ \underbrace{(|0-0| + |1-1| + |0-0| + \cdots + |1-1|)}_{mn-k \text{ tines}}) = \frac{1}{m}(k+0) = \frac{k}{m}$$

Thus it follows that $D^s(F_1, F_2) \in \left\{\frac{k}{m} | k = 0, 1, 2, \ldots, mn\right\}$
(2) Note that $L^s(F_1, F_2) = \frac{1}{n}D^s(F_1, F_2)$. The result now follows from (1).
(3), (4) Follow immediately from the fact that the quantity $\left|F_1(e_i)(x_j) - F_2(e_i)(x_j)\right|$ is either 0 or 1, only and from (1) and (2). □

Corollary 1.2 *Let $m = |E|$, $n = |U|$. Then for any two soft sets (F_1, E) and (F_2, E) we have*
(1) $D^s(F_1, F_2) \le n$
(2) $L^s(F_1, F_2) \le 1$
(3) $E^s(F_1, F_2) \le \sqrt{n}$
(4) $Q^s(F_1, F_2) \le 1$

From the definitions of the four distances defined for soft sets one can easily verify the following:

Theorem 1.29 *All the functions $D^S, L^s, E^s, Q^s : S(U)_E \to R^+$ defined are metrics, where R^+ is the set of all non-negative real numbers.*

1.5.3 Distance Based Similarity Measure of Soft Sets

Based on the distances defined in Sect. 1.5.2, the following similarity measures for soft sets (F_1, E_1) and (F_2, E_2) defined over the same finite universe U can be defined:

Definition 1.52 (1) Based on Hamming distance, a similarity measure can be defined as $S'(F_1, F_2) = \frac{1}{1+D^S(F_1,F_2)}$

(2) Following Williams and Steele [150], one can define a similarity measure as $S'(F_1, F_2) = e^{-\alpha . D^S(F_1,F_2)}$ where α is a positive real number (parameter) called the steepness measure.

(3) Using Euclidean distance, similarity measure can be defined as $S'(F_1, F_2) = \frac{1}{1+E^S(F_1,F_2)}$

(4) Following Williams and Steele [150], another similarity measure is defined as: $S'(F_1, F_2) = e^{-\alpha.E^s(F_1,F_2)}$ where α is a positive real number (parameter) called the steepness measure.

It is clear that for distance-based similarity measures, the similarity between a soft set and its complement is never zero.

Lemma 1.6 *For any two soft sets* (F_1, E) *and* (F_2, E), *the following holds: (1)* $0 \leq S'(F_1, F_2) \leq 1$ *(2)* $S'(F_1, F_2) = S'(F_2, F_1)$ *(3)* $S'(F_1, F_2) = 1 \Leftrightarrow (F_1, E_1) = (F_2, E_2)$

Proof Follows from definition clearly. □

The following Table compares the properties of the two measures of similarity of soft sets discussed above. Here $X_{A,B}$ denote the similarity measure between two soft sets (F_1, E_1) and (F_2, E_2) whose membership matrices are A and B respectively (Table 1.2).

Table 1.2 Comparison of similarity measures

Property	S	S'
$X_{A,B} = X_{B,A}$	Y	Y
$0 \leq X_{A,B} \leq 1$	Y	Y
$A = B \Rightarrow X_{A,B} = 1$	Y	Y
$X_{A,B} = 1 \Rightarrow A = B$	N	Y
$E_1 \cap E_2 = \phi \Rightarrow X_{A,B} = 0$	Y	N
$X_{A,A^C} = 0$	Y	N

Y = Yes, N = No

1.6 Softness of Soft Sets

The study of uncertainty measures is an important topic for the theories to deal with uncertainty. Majumdar and Samanta [81, 82] initiated the study of uncertainty measures of soft sets. Majumdar [83] pointed out that the associated uncertainty of a soft set arises from the parameterized classification of objects in the universe. The entropy, i.e., the associated uncertainty, of a soft set should be maximum if either the objects cannot be classified at all w.r.t the parameters or every object of the universe can be classified by every parameter. Again the entropy of a soft set is minimum if each object can be classified just by one parameter. Thus, the entropies of null soft set and absolute soft set are maximum, whereas the entropy of deterministic soft set is minimum. Furthermore, for a superset of a soft set the uncertainty ultimately increases in comparison with its subset, as new objects being introduced in the set which shares same parameters with other objects. Based on this observation, Majumdar [83] proposed the notion of equivalent soft sets and the axiomatic definition of soft set entropy.

Entropy as a measure of fuzziness was first mentioned by Zadeh [161] in 1965. Later De Luca-Termini [75] axiomatized the non-probabilistic entropy. According to them the entropy E of a fuzzy set A need to satisfy certain axioms. In a similar line entropy of a soft is defined by Majumdar [83]. For that purpose he give two new definitions, namely deterministic soft set and equivalent soft sets. The definitions, results and examples given in this section are taken from [83, 110].

Definition 1.53 A soft set (F, A) is said to be a deterministic soft set over U if the following holds:
(i) $\cup_{e\in A} F(e) = U$
(ii) $F(e) \cap F(f) = \phi$, where $e, f \in A$

Definition 1.54 Let (F, A) be any soft set. Then another soft set (F^*, A) is said to be equivalent to (F, A) if there exists a bijective mapping σ from A to A defined as: $\sigma(F_x) = F_x^*$, where $F_x = \{e : x \in F(e)\}$ and $F_x^* = \{e : x \in F^*(e)\}$.

We denote the collection of all soft sets which are equivalent with (F, A) as $C(F)$.

Example 13 Let the universe and parameter set be $U = \{x_1, x_2, x_3, x_4\}$ $\&A = \{e_1, e_2, e_3\}$ respectively. Then let us consider the following soft set (F, A) as follows: $F(e_1) = \{x_1, x_2\}, F(e_2) = \{x_2, x_3\}, F(e_3) = \{x_1, x_4\}$ $F(e_1) = \{x_1, x_2\}, F(e_2) = \{x_2, x_3\}, F(e_3) = \{x_1, x_4\}$ Then the following soft set (G, A) is equivalent to the soft set (F, A) where $G(e_1) = \{x_1, x_4\}, G(e_2) = \{x_1, x_2\}, G(e_3) = \{x_2, x_3\}$. This is because there is a bijective mapping σ on A such that $\sigma(e_1) = e_2, \sigma(e_2) = e_3 \& \sigma(e_3) = e_1$ and $\sigma(F_x) = G_x \forall x \in U$. The softness of a soft set and its equivalent soft sets need to be same because the amount of imperfectness or ambiguity of information is same in both cases. Based on these we have the following.

Definition 1.55 A mapping $S : S(U)_E \to [0, 1]$ is said to be soft set entropy or softness measure if S satisfies the following properties:
(S1) $S(\widetilde{\Phi}_E) = 1, S(\widetilde{U}_E) = 1$
(S2) $S(F, A) = 0$, if (F, A) is deterministic soft set
(S3) $S(F, A) \leq S(G, A)$ if $(F, A)(\neq \widetilde{\Phi}_A) \subseteq (G, A)$
(S4) $S(F^*, A) = S(F, A)$, where $(F^*, A) \in C(F)$

According to this definition an ordinary set has softness zero, as an ordinary set can be thought of as a soft set with a single parameter and thus is a deterministic soft set. For simplicity, we denote $S(F, A)$ by $S(F)$

Theorem 1.30 *The function* $S : S(U)_E \to [0, 1]$ *defined as*
$$S(F) = 1 - \frac{|U|}{\sum_{x\in U} |\{e : x \in F(e)\}|}, \text{ if } F \neq \tilde{\Phi}_A \text{ or } \tilde{U}_A$$
$$= 1, \text{ if } F = \widetilde{\Phi}_A \text{ or } \widetilde{U}_A$$
is an entropy (or measure of softness) of a soft set.

Proof Here (S1) holds obviously from construction.
(S2) For a deterministic soft set $\sum_{x\in U}|\{e : x \in F(e)| = |U|$ because each element is attached with exactly one parameter$\Rightarrow |\{e : x \in F(e)| = 1\forall x \in U$ Hence $S(F) = 0$ Thus (S2) holds.
Next let $F\&G$ be two soft sets such that $F(\neq \tilde{\Phi}_A) \subseteq G \Rightarrow \forall e \in E, F(e) \subseteq G(e) \Rightarrow \{e : x \in F(e)\} \subseteq \{e : x \in G(e)\}.\therefore \sum_{x\in U}|\{e : x \in F(e)\}| \leq \sum_{x\in U}|\{e : x \in G(e)\}|$.
$\Rightarrow \frac{|U|}{\sum_{x\in U}|\{e:x\in F(e)\}|} \geq \frac{|U|}{\sum_{x\in U}|\{e:x\in G(e)\}|} \Rightarrow S(F) \leq S(G)$ Thus (S3) also holds.
(S4) Let $F^* \in C(F)$ Then $\sum_{x\in U}|\{e : x \in F(e)\}| = \sum_{x\in U}|\{e : x \in F^*(e)\}|$. Therefore $S(F) = S(F^*)$. Hence the theorem. □

Example 14 Consider a collection $U = \{h_1, h_2, \ldots, h_5, h_6\}$ of six houses. Some parameters of a good house are collected and expressed as $E = \{e_1 =$cheap,$e_2 =$ well constructed,$e_3 =$ costly,$e_4 =$ in good neighbourhood,$e_5 =$ good location$\}$. Define the soft set $F : E \to P(U)$ as follows: Let $F(e_1) = \{h_1, h_4\}, F(e_2) = \{h_2, h_3, h_5\}, F(e_3) = \{h_2, h_3\}, F(e_4) = \{h_1, h_2, h_3\}, F(e_5) = \{h_4\}$. Here $S(F) = 1 - \frac{|U|}{\sum_{x\in U}|\{ex\in F(e)\}|} = 1 - \frac{6}{11} \approx 0.45$.

The notion of entropy can also be defined using distance between two soft sets. For this we denote the range of the function F in (F, A) by R_T.

Definition 1.56 Let (F, A) be any soft set with the universe U. Then the nearest soft set of (F, A) is the soft set (F_{near}, A) which is a deterministic over soft set over R_T and is obtained from (F, A) by eliminating the least number of elements.

Definition 1.57 Let (F, A) be any soft set with the universe U. Then the farthest soft set of (F, A) is the soft set (F_{far}, A) which is a deterministic over soft set over R_T and is obtained from (F, A) by including additional elements in (F, A) such that $F_{\text{far}}(e) = R_T \forall e \in A$.

Definition 1.58 Let d_1 and d_2 denote the mean hamming distances between a non null non absolute soft set (F, A) from (F_{near}, A) and and (F_{far}, A) respectively.(F, A) is said to be a soft set of Type I if $d_1 \leq d_2$ and (F, A) is said to be a soft set of Type II if $d_1 > d_2$.

Definition 1.59 The distance based softness measure or soft entropy of (F, A) is defined as
$$E(F, A) = \begin{cases} d_1/d_2, & \text{if } (F, A) \neq \tilde{\Phi}_A, (F, A) \neq \tilde{U}_A, d_1 \leq d_2 \\ d_2/d_2, & \text{if } (F, A) \neq \tilde{\Phi}_A, (F, A) \neq \tilde{U}_A, d_2 \leq d_1 \\ 1, & \text{if } (F, A) = \tilde{\Phi}_A \text{ or } (F, A) = \tilde{U}_A \end{cases}$$

It is easily verifiable that all the properties of soft set entropy are satisfied by $E(F, A)$ for Type-I Soft sets.

1.7 Representations of Fuzzy Sets, Rough Sets and Topological Spaces as Soft Sets

Soft sets being a a generalized tool, can be used for representing many of the already existing and well known structures in mathematical modelling like fuzzy sets [160], rough sets [97], topological spaces [69] etc. In this section, we give the basic definitions of these structures and see how each can be identified with a suitable kind of soft set. Molodtsov [88] gives representation of fuzzy sets and topological spaces as soft sets.

1.7.1 Fuzzy Sets as Soft Sets

Definition 1.60 Let X be a nonempty set. A fuzzy set A of X is a mapping $A : X \to [0, 1]$, that is, $A = \{(x, \mu_A(x)) : x \in X\}$ where $\mu_A(x)$ is the membership grade of x in A.

Definition 1.61 α-cut or α-level set of a fuzzy set A is defined as $A_\alpha = \{x : \mu_A(x) \geq \alpha\}$. That is, α-cuts are crisp sets that contain elements of the domain associated with membership grades greater than or equal to a certain level α.

Theorem 1.31 (Decomposition theorem for fuzzy sets [66]) *Let A be a fuzzy set and define a fuzzy set ${}_\alpha A$ for each α-cut of A as ${}_\alpha A = \{(x, \alpha) : x \in A_\alpha\}$, Then $A = \cup_{\alpha \in [0,1]} {}_\alpha A$.*

Molodtsov [88] represented fuzzy sets as a special type of soft sets in the following manner.

Suppose A is a fuzzy set of the universe U, take the parameter set $E = [0, 1]$, and define the mapping $F : E \to P(U)$ as follows: $F(a) = \{x \in U : \mu_A(x) \geq a\}$, $a \in E$. In other words, $F(a)$ is the a-level set of A. According to this manner and by using the decomposition theorem of fuzzy sets given above, every fuzzy set can be represented as a soft set in a unique manner.

1.7.2 Topological Spaces as Soft Sets

Definition 1.62 Let X be a set and $P(X)$ its power set. A collection T in $P(X)$ is called a topology on X if T has the following properties:

(i) $x, \phi \in T$
(ii) Any union of elements of T belongs to T.
(iii) Finite intersection of elements of T belongs to T.

(X, T) is called a topological space. Members of T are called open sets and for any $x \in X$, an open set containing x is called a neighborhood of x.

The representation of topological spaces as soft sets is also given by Molodtsov [88]. For a topological space (X, T), let $F(x)$ denote the collection of all neighborhoods of a point x in X. i.e., define $F : X \to P(P(X))$ as $F(x) = \{V \in T : x \in V\}$. Now the ordered pair (F, X) is clearly a soft set over the universe $U = P(X)$ with parameter set $E = X$.

1.7.3 Rough Sets as Soft Sets

Let U be an initial universe and $R(R \subset U \times U)$ is an equivalence relation on U. In Pawlak's [97] sense, the ordered pair (U, R) is called an approximation space. Similar to the case of fuzzy sets and topological spaces, rough sets can also be put into the framework of soft sets. For each subset A of U, the approximation mapping $\underline{apr}_R$ maps A to the its lower approximation $\underline{apr}_R(A)$ and the approximation mapping $\overline{apr}_R$ maps A to its upper approximation $\overline{apr}_R(A)$, where $\underline{apr}_R(A) = \{x \in U : [X]_R \subseteq A\}$ and $\overline{apr}_R(A) = \{x \in U : [x]_R \cap A \neq \phi\}$ with $[x]_R$ being the equivalence class of x with respect to R. Therefore, the rough set model (U, R) can be seen as two soft sets $(\underline{apr}_R, P(U))$ and $(\overline{apr}_R, P(U))$ over U.

Chapter 2
Algebraic Structures of Soft Sets

The algebraic structure of set theories dealing with uncertainties has been studied by many authors. For example, see Rosenfeld [114] for fuzzy groups and Biswas et al. [22] and others [23, 56] for rough groups. In this chapter we deal with algebraic structures of soft sets particularly soft groups, soft ideals and BCK/BCI algebras, soft rings and semi rings, and soft modules. We also discuss lattice structure of soft sets.

2.1 Soft Groups

In this section we introduce a basic version of soft group theory, which extends the notion of a group to include the algebraic structures of soft sets. The definitions, results and examples discussed in this section are mainly due to [5, 40, 124].

Unlike other versions of generalized group structures, a soft group is a parameterized family of subgroups. Further, fuzzy groups may be considered as a special case of the soft groups. Throughout this section, G is a group and A is any nonempty set. R will refer to an arbitrary binary relation between an element of A and an element of G. A set-valued function $F : A \to P(G)$ can be defined as $F(x) = \{y \in G : (x, y) \in R, x \in A \text{ and } y \in G\}$. The pair (F, A) is then a soft set over G. Defining a set-valued function from A to G also defines a binary relation R on $A \times G$, given by $R = \{(x, y) \in A \times G : y \in F(x)\}$.The triplet (A, G, R) is referred to as an approximation set.

Definition 2.1 Let (F, A) be a soft set over G. Then (F, A) is said to be a soft group over G if and only if $F(x)$ is a subgroup of G ($F(x) < G$) for all $x \in A$.

Example 15 Suppose that $G = A = S_3 = \{e, (12), (13), (23), (123), (122)\}$, and that we define the set-valued function $F(x) = \{y \in G : xRy \iff y = x^n, n \in N\}$.

S. J. John, *Soft Sets*, Studies in Fuzziness and Soft Computing 400, https://doi.org/10.1007/978-3-030-57654-7_2

Table 2.1 Tabular Representation of the soft group in Example 15

$x \downarrow$	y = (e)	(12)	(13)	(23)	(123)	(132)
e	1	0	0	0	0	0
(12)	1	1	0	0	0	0
(13)	1	0	1	0	0	0
(23)	1	0	0	1	0	0
(123)	1	0	0	0	1	1
(132)	1	0	0	0	1	1

Then the soft group (F, A) is a parameterized family $\{F(x) : x \in A\}$ of subsets, which gives us a collection of subgroups of G. Now consider the particular mapping F defined above, which is also a subgroup of G. In this we can view the soft group (F, A) as the collection of subgroups of G given below:

$F(e) = \{e\}, \quad F(12) = \{e, (12)\}, \quad F(13) = \{e, (13)\}, \quad F(23) = \{e, (23)\}$ $F(123) = F(132) = \{e, (123), (132)\}$. An equivalent tabular representation of this soft group can be given as (Table 2.1)

If $y = x^n$ and $n \in N$, then (x, y) is indicated by the number 1. If $y \neq x^n$, then (x, y) is indicated by the number 0. If G_i is a collection y whose (x, y) are all assigned the number i in the table above, then G_i is a subgroup of G. For computing purposes, any soft group can be represented numerically in the form of a table. Rosenfeld [114] defined fuzzy groups as follows:

Definition 2.2 Let X be a group, a fuzzy subset A of X will be called a fuzzy subgroup of X if

(i) $\mu_A(xy) \geqslant \min(\mu_A(x), \mu_A(y))$ for all $x, y \in X$
(ii) $\mu_A\left(x^{-1}\right) \geqslant \mu_A(x)$ for all $x \in X$

Theorem 2.1 *Fuzzy groups are special types of soft groups.*

Proof Let G be a given group and A be a fuzzy subgroup of G with membership μ_A; i.e., μ_A is a mapping of G into $[0, 1]$. Now consider the family of α -level subgroups of G for the function μ_A given by $F(\alpha) = \{x \in G : \mu_A(x) \geqslant \alpha\}, \alpha \in [0, 1]$. If we know the family F, then we can find the functions $\mu_A(x)$ by means of the formula $\mu_A(x) = \sup\{\alpha : x \in F(\alpha)\}$. Thus, each of Rosenfeld's fuzzy groups A is equivalent to the soft group $(F, [0, 1])$. □

The concept of support set is defined for both fuzzy sets and formal power series in the literature. Similar notion can be defined for soft sets also.

Definition 2.3 For a soft set (F, A) the set $\mathrm{Supp}(F, A) = \{x \in A | F(x) \neq \emptyset\}$ is called the support of the soft set (F, A).

Clearly a null soft set is indeed a soft set with an empty support, and we say that a soft set (F, A) is non-null if Supp $(F, A) \neq \emptyset$.

Theorem 2.2 *Let (F, A), (Q, A) and (T, B) be soft groups over G. Then the following hold.*

(a) If it is non-null, then the extended intersection $(F, A) \cap_E (T, B)$ is a soft group over G.

(b) If it is non-null,then the intersection $(F, A)\widetilde{\cap}(Q, A)$ is a soft group over G.

Proof (a) Let $(F, A) \cap_E (T, B) = (H, C)$, where $C = A \cup B$ and

$$H(x) = \begin{cases} F(x) & \text{if } x \in A - B \\ T(x) & \text{if } x \in B - A \text{ for all } x \in C. \\ F(x) \cap T(x) & \text{if } x \in A \cap B \end{cases}$$

Suppose that (H, C) is a non-null soft set over G. Let $x \in \mathrm{Supp}(H, C)$. If $x \in A - B$, then $H(x) = F(x) \neq \emptyset$ is a subgroup of G; if $x \in B - A$, then $H(x) = T(x) \neq \emptyset$ is a subgroup of G; and if $x \in A \cap B$, $H(x) = F(x) \cap T(x) \neq \emptyset$. Thus $\emptyset \neq F(x)$ and $\emptyset \neq T(x)$ are both subgroups of G, and so is their intersection. It follows that (H, C) is a soft group over G.

(b) Let $(F, A)\widetilde{\cap}(Q, A) = (K, A)$, where $K(x) = F(x) \cap Q(x)$ for all $x \in A$. Suppose that (K, A) is a non-null soft set over G. If $x \in \mathrm{Supp}(K, A)$, then $K(x) = F(x) \cap Q(x) \neq \emptyset$. Thus $\emptyset \neq F(x)$ and $\emptyset \neq Q(x)$ are both groups of G. Hence, $K(x)$ is a subgroup of G for all $x \in \mathrm{Supp}(K, A)$. Therefore (K, A) is a soft group over G, as required. □

Theorem 2.3 *Let (F, A) and (H, B) be two soft groups over G. If $A \cap B = \emptyset$,then $(F, A)\widetilde{\cup}(H, B)$ is a soft group over G.*

Proof Since $A \cap B = \emptyset$, it follows that either $x \in A - B$ or $x \in B - A$ for all $x \in C$. Let $(F, A)\widetilde{\cup}(H, B) = (U, C)$. If $x \in A - B$ then $U(x) = F(x) < G$, and if $x \in B - A$ then $U(x) = H(x) < G$. Thus, $(F, A)\widetilde{\cup}(H, B)$ is a soft group over G. □

Theorem 2.4 *Let (F, A) and (H, B) be two soft groups over G. Then $(F, A) \wedge (H, B)$ is soft group over G.*

Proof Let $(F, A) \wedge (H, B) = (U, A \times B)$. Now $U(\alpha, \beta) = F(\alpha) \cap H(\beta)$ for all $(\alpha, \beta) \in A \times B$. As $F(\alpha)$ and $H(\beta)$ are subgroups of G, $F(\alpha) \cap H(\beta)$ is a subgroup of G. Therefore $U(\alpha, \beta)$ is a subgroup of G for all $(\alpha, \beta) \in A \times B$. Hence it follows that $(F, A) \wedge (H, B)$ is a soft group over G. □

Definition 2.4 Let (F, A) be a soft group over G. Then

(1) (F, A) is said to be an identity soft group over G if $F(x) = \{e\}$ for all $x \in A$, where e is the identity element of G ; and

(2) (F, A) is said to be an absolute soft group over G if $F(x) = G$ for all $x \in A$

Theorem 2.5 *(1) Let (F, A) be a soft group over G and f be a homomorphism from G to K. If $F(x) = Kerf$ for all $x \in A$, then $(f(F), A)$ is the identity soft group over K.*

(2) Let (F, A) be an absolute soft group over G, and let f be a homomorphism from G on to K. Then $(f(F), A)$ is an absolute soft group over K.

Proof (1) For $x \in A$, $f(F(x)) = e_K$ where e_K is identity element of K. Now by definition, $(f(F), A)$ is an identity soft group over K

(2) $F(x) = G$ for all $x \in A$, since (F, A) is an absolute soft group over G. It follows that $f(F(x)) = f(G) = K$ for all $x \in A$. Hence, $(f(F), A)$ is an absolute soft group over K. □

Definition 2.5 Let (F, A) and (H, K) be two soft groups over G. Then (H, K) is a soft subgroup of (F, A), written $(H, K)\tilde{<}(F, A)$, if

(1) $K \subset A$
(2) $H(x) < F(x)$ for all $x \in K$

Example 16 Let $G = S_3$, $A = S_3$, and $K = A_3$. If we define the functions $F(x) = \{y \in S_3 : xRy \Longleftrightarrow y = x^n, n \in N\}$ and $H(x) = \{y \in A_3 : xRy \Longleftrightarrow y \in \langle x \rangle\}$, then $(H, K)\tilde{<}(F, A)$ since $A_3 < S_3$ and $H(x) < F(x)$ for all $x \in A_3$.

Many properties of soft subgroups are similar to properties of classical subgroups. The proofs are all straightforward.

Theorem 2.6 *Let (F, A) and (H, A) be two soft groups over G*

(1) If $F(x) \subseteq H(x)$ for all $x \in A$, then (F, A) is a soft subgroup of (H, A)
(2) If $E = \{e\}$ and (U, E), (F, G) are both soft groups over G, then (U, E) is a soft subgroup of (F, G).

Corollary 2.1 *If (F, G) is a soft group over G, then (F, G) and (F, E) are both soft subgroups of (F, G).*

Theorem 2.7 *Let (F, A) be a soft group over G and $(F_i, A_i)_{i \in I}$ be a nonempty family of soft subgroups of (F, A). Then we have the following.*

(1) $\tilde{\bigcap}_{i \in I} (F_i, A_i)$ is a soft subgroup of (F, A), if it is non-null.
(2) $\tilde{\bigwedge}_{i \in I} (F_i, A_i)$ is a soft subgroup of $\tilde{\bigwedge}_{i \in I}(F, A)$, if it is non-null.
(3) If $\{A_i | i \in I\}$ are pairwise disjoint, then $\tilde{\bigcup}_{i \in I} (F_i, A_i)$ is a soft subgroup over (F, A).

Proof (1) Let $\tilde{\bigcap}_{i \in I} (F_i, A_i) = (G, B)$, where $B = \bigcap_{i \in I} A_i \neq \emptyset$ and $G(x) = \bigcap_{i \in I} F_i(x)$ for all $x \in B$. First, we check that $B = \bigcap_{i \in I} A_i$, which is the parameter set of $\tilde{\bigcap}_{i \in I} (F_i, A_i)$, is a subset of A. Suppose that the soft set (G, B) is non-null. If $x \in \text{Supp}(G, B)$, then $G(x) = \bigcap_{i \in I} F_i(x) \neq \emptyset$. It follows that, for all $i \in I$, the nonempty set $F_i(x)$ is a subgroup of $F(x)$, since (F_i, A_i) is a family of soft subgroups of (F, A). Hence, $G(x)$ is a subgroup of $F(x)$ for all $x \in \text{Supp}(G, B)$. This completes the proof.

(2) Let $\tilde{\wedge}_{i\in I}(F_i, A_i) = (G, B)$, where $B = \prod_{i\in I} A_i$ and $G(x) = \bigcap_{i\in I} F_i(x_i)$ for all $x = (x_i)_{i\in I} \in B$. Since $B = \prod_{i\in I} A_i \subseteq \prod_{i\in I} A$, the first condition in the definition of soft subgroup is satisfied. Suppose that the soft set (G, B) is non-null. If $x = (x_i)_{i\in I} \in \mathrm{Supp}(G, B)$, then $G(x) = \bigcap_{\mathrm{iisl}} F_i(x_i) \neq \emptyset$. Thus the nonempty set $F_i(x_i)$ is a subgroup of $F(x)$, since (F_i, A_i) is a family of soft subgroups of (F, A) for all $i \in I$. Hence, $G(x)$ is a subgroup of $F(x)$ for all $x \in \mathrm{Supp}(G, B)$, which completes the proof.

(3) let $\tilde{\bigcup}_{i\in I}(F_i, A_i) = (G, B)$. Then $B = \bigcup_{i\in I} A_i$ and for all $x \in B, G(x) = \bigcup_{i\in I} F_i(x)$, where $I(x) = \{i \in I | x \in A_i\}$. Since $B = \bigcup_{i\in I} A_i$, which is the parameter set of $\tilde{\bigcup}_{i\in I}(F_i, A_i)$, is a subset of A, the first condition of definition of soft subgroup is satisfied. Note first that (G, B) is non-null, since $\mathrm{Supp}\,(G, B) = \bigcup_{i\in I} \mathrm{Supp}\,(F_i, A_i) \neq \emptyset$. Let $x \in \mathrm{Supp}(G, B)$. Then $G(x) = \bigcup_{i\in I} F_i(x) \neq \emptyset$, and so we have $F_{i_0}(x) \neq \emptyset$ for some $i_0 \in I(x)$. Yet, from the hypothesis, we know that $\{A_i | i \in I\}$ are pairwise disjoint. Hence, the above i_0 is in fact unique. Therefore, $G(x)$ coincides with $F_{i_0}(x)$. Furthermore, since $\left(F_{i_0}, A_{i_0}\right)$ is a soft subgroup of (F, A), the nonempty set $F_{i_0}(x)$ is a subgroup of $F(x)$ for all $x \in \mathrm{Supp}(G, B)$. This completes the proof.

Theorem 2.8 *Let (F, A) and (H, B) be two soft groups over G, and (F, A) be a soft subgroup of (H, B). If f is a homomorphism from G to K, then $(f(F), A)$ and $(f(H), B)$ are both soft subgroups over K and $(f(F), A)$ is a soft subgroup of $(f(H), B)$.*

Proof Since f is a homomorphism from G to K, $f(F(x))$ and $f(H(y))$ are subgroups of K for all $x \in A$ and for all $y \in B$. $(f(F), A)$ and $(f(H), B)$ are therefore soft groups over K. If (F, A) is a soft subgroup of (H, B), then $F(x)$ is a subgroup of $H(x)$ and $f(F(x))$ is a subgroup of $f(H(x))$ for all $x \in A$. Thus $(f(F), A)$ is a soft subgroup of $(f(H), B)$. □

Definition 2.6 Let (F, A) and (H, B) be two soft groups over G and K respectively, and let $f : G \to K$ and $g : A \to B$ be two functions. Then we say that (f, g) is a soft homomorphism, and that (F, A) is soft homomorphic to (H, B). The latter is written as $(F, A) \sim (H, B)$, if the following conditions are satisfied:

(1) f is a homomorphism from G onto K
(2) g is a mapping from A onto B , and
(3) $f(F(x)) = H(g(x))$ for all $x \in A$.

In Definition 2.6, if f is an isomorphism from G to K and g is a one-to-one mapping from A onto B, then we say that (f, g) is a soft isomorphism and that (F, A) is soft isomorphic to (H, B). The latter is denoted by $(F, A) \simeq (H, B)$.

Example 17 Consider the groups $(Z, +)$ and $(Z_m, \oplus)$. We define a homomorphism from Z onto Z_m such as $f(k) = \overline{k}$ for $k \in Z$, and a mapping g from Z^+ onto Z_m as $g(k) = \overline{k}$ for $k \in Z^+$. Let $F : Z^+ \to P(Z)$ and $F(x) = \{y \in Z : y = 5kx, k \in Z\}$; let $H : Z_m \to P(Z_m)$ and $H(u) = \{\overline{y} \in Z_m : y = uk, k \in 5Z\}$. Then we obtain $F(x) = 5xZ$ and $H(u) = \{\overline{ku} : k \in 5Z\}$. It is clear that $\left(F, Z^+\right)$ and (H, Z_m) are soft groups over Z and Z_m, respectively. Since $f(F(x)) = \{\overline{5xk} : k \in Z\}$ and $H(g(x)) = \{\overline{xs} : s \in 5Z\}$, we get $f(F(x)) = H(g(x))$. Hence (f, g) is a soft homomorphism, and $\left(F, Z^+\right)$ is soft homomorphic to (H, Z_m).

Definition 2.7 Let (F, A) be a soft group over G and (H, B) be a soft subgroup of (F, A). Then we say that (H, B) is a normal soft subgroup of (F, A), if $H(x)$ is a normal subgroup of $F(x)$ for all $x \in B$.

Definition 2.8 Let $(F_i, A_i)_{i\in I}$ be a nonempty family of soft subgroups (normal soft subgroups) over a common abelian group $(G, +)$. The sum of the nonempty family of soft subgroups (normal soft subgroups) $(F_i, A_i)_{i\in I}$ over G is defined as the soft set $(H, B) = \widetilde{\Sigma}_{i\in I}(F_i, A_i)$, where $B = \prod_{i\in I} A_i$ and $H(x) = \Sigma_{i\in I} F_i(x_i)$ for all $x = (x_i)_{i\in I} \in B$. If $A_i = A$ and $F_i = F$ for all $i \in I$, then $\widetilde{\Sigma}_{i\in I}(F_i, A_i)$ is denoted by $\widetilde{\Sigma}_{i\in I}(F, A)$. In this case, $\prod_{i\in I} A_i = \prod_{i\in I} A$ means the direct power A^I.

Theorem 2.9 *Let (F, A) be a soft group over G and $(F_i, A_i)_{i\in I}$ be a nonempty family of soft subgroups of (F, A). Then we have the following:*

(i) $\bigcap_{Ei\in I}(F_i, A_i)$ is a soft subgroup of (F, A), if it is non-null.
(ii) If $F_i(x_i) \subseteq F_j(x_j)$ or $F_j(x_j) \subseteq F_i(x_i)$ for all $i, j \in I$ and $x_i \in I$ then $\bigcup_{R_{i\in I}}(F_i, A_i)$ is a soft subgroup of (F, A), whenever it is non- null.
(iii) If $F_i(x_i) \subseteq F_j(x_j)$ or $F_j(x_j) \subseteq F_i(x_i)$ for all $i, j \in I$ and $x_i \in I$ then $\widetilde{\bigvee}_{i\in I}(F_i, A_i)$ is a soft subgroup of $\widetilde{\bigvee}_{i\in I}(F, A)$, whenever it is non- null.
(iv) $\widetilde{\prod}_{i\in I}(F_i, A_i)$ is a soft subgroup of $\widetilde{\prod}_{i\in I}(F, A)$, whenever it is non-null.
(v) If $(G, +)$ is abelian, then $\widetilde{\sum}_{i\in I}(F_i, A_i)$ is a soft subgroup of $\widetilde{\sum}_{i\in i}(F, A)$, whenever it is non-null.

Proof (i) Assume that $(F_i, A_i)_{i\in I}$ is a nonempty family of soft subgroups of (F, A). Let $\bigcap_{Ei\in I}(F_i, A_i) = (H, B)$, where $B = \bigcup_{i\in I} A_i$ and $H(x) = \bigcup_{i\in I(x)} F_i(x)$, and $I(x) = \{i \in I | x \in A_i\}$ for all $x \in B$. Now $B = \bigcup_{i\in I} A_i$ is a subset of A. Suppose that the soft set (H, B) is non-null. If $x \in \mathrm{Supp}(H, B)$, then $H(x) = \bigcap_{i\in I} F_i(x) \neq \emptyset$. It follows that, for all $i \in I$, the nonempty set $F_i(x)$ is a subgroup of $F(x)$, since (F_i, A_i) is a family of soft subgroups of (F, A). Hence, $H(x)$ is a subgroup of $F(x)$ for all $x \in \mathrm{Supp}(H, B)$. This completes the proof.

(ii) Assume that $(F_i, A_i)_{i\in I}$ is a nonempty family of soft subgroups of (F, A). Let $(H, B) = \bigcup_{Ri\in I}(F_i, A_i)$, where $B = \bigcap_{i\in I} A_i \neq \emptyset$ and $H(x) = \bigcup_{i\in I} F_i(x)$ for all $x \in B$. Let$B = \bigcap_{i\in I} A_i$ is a subset of A. Let $x \in \mathrm{Supp}(H, B)$. Since Supp $(H, B) = \bigcup_{i\in I}\mathrm{Supp}(F_i, A_i) \neq \emptyset$, we have $F_{i_0}(x) \neq \emptyset$ for some $i_0 \in I$. By assumption, $\bigcup_{i\in I} F_i(x_i)$ is a subgroup of $F(x)$ for all $x \in \mathrm{Supp}(H, B)$, since (F_i, A_i) is a family of soft subgroups of (F, A). Hence, $H(x)$ is a subgroup of $F(x)$ for all $x \in \mathrm{Supp}(H, B)$. This completes the proof.

(iii) Assume that $(F_i, A_i)_{i\in I}$ is a nonempty family of soft subgroups of (F, A). Let $(H, B) = \tilde{\bigvee}_{i\in I} (F_i, A_i)$, where $B = \prod_{i\in I} A_i$ and $H(x) = \bigcup_{i\in I} F_i(x)$ for all $x = (x_i)_{i\in I} \in B$. Since $B = \prod_{i\in I} A_i \subseteq \prod_{i\in I} A$, the first condition in the definition of soft subgroup is satisfied. Let $x = (x_i)_{i\in I} \in \text{Supp}(H, B)$. Then $H(x) = \bigcup_{i\in I} F_i(x_i) \neq \emptyset$, so we have $F_{i_0}(x_{i_0}) \neq \emptyset$ for some $i_0 \in I$. By assumption, $\bigcup_{i\in I} F_i(x_i)$ is a subgroup of $F(x)$ for all $x = (x_i)_{i\in I} \in B$. Hence, $H(x)$ is a subgroup of $F(x)$ for all $x \in \text{Supp}(H, B)$. This completes the proof.

(iv) Assume that $(F_i, A_i)_{i\in I}$ is a nonempty family of soft subgroups of (F, A). Let $(H, B) = \tilde{\prod}_{i\in I} (F_i, A_i)$, where $B = \prod_{i\in I} A_i$ and $H(x) = \prod_{i\in I} F_i(x_i)$ for all $x = (x_i)_{i\in I} \in B$. Let $x = (x_i)_{i\in I} \in \text{Supp}(H, B)$. Then $H(x) = \prod_{i\in I} F_i(x_i) \neq \emptyset$, so we have $F_i(x_i) \neq \emptyset$ for all $i \in I$. Since (F_i, A_i) is a family of soft subgroups of (F, A), we have that $F_i(x_i)$ is a subgroup of $F(x_i)$ for all $x = (x_i)_{i\in I} \in B$. That is, $\prod_{i\in I} F_i(x_i)$ is a subgroup of $\prod_{i\in I} F(x_i)$. Hence, $H(x)$ is a subgroup of $F(x)$ for all $x \in \text{Supp}(H, B)$. This completes the proof.

(v) Assume that $(F_i, A_i)_{i\in I}$ is a nonempty family of soft subgroups of (F, A). Let $(H, B) = \tilde{\Sigma}_{i\in I} (F_i, A_i)$, where $B = \prod_{i\in I} A_i$ and $H(x) = \Sigma_{i\in I} F_i(x_i)$ for all $x = (x_i)_{i\in I} \in B$. Let $x = (x_i)_{i\in I} \in \text{Supp}(H, B)$. Then $H(x) = \Sigma_{i\in I} F_i(x_i) \neq \emptyset$, so we have $F_i(x_i) \neq \emptyset$ for all $i \in I$. Since (F_i, A_i) is a family of soft subgroups of (F, A) and G is abelian, we have that $F_i(x_i)$ is a subgroup of $F(x_i)$ for all $x = (x_i)_{i\in I} \in B$. That is, $\tilde{\sum}_{i\in I} F_i(x_i)$ is a subgroup of $\tilde{\sum}_{i\in I} F(x_i)$. Hence, $H(x)$ is a subgroup of $F(x)$ for all $x \in \text{Supp}(H, B)$. This completes the proof. □

Proposition 2.1 *Let (F, A) be a soft group over G and $(F_i, A_i)_{i\in I}$ be a nonempty family of soft subgroups of (F, A). Then $\tilde{\bigcap}_{i\in I} (F_i, A_i)$ is a soft subgroup of (F_i, A_i) for each $i \in I$, if it is non-null.*

Proof Let $\tilde{\bigcap}_{i\in I} (F_i, A_i) = (H, C)$, where $C = \bigcap_{i\in I} A_i \neq \emptyset$ and $H(x) = \bigcap_{i\in I} F_i(x)$ for all $x \in C$. First,we check the parameter sets. $\bigcap_{i\in I} A_i$, which is the parameter set of $\tilde{\bigcap}_{i\in I} (F_i, A_i)$, is a subset of the parameter set of (F_i, A_i) for each $i \in I$. Suppose that (H, C) is a non-null soft set over G. If $x \in \text{Supp}(H, C)$, then $H(x) = \bigcap_{i\in I} F_i(x) \neq \emptyset$. Thus $\emptyset \neq F_i(x)$ are subgroups of G for all $i \in I$. Therefore, $H(x) = \bigcap_{i\in I} F_i(x)$ is a subgroup of G. Moreover, since $\bigcap_{i\in I} F_i(x) \subseteq F_i(x)$, for all $i \in I$ and for all $x \in \bigcap_{i\in I} A_i$, the rest of the proof is obvious from Theorem 2.6(1). □

The following proposition and theorem are a direct consequent of the relevant definitions and Theorem 2.7.

Proposition 2.2 *Let (F, A) and (T, A) be soft groups over G. Then $(F, A) \cap_E (T, A)$ is a soft subgroup of both (F, A) and (T, A), if it is non null.*

Theorem 2.10 *Let (F, A) be a soft group over G and $(F_i, A_i)_{i\in I}$ be a nonempty family of normal soft subgroups of (F, A). Then we have the following:*

(i) $\widetilde{\bigcap}_{i\in I}(F_i, A_i)$ *is a normal soft subgroup of* (F, A)*, if it is non-null.*
(ii) $\tilde{\wedge}_{i\in I}(F_i, A_i)$ *is a normal soft subgroup of* $\tilde{\wedge}_{i\in I}(F, A)$*, if it is non-null.*
(iii) If $\{A_i | i \in I\}$ *are pairwise disjoint, then* $\bigcup_{i\in I}(F_i, A_i)$ *is a normal soft subgroup over* G.

Remark 2.1 Results similar to Theorem 2.9 holds for normal soft subgroups also.

2.1.1 Normalistic Soft Groups

Definition 2.9 Let G be a group and (F, A) be a non-null soft set over G. Then (F, A) is called a normalistic soft group over G if $F(x)$ is a normal subgroup of G for all $x \in \text{Supp}(F, A)$

Example 18 Let $G = A_3 = \{e, (123), (132)\}$ be alternating groups of S_3 and the soft set (F, A) over G , where $F : A \to P(G)$ is set-valued function defined by

$$F(x) = \{y \in A_3 | xRy \Leftrightarrow y \in \langle x\rangle\}$$

for all $x \in A = G$. Then $F(e) = \{e\}$, $F(123) = F(132) = \{e, (123), (132)\}$. Since $F(x)$ is a normal subgroup of A_3 for all $x \in \text{Supp}\,(F, A_3)$, (F, A_3) is a normalistic soft group over A_3.

Clearly every normalistic soft group over G is a soft group over G. However, the converse is not true in general.

Example 19 Let $S_3 = \{e, (12), (13), (23), (123), (132)\}$. Consider the function defined by

$$F(x) = \left\{y \in S_3 | xRy \Leftrightarrow y = x^n, n \in \mathbb{N}\right\}$$

for all $x \in S_3$. Then $F(e) = \{e\}$, $F(12) = \{e, (12)\}$, $F(13) = \{e, (13)\}$, $F(23) = \{e, (23)\}$, $F(123) = F(132) = \{e, (123), (132)\}$, which are all subgroups of S_3. Hence, (F, S_3) is a soft group over S_3. Nevertheless, $F(e) = \{e\}$, $F(12) = \{e, (12)\}$ $F(13) = \{e, (13)\}$ are not normal subgroups of S_3. Therefore, (F, S_3) is not a normalistic soft group over S_3.

The following proposition holds clearly.

Proposition 2.3 *Let* G *be a group,* (F, A) *be a soft set over* G*, and* $B \subset A$*. If* (F, A) *is a normalistic soft group over* G*, then so is* (F, B)*, whenever* (F, B) *is non-null.*

Theorem 2.11 *Let* (F, A) *and* (T, B) *be normalistic soft groups over* G*. Then the following hold:*

(i) $(F, A)\widetilde{\cap}(T, B)$ *is a normalistic soft group over* G*, if it is non-null.*
(ii) $(F, A) \cap_E (T, B)$ *is a normalistic soft group over* G*, if it is non-null.*
(iii) $(F, A)\widetilde{\wedge}(T, B)$ *is a normalistic soft group over* G*, if is non-null.*
(iv) *If* $F(x)$ *and* $T(x)$ *are ordered by inclusion for all* $x \in \mathrm{Supp}\,((F, A) \cup_R (T, B))$*, then* $(F, A) \cup_R (T, B)$ *is a normalistic soft group over* G*, whenever it is non-null.*
(v) *If it is non null, the soft set* $(F, A)\widetilde{\vee}(T, B) = (N, A \times B)$ *is a normalistic soft group over* G*, whenever* $F(x)$ *and* $T(y)$ *are ordered by inclusion for all* $(x, y) \in$ *Supp* $(N, A \times B)$.

Proof (i) Let $(F, A)\widetilde{\cap}(T, B) = (H, C)$, where $C = A \cap B \neq \emptyset$ and $H(x) = F(x) \cap T(x)$ for all $x \in C$. Assume that (H, C) is a non-null soft set over G. If $x \in \mathrm{Supp}(H, C)$, then $H(x) = F(x) \cap T(x) \neq \emptyset$. Therefore, the nonempty sets $F(x)$ and $T(x)$ are both normal subgroups of G. If follows that $H(x)$ is a normal subgroup of G for all $x \in \mathrm{Supp}\,(H, C)$. Thus (H, C) is a normalistic soft group over G.

(ii) Let $(F, A) \cap_E (T, B) = (H, C)$, where $C = A \cup B$ and

$$H(x) = \begin{cases} F(x) & \text{if } x \in A - B \\ T(x) & \text{if } x \in B - A \\ F(x) \cap T(x) & \text{if } x \in A \cap B \end{cases}$$

for all $x \in C$. Suppose that (H, C) is a non-null soft set over G. Let $x \in \mathrm{Supp}(H, C)$. If $x \in A - B$, then $H(x) = F(x) \neq \emptyset$ is a normal subgroup of G; if $x \in B - A$, then $H(x) \neq \emptyset$ is a normal subgroup of G ; and if $x \in A \cap B$, $H(x) = F(x) \cap T(x) \neq \emptyset$. Thus $\emptyset \neq F(x)$ and $\emptyset \neq T(x)$ are both normal subgroups of G, and so is their intersection. It follows that (H, C) is a normalistic soft group over G.

(iii) Let $(F, A)\widetilde{\wedge}(T, B) = (H, A \times B)$, where $H(x, y) = F(x) \cap T(y)$ for all $(x, y) \in A \times B$. Assume that (H, C) is non-null soft set over G. If $(x, y) \in \mathrm{Supp}(H, C)$, then $H(x, y) = F(x) \cap T(y) \neq \emptyset$. Since (F, A) and (T, B) are normalistic soft groups over G, we know that the nonempty sets $F(x)$ and $T(y)$ are both normal subgroups of G. Therefore, $H(x, y)$ is a normal subgroup of G for all $(x, y) \in \mathrm{Supp}(H, C)$. Thus we can deduce that $(F, A)\widetilde{\wedge}(T, B) = (H, C)$ is a normalistic soft group over G.

(iv) Let $(F, A) \cup_R (T, B) = (S, A \cap B)$, where $S(x) = F(x) \cup T(x)$ for all $x \in A \cap B \neq \emptyset$. Then, by hypothesis, $(S, A \cap B)$ is a non-null soft set over G. If $x \in \mathrm{Supp}(S, A \cap B)$, then $S(x) = F(x) \cup T(x) \neq \emptyset$. Since $F(x)$ and $T(x)$ are ordered by inclusion for all $x \in \mathrm{Supp}(S, A \cap B)$, $F(x) \cup T(x) = F(x)$ or $F(x) \cup T(x) = T(x)$. Since $\emptyset \neq F(x)$ and $\emptyset \neq T(x)$ are both normal subgroups of G, $S(x)$ is a normal subgroup of G for all $x \in \mathrm{Supp}(S, A \cap B)$. Therefore, $(S, A \cap B)$ is a normalistic soft group over C.

(v) Let $(F, A)\widetilde{\vee}(T, B) = (N, A \times B)$, where $N(x, y) = F(x) \cup T(y)$ for all $(x, y) \in A \times B$. Then, by hypothesis, $(N, A \times B)$ is a non-null soft set over G. If $(x, y) \in \mathrm{Supp}(N, A \times B)$, then $N(x, y) = F(x) \cup T(y) \neq \emptyset$. since $F(x)$

Table 2.2 Operation table for group in Example 20

$+$	0	a	b	c
0	0	a	b	c
a	a	0	c	b
b	b	c	0	a
c	c	b	a	0

and $T(y)$ are ordered by inclusion for all $(x, y) \in \text{Supp}(N, A \times B)$, $F(x) \cup T(y) = F(x)$ or $F(x) \cup T(y) = T(y)$. Since $\emptyset \neq F(x)$ and $\emptyset \neq T(y)$ are both normal subgroups of G, $N(x, y)$ is a normal subgroup of G for all $(x, y) \in \text{Supp}(N, A \times B)$. Therefore, $(N, A \times B)$ is a normalistic soft group over G.

□

Proof of the following theorem is straightforward.

Theorem 2.12 *Let (F, A) and (T, B) be normalistic soft groups over G. If A and B are disjoint, then the union $(F, A)\widetilde{\cup}(T, B)$ is a normalistic soft group over G, if is non-null.*

Note that, if A and B are not disjoint, Theorem 2.12, does not hold in general, as can be seen from the following example.

Example 20 Let $G = \{0, a, b, c\}$ be the Klein- 4 group with the operation table given in (Table 2.2).

Let $A = G$ and the soft set (F, A) over G, where $F : A \to P(G)$ is a set-valued function defined by

$$F(x) = \{y \in G | x\alpha y \Leftrightarrow y = nx \text{ for some } n \in \mathbb{N}\}$$

for all $x \in A$. Here, $nx = x + x \ldots + x$ means the n -fold sum of x and $0x = 0$. Then $F(0) = \{0\}$, $F(a) = \{0, a\}$, $F(b) = \{0, b\}$ $F(c) = \{0, c\}$, which are all subgroups of G. Hence, (F, A) is a soft group over G.

Let $B = G$ and the soft set (H, B) over G, where $H : B \to P(G)$ is defined by

$$H(x) = \{0\} \cup \{y \in G | x\alpha y \Leftrightarrow x + y = b\}$$

Then $H(0) = \{0, b\}$, $H(a) = \{0, c\}$, $H(b) = \{0\}$, $H(c) = \{0, a\}$, which are all subgroups of G. Hence, (H, B) is a soft group over G. It is seen that (F, A) and (H, B) are both normalistic soft groups over G. Consider $(F, A)\widetilde{\cup}(H, B) = (K, C)$, where $C = A \cup B$. Since $K(a) = F(a) \cup H(a) = \{0, a, c\}$ is not a normal subgroup of G, (K, C) is not a normalistic soft group over G.

Definition 2.10 Let (F, A) and (H, B) be two normalistic soft groups over G_1 and G_2 respectively. The product of normalistic soft groups (F, A) and (H, B) is defined as $(F, A) \times (H, B) = (U, A \times B)$, where $U(x, y) = F(x) \times H(y)$ for all $(x, y) \in A \times B$.

Theorem 2.13 *Let (F, A) and (H, B) be two normalistic soft groups over G_1 and G_2, respectively. If it is non-null, then the product $(F, A) \times (H, B)$ is a normalistic soft group over $G_1 \times G_2$*

Proof Let $(F, A) \times (H, B) = (U, A \times B)$, where $U(x, y) = F(x) \times H(y)$ for all $(x, y) \in A \times B$. Then, by hypothesis, $(U, A \times B)$ is a non-null soft set over $G_1 \times G_2$. If $(x, y) \in \text{Supp}(U, A \times B)$, then $U(x, y) = F(x) \times H(y) \neq \emptyset$. Since $\emptyset \neq F(x)$ is a normal subgroup of G_1 and $\emptyset \neq H(y)$ is a normal subgroup of G_2 , it follows that $U(x, y)$ is a normal subgroup of $G_1 \times G_2$ for all $(x, y) \in \text{Supp}(U, A \times B)$. Therefore $(U, A \times B)$ is a normalistic soft group over $G_1 \times G_2$. □

Note that if N_1 and N_2 are two normal subgroups of a group $(G, +)$, then the sum of these two normal subgroups is defined as the following: $N_1 + N_2 = \{n_1 + n_2 | n_1 \in N_1 \wedge n_2 \in N_2\}$.

Definition 2.11 Let (F, N_1) and (H, N_2) be two soft normalistic soft groups over the abelian group $(G, +)$. The sum of normalistic soft groups (F, N_1) and (H, N_2) is defined as $(F, N_2) + (H, N_2) = (T, N_1 \times N_2)$, where $T(x, y) = F(x) + H(y)$ for all $(x, y) \in N_1 \times N_2$. Recall that, if $(F_i, A_i)_{i \in I}$ is a nonempty family of normalistic soft groups over a common abelian group G, then the sum of the nonempty family of normalistic soft groups $(F_i, A_i)_{i \in I}$ over G, $\widetilde{\sum}_{i \in I} (F_i, A_i)$, is defined similar to that of soft subgroups and normal soft subgroups as in Definition 2.8.

For every abelian group $(G, +)$, we have the following theorem.

Theorem 2.14 *Let (F, N_1) and (H, N_2) be normalistic soft groups over the abelian group $(G, +)$. Then, if it is non-null, the sum $(F, N_1) + (H, N_2)$ is a normalistic soft group over G.*

Proof Let $(F, N_1) + (H, N_2) = (T, N_1 \times N_2)$, where $T(x, y) = F(x) + H(y)$ for all $(x, y) \in N_1 \times N_2$. Then by hypothesis, $(T, N_1 \times N_2)$ is a non-null soft set over G. If $(x, y) \in \text{Supp}\,(T, N_1 \times N_2)$, then $T(x, y) = F(x) + H(y) \neq \emptyset$. Since $\emptyset \neq F(x)$ is a normal subgroup of G and $\emptyset \neq H(y)$ is a normal subgroup of G, it follows that $T(x, y)$ is a normal subgroup of G for all $(x, y) \in \text{Supp}\,(T, N_1 \times N_2)$. Therefore $(T, N_1 \times N_2)$ is a normalistic soft group over G. □

Theorem 2.15 *Let $(F_i, A_i)\, i \in I$ be a nonempty family of normalistic soft group G. Then we have the following.*

(i) $\widetilde{\wedge}_{i \in I} (F_i, A_i)$ *is a normalistic soft group over G, if is non-null.*
(ii) $\widetilde{\cap} (F_i, A_i)$ *is a normalistic soft group over G, if it is non-null.*
(iii) $\cap_E (F_i, A_i)$ *is a normalistic soft group* $\underset{\smile}{o}$*ver G, if is non-null.*
(iv) If $\{A_i | i \in I\}$ are pairwise disjoint, then $\bigcup_{i \in I} (F_i, A_i)$ *is a normalistic soft group over G.*

(v) If $F_i(x_i) \subseteq F_j(x_j)$ or $F_j(x_j) \subseteq F_i(x_i)$ for all $i, j \in I$ and $x_i \in I$, then $\widetilde{\cup}_{\Re i \in I}(F_i, A_i)$ is a normalistic soft group over G, whenever it is non-null.

(vi) If $F_i(x_i) \subseteq F_j(x_j)$ or $F_j(x_j) \subseteq F_i(x_i)$ for all $i, j \in I$ and $x_i \in I$, then $\widetilde{\vee}_{i \in I}(F_i, A_i)$ is a normalistic soft group over G, whenever it is non-null.

(vii) If G is an abelian group, then $\sum_{i \in I}(F_i, A_i)$ is a normalistic soft group over G, whenever it is non-null.

Proof Proof follows in similar lines with that of Theorems 2.7, 2.8 and 2.11. Hence is omitted. □

Proposition 2.4 *Let $(F_i, A_i)_{i \in I}$ be a nonempty family of normalistic soft groups of G_i. If it is non-null, $\widetilde{\prod}_{i \in I}(F_i, A_i)$ is a normalistic soft group over $\prod_{i \in I} G_i$.*

Proof Let $(H, B) = \widetilde{\prod}_{i \in I}(F_i, A_i)$, where $B = \prod_{i \in I} A_i$ and $H(x) = \prod_{i \in I} F_i(x_i)$ for all $x = (x_i)_{i \in I} \in B$. Let $x = (x_i)_{i \in I} \in \mathrm{Supp}(H, B)$. Then $H(x) = \prod_{i \in I} F_i(x_i) \neq \emptyset$, so we have $F_i(x_i) \neq \emptyset$ for all $i \in I$. Since (F_i, A_i) is a family of normalistic soft groups of G_i for all $i \in I$, we have that $\prod_{i \in I} F_i(x_i)$ is a normal subgroup of $\prod_{i \in I} G_i$ for all $x = (x_i)_{i \in I} \in B$. That is, the Cartesian product $\widetilde{\prod}_{i \in I}(F_i, A_i)$ is a normalistic soft group over $\prod_{i \in I} G_i$. □

Definition 2.12 Let (F, A) be a normalistic soft group over G. Then we have the following. (a) (F, A) is said to be trivial normalistic soft group if $F(x) = \{e_G\}$ for all $x \in \mathrm{Supp}(F, A)$. (b) (F, A) is said to be whole normalistic soft group if $F(x) = G$ for all $x \in \mathrm{Supp}(F, A)$.

Example 21 Let $G = \{1, -1, i, -i\}$ be the Klein-4 group with the operation table given below (Table 2.3).

Let $A = \{1, -1\}$ and the soft set (F, A) over G, where $F : A \to P(G)$ is a set-valued function defined by

$$F(x) = \{y \in G | x\alpha y \Leftrightarrow y = x^2\}$$

Then $F(1) = F(-1) = \{1\} = \{e_G\}$ for all $x \in \mathrm{Supp}(F, A)$, (F, A) is a trivial normalistic soft group over G. Let the soft set (H, B) over G where $H : B \to P(G)$ is defined by $H(x) = \{y \in G | x\alpha y \Leftrightarrow y = nx \text{ for some } n \in \mathbb{N}\}$ for all $x \in B = \{i, -i\}$. Then $H(i) = H(-i) = \{1, -1, i, -i\}$. Since $H(x) = G$ for all $x \in \mathrm{Supp}(H, B)$, (H, B) is a whole normalistic soft group over G.

Table 2.3 Operation table for group in Example 21

$*$	1	-1	i	$-i$
1	1	-1	i	$-i$
-1	-1	1	$-i$	i
i	i	$-i$	-1	1
$-i$	$-i$	i	1	-1

Proposition 2.5 *Let (F, A) and (G, B) be normalistic soft groups over G Then the following hold.*

(i) If (F, A) and (G, B) are trivial normalistic soft groups over G, then $(F, A)\tilde{\cap}(G, B)$ is a trivial normalistic soft group over G.
(ii) If (F, A) and (G, B) are whole normalistic soft groups over G, then $(F, A)\tilde{\cap}(G, B)$ is a whole normalistic soft group over G.
(iii) If (F, A) is a trivial normalistic soft group over G and (G, A) is a whole normalistic soft group over G, then $(F, A)\tilde{\cap}(G, B)$ is a trivial normalistic soft group over G.
(iv) If (F, A) and (G, B) are trivial normalistic soft groups over G, where G is abelian, then $(F, A) + (G, B)$ is a trivial normalistic soft group over G.
(v) If (F, A) and (G, B) are whole normalistic soft groups over G, where G is abelian, then $(F, A) + (G, B)$ is a whole normalistic soft group over G.
(vi) If (F, A) is a trivial normalistic soft group over G and (G, B) is a whole normalistic soft group over G, where G is abelian, then $(F, A) + (G, B)$ is a whole normalistic soft group over G

Proof Proof follows from relevant Definitions and Theorems 2.11 and 2.14. □

Proposition 2.6 *Let (F, N_1) and (G, N_2) be two normalistic soft groups over G_1 and G_2, respectively. The following hold:*

(i) If (F, N_1) and (G, N_2) are trivial normalistic soft groups over G_1 and C_2, respectively, then $(F, N_1) \times (G, N_2)$ is a trivial normalistic soft group over $G_1 \times G_2$.
(ii) If (F, N_1) and (G, N_2) are whole normalistic soft groups over G_1 and C_2, respectively, then $(F, N_1) \times (G, N_2)$ is a whole normalistic soft group over $G_1 \times G_2$.

Proof Proof follows from Definitions 2.10, 2.12 and Theorem 2.13. □

Let G_1 and G_2 be two groups, (F, A) and (H, B) be soft sets over G_1 and G_2, respectively, and $f : G_1 \to G_2$ be a mapping of groups. Then the soft set $(f(F),\ \text{Supp(F, A)})$ over G_2 can be defined, where

$$f(F) : \text{Supp}(F, A) \to P(G_2)$$

is given by $f(F)(x) = f(F(x))$ for all $x \in \text{Supp}(F, A)$. It is also worth noting that $\text{Supp}(F, A) = \text{Supp}(f(F), Supp(F, A)))$. Moreover, if f is a bijective mapping, then $\left(f^{-1}(H), \text{Supp}(H, B)\right)$ is a soft set over G_1, where

$$f^{-1}(H) : \text{Supp}(H, B) \to P(G_1)$$

is given by $f^{-1}(H)(y) = f^{-1}(H(y))$ for all $y \in \text{Supp}(H, B)$. Similarly, $\text{Supp}(H, B)$ $=\text{Supp}\left(f^{-1}(G), Supp(H, B)\right)$.

Proposition 2.7 *Let $f : G_1 \to G_2$ be a group epimorphism. If (F, A) is a normalistic soft group over G_1, then $(f(F),\ Supp\ (F, A))$ is a normalistic soft group over G_2.*

Proof Note first that, since (F, A) is a normalistic soft group over G_1 , then it has to be non-null; therefore $(f(F), \text{Supp}(F, A))$ is non-null over G_2, too. We have $f(F)(x) = f(F(x)) \neq \emptyset$ for all $x \in \text{Supp}(f(F), Supp(F, A))$. Because of the fact that (F, A) is a normalistic soft group over G_1, the nonempty set $F(x)$ is a normal subgroup of G_1 for all $x \in \text{Supp}(F, A)$. Thus, we can conclude that its homomorphic image $f(F(x))$ is a normal subgroup of G_2. So, $f(F(x))$ is a normal subgroup of G_2 for all $x \in \text{Supp}(f(F), \text{Supp}(F, A))$. This means that $(f(F), \text{Supp}(F, A))$ is a normalistic soft group over G_2. □

Proposition 2.8 *Let $f : G_1 \to G_2$ be a group isomorphism. If (H, B) is a normalistic soft group over G_2, then $\left(f^{-1}(H), \text{Supp}(H, B)\right)$ is a normalistic soft group over G_1.*

Proof Note first that, since (H, B) is a normalistic soft group over G_2, it has to be non-null; then so does $\left(f^{-1}(H), \text{Supp}(H, B)\right)$ over G_1. We have $f^{-1}(H)(y) = f^{-1}(H(y)) \neq \emptyset$ for all $y \in \text{Supp}\left(f^{-1}(H), \text{Supp}(H, B)\right)$. Because of the fact that (H, B) is a normalistic soft group over G_2, the nonempty set $H(y)$ is a normal subgroup of G_2 for all $y \in \text{Supp}\left(f^{-1}(H), \text{Supp}(H, B)\right)$. Thus, we can conclude that $f^{-1}(H(y))$ is a normal subgroup of G_1 for all $y \in \text{Supp}\left(f^{-1}(H), \text{Supp}(H, B)\right)$. This means that $\left(f^{-1}(H), \text{Supp}(H, B)\right)$ is a normalistic soft group over G_1. □

Theorem 2.16 *Let (F, A) be a normalistic soft group over G_1 and let $f : G_1 \to G_2$ be an epimorphism of groups. Then we have the following:*

(i) If $F(x) = Kerf$ for all $x \in \text{Supp}(F, A)$, then $(f(F), \text{Supp}(F, A))$ is a trivial normalistic soft group over G_2.
(ii) If (F, A) is whole, then $(f(F), \text{Supp}(F, A))$ is a whole normalistic soft group over G_2.
(iii) If f is injective and (H, B) is trivial, then $\left(f^{-1}(H), \text{Supp}(H, B)\right)$ is a trivial normalistic soft group over G_1.
(iv) If f is injective and $H(y) = f(G_1)$ for all $y \in \text{Supp}(H, B)$, then $\left(f^{-1}(H), \text{Supp}(H, B)\right)$ is a whole normalistic soft group over G_1.

Proof (i) Assume that $F(x) = \text{Kerf}$ for all $x \in \text{Supp}(F, A)$. Then $f(F)(x) = f(F(x)) = \left\{0_{G_2}\right\}$ for all $x \in \text{Supp}(F, A)$. That is, $(f(F), \text{Supp}(F, A))$ is a trivial normalistic soft group over G_2 by Proposition 2.7.
(ii) Suppose that (F, A) is whole. Then $F(x) = G_1$ for all $x \in \text{Supp}(F, A)$. It follows that $f(F)(x) = f(F(x)) = f(G_1) = G_2$ for all $x \in \text{Supp}(F, A)$, which means that $(f(F), \text{Supp}(F, A))$ is a whole normalistic soft group over G_2 by Proposition 2.7.
(iii) Assume that f is injective and (H, B) is trivial. Then $H(y) = \left\{0_{G_2}\right\}$ for all $y \in \text{Supp}(H, B)$. Thus, $f^{-1}(H)(y) = f^{-1}(H(y)) = f^{-1}\left(0_{G_2}\right) = \text{Kerf} = \left\{0_{G_1}\right\}$ for all $y \in \text{Supp}(G, B)$ since f is injective. It follows that $\left(f^{-1}(H), \text{Supp}(H, B)\right)$ is a trivial normalistic soft group over G_1, by Proposition 2.8.

(iv) Let $H(y) = f(G_1)$ for all $y \in \text{Supp}(H, B)$. Then $f^{-1}(H)(y) = f^{-1}(H(y)) = f^{-1}(f(G_1)) = G_1$ for all $y \in \text{Supp}(H, B)$. That is to say, $\left(f^{-1}(H), \text{Supp}(H, B)\right)$ is a whole normalistic soft group over G_1 by Proposition 2.8. □

Definition 2.13 A group G is said to satisfy condition (C_N) if $H \lessdot K \lessdot G$, then $H \lessdot G$.($\lessdot$ means normal subgroup of)

It is easily seen that group S_3 satisfies the condition (C_N); nevertheless, the dihedral group D_4 does not satisfy this condition.

Proposition 2.9 *Let G be a group satisfying condition (C_N) and let (F, A) be a normalistic soft group over G. If (H, B) is a normal soft subgroup of (F, A), then (H, B) is also a normalistic soft group over G.*

Proof If (H, B) is a normal soft subgroup of (F, A), then, for all $x \in Supp(H, B)$, $H(x) \lessdot F(x)$. Since (F, A) is a normalistic soft group over G, $F(x) \lessdot G$ for all $x \in \text{Supp}(F, A)$. Thus we have that $H(x) \lessdot F(x) \lessdot G$ for all $x \in \text{Supp}(H, B)$. Since G satisfies the condition C_N, $H(x) \lessdot G$ for all $x \in \text{Supp}(H, B)$. Hence, (H, B) is a normalistic soft group over G. □

Now we define normalistic soft group homomorphisms.

Definition 2.14 Let (F, A) and (H, B) be normalistic soft groups over G_1 and G_2, respectively. Let $f : G_1 \to G_2$ and $g : A \to B$ be two mappings. Then the pair (f, g) is called a soft homomorphism from (F, A) to (H, B) if the following conditions are satisfied:

(i) f is a group homomorphism.
(ii) $f(F(x)) = H(g(x))$ for all $x \in A$.

If (f, g) is a soft homomorphism and f and g are both surjective, then we say that (F, A) is normalistic softly homomorphic to (H, B) under the soft homomorphism (f, g), which is denoted by $(F, A) \sim (H, B)$, and then (f, g) is called a normalistic soft group homomorphism. Furthermore, if f is an isomorphism of groups and g is a bijective mapping, then (f, g) is said to be a normalistic soft group isomorphism. In this case, we say that (F, A) is normalistic softly isomorphic to (H, B), which is denoted by $(F, A) \simeq_N (H, B)$.

Theorem 2.17 *Let G_1, G_2, and G_3 be groups and (F, A), (H, B), and (T, C) be normalistic soft groups over G_1, G_2, and G_3 respectively. Let (f, g) from (F, A) to (H, B) and (f^*, g^*) from (H, B) to (T, C) be a soft homomorphisms. Then $(f^* \circ f, g^* \circ g)$ from (F, A) to (H, C) is a soft homomorphism from G_1 to G_3.*

Proof Let (f, g) from G_1 to G_2 be a soft homomorphism from (F, A) to (H, B). Then there exists a group homomorphism f such that $f : G_1 \to G_2$ and a mapping g such that $g : A \to B$ which satisfy $f(F(x)) = H(g(x))$ for all $x \in A$. Let (f^*, g^*) from G_2 to G_3 be a soft homomorphism from (H, B) to (T, C), then there exists a group homomorphism f^* such that $f^* : G_2 \to G_3$ and a mapping g^* such that

$g : B \to C$ which satisfy $f^*(H(x)) = T(g^*(x))$ for all $x \in B$. We need to show that $(f^* \circ f)(F(x)) = T((g^* \circ g)(x))$ for all $x \in A$. Let $x \in A$. Then

$$\begin{aligned}(f^* \circ f)(F(x)) &= f^*(f(F(x))) \\ &= f^*(H(g(x))) \\ &= T\left(g^*(g(x))\right) \\ &= T\left(\left(g^* \circ g\right)(x)\right)\end{aligned}$$

Therefore, the proof is completed. □

Theorem 2.18 *The relation $\simeq_N$ is an equivalence relation on normalistic soft groups.*

Proof (i) Reflexive: Let (F, A) be a normalistic soft group over G. Then $(F, A) \simeq_N (F, A)$ under the normalistic soft group isomorphism (I_A, I_A), where I_A is the identity function of A.
(ii) Symmetric: Let G_1 and G_2 be groups and (F, A) and (H, B) be normalistic soft groups over G_1 and G_2, respectively. Assume that $(F, A) \simeq_N (H, B)$. Then there exists an isomorphism f such that $f : G_1 \to G_2$ and a bijective mapping g such that $g : A \to B$ which satisfy $f(F(x)) = H(g(x))$ for all $x \in A$. One can easily say that $(H, B) \simeq_N (F, A)$ under the normalistic soft group isomorphism $\left(f^{-1}, g^{-1}\right)$, since $f^{-1} : G_2 \to G_1$ is an isomorphism and $g^{-1} : B \to A$ is a bijective mapping. Moreover, since

$$\begin{aligned}f(F(x)) = H(g(x)) &\Rightarrow f^{-1}(f(F(x))) = f^{-1}(H(g(x))) \\ &\Rightarrow F(x) = f^{-1}(H(g(x))) \\ &\Rightarrow F\left(g^{-1}(x)\right) = f^{-1}\left(H\left(g\left(g^{-1}(x)\right)\right)\right) \\ &\Rightarrow F\left(g^{-1}(x)\right) = f^{-1}(H(x))\end{aligned}$$

for all $x \in B$, $f^{-1}(H(x)) = F\left(g^{-1}(x)\right)$ is satisfied for all $x \in B$.
(iii) Transitive: Considering the fact that the composition of two isomorphism is an isomorphism and the composition of two bijective mapping is a bijective mapping, the transitive property is obvious. □

Proposition 2.10 *Let G_1 and G_2 be groups and (F, A) and (H, B) be soft sets over G_1 and G_2, respectively. If (F, A) is a normalistic soft group over G_1 and $(F, A) \simeq_N (H, B)$, then (H, B) is a normalistic soft group over G_2.*

Proof We need to show that $H(y)$ is a normal subgroup of G_2 for all $y \in \mathrm{Supp}(H, B)$. Since $(F, A) \simeq_N (H, B)$, there exists an isomorphism f from G_1 to G_2 and a bijective mapping g from A to B which satisfy $f(F(x)) = H(g(x))$ for all $x \in A$. Assume that (F, A) is a normalistic soft group over G_1. Then $F(x)$ is a normal subgroup of G_1 for all $x \in \mathrm{Supp}(F, A)$, therefore $f(F(x))$ is a normal subgroup of G_2 for all $x \in \mathrm{Supp}(F, A)$. Since g is a bijective mapping, for all $y \in \mathrm{Supp}(H, B) \subseteq B$, there exists an $x \in A$ such that $y = g(x)$. Hence, $H(y)$ is a normal subgroup of G_2 for all $y \in \mathrm{Supp}(H, B)$ since $f(F(x)) = H(y)$. □

Theorem 2.19 *Let $f : G_1 \to G_2$ be an epimorphism of groups and (F, A) and (H, B) be two normalistic soft groups over G_1 and G_2, respectively.*

(i) The map (f, I_A) from (F, A) to (K, A) is a normalistic soft group homomorphism from G_1 to G_2, where $I_A : A \to A$ is the identity mapping and the set-valued function $K : A \to P(G_2)$ is defined by $K(x) = f(F(x))$ for all $x \in A$.

(ii) If $f : G_1 \to G_2$ is an isomorphism, then the map $\left(f^{-1}, I_B\right)$ from (H, B) to (T, B) is a normalistic soft group homomorphism from G_1 to G_2, where $I_B : B \to B$ is the identity mapping and the set-valued function $T : B \to P(G_1)$ is defined by $T(x) = f^{-1}(H(x))$ for all $x \in B$.

Proof The proof follows from Definition 2.14 and is therefore omitted. □

2.2 Soft BCK/BCI-Algebras

A BCK/BCI-algebra is an algebraic structure introduced by K. Iseki, Y. Imai, and S. Tanaka in 1966, that describes fragments of the propositional calculus involving implication known as BCK/BCI logics and was extensively investigated by several researchers. For a detailed review, refer [52–55]. Soft BCK/BCI algebra as introduced by Y.B. Jun in 2008 will be discussed in this section. Definitions, results and examples given are taken from [58].

Definition 2.15 An algebra $(X; *, 0)$ of type $(2, 0)$ is called a BCI-algebra if it satisfies the following conditions:

(I) $(\forall x, y, z \in X)(((x * y) * (x * z)) * (z * y) = 0)$
(II) $(\forall x, y \in X)((x * y)) * y = 0)$
(III) $(\forall x \in X)(x * x = 0)$
(IV) $(\forall x, y \in X)(x * y = 0, y * x = 0 \Rightarrow x = y)$

If a BCl-algebra X satisfies the following identity:

(V) $(\forall x \in X)(0 * x = 0)$

then X is called a BCK-algebra.

Any BCK-algebra X satisfies the following axioms:

(a1) $(\forall x \in X)(x * 0 = x)$,
(a2) $(\forall x, y, z \in X)(x \leq y \Rightarrow x * z \leq y * z, z * y \leq z * x)$,
(a3) $(\forall x, y, z \in X)((x * y) * z = (x * z) * y)$,
(a4) $(\forall x, y, z \in X)((x * z) * (y * z) \leq x * y)$.

where $x \leq y$ if and only if $x * y = 0$. A BCK-algebra X is s said to be commutative if $x \wedge y = y \wedge x$ for all $x, y \in X$ where $x \wedge y = y * (y * x)$. A commutative BCK-algebra will be written by cBCK-algebra for short. A nonempty subset S of a BCK/BCIalgebra X is called a BCK/BCI-subalgebra of X if $x * y \in S$ for all $x, y \in S$. A mapping $f : X \to Y$ of BCK/BCI-algebras is called a homomorphism if $f(x * y) = f(x) * f(y)$ for all $x, y \in X$. For a homomorphism $f : X \to Y$ of BCK/BCI-algebras, the *kernel* of f, denoted by ker(f) is defined to be the set

Table 2.4 Cayley table for BCK-algebra in Example 22

$*$	0	a	b	c	d
0	0	0	0	0	0
a	a	0	a	a	a
b	b	b	0	b	b
c	c	c	c	0	c
d	d	d	d	d	0

$$\ker(f) = \{x \in X \mid f(x) = 0\}.$$

Let X be a BCK/BCI-algebra. A fuzzy set $\mu : X \to [0, 1]$ is called a *fuzzy subalgebra* of X if $\mu(x * y) \geq \min\{\mu(x), \mu(y)\}$ for all $x, y \in X$.

In what follows, let X and A be a BCK/BCI-algebra and a nonempty set, respectively, and R will refer to an arbitrary binary relation between an element of A and an element of X, that is, R is a subset of $A \times X$ unless otherwise specified. A set-valued function $F : A \to \mathcal{P}(X)$ can be defined as $F(x) = \{y \in X \mid xRy\}$ for all $x \in A$. The pair (F, A) is then a soft set over X. For any element x of a BCI-algebra X, we define the order of x, denoted by $o(x)$, as

$$o(x) = \min\{n \in \mathbb{N} \mid 0 * x^n = 0\}$$

Definition 2.16 Let (F, A) be a soft set over X. Then (F, A) is called a soft BCK/BCI-algebra over X if $F(x)$ is a BCK/BCI-subalgebra of X for all $x \in A$.

Let us illustrate this definition using the following examples.

Example 22 Let $X = \{0, a, b, c, d\}$ be a BCK-algebra with Cayley table (Table 2.4):

Let (F, A) be a soft set over X, where $A = X$ and $F : A \to \mathcal{P}(X)$ is a set-valued function defined by

$F(x) = \{y \in X \mid xRy \Leftrightarrow y \in x^{-1}I\}$

for all $x \in A$ where $I = \{0, a\}$ and $x^{-1}I = \{y \in X \mid x \wedge y \in I\}$. Then $F(0) = F(a) = X$, $F(b) = \{0, a, c, d\}$, $F(c) = \{0, a, b, d\}$ and $F(d) = \{0, a, b, c\}$ are BCK-subalgebras of X. Therefore (F, A) is a soft BCK-algebra over X.

Example 23 Consider a BCI-algebra $X = \{0, a, b, c\}$ with the Cayley table (Table 2.5):

Then, $(X; ., 0)$ is a BCI-algebra. Let (F, A) be a soft set over X, where $A = X$ and $F : A \to \mathcal{P}(X)$ is a set-valued function defined as follows:

$F(x) = \{0\} \cup \{y \in X \mid xRy \Leftrightarrow o(x) = o(y)\}$

for all $x \in A$. Then $F(0) = F(a) = F(b) = F(c) = \{0, a, b, c\}$ is a BCI-sub algebra of X, but $F(d) = F(e) = F(f) = F(g) = \{0, d, e, f, g\}$ is not a BCI-subalgebras of X. Hence (F, A) is not a soft BCI-algebra over X. If we take $B = \{a, b, c\}$ and define a set-valued function $G : B \to \mathcal{P}(X)$ by

$$G(x) = \{y \in X \mid xRy \Leftrightarrow o(x) = o(y)\}$$

Table 2.5 Cayley table for BCK-algebra in Example 23

$*$	0	a	b	c	d	e	f	g
0	0	0	0	0	d	d	d	d
a	a	0	0	0	e	d	d	d
b	b	b	0	0	f	f	d	d
c	c	b	a	0	g	f	e	d
d	d	d	d	d	0	0	0	0
e	e	d	d	d	a	0	0	0
f	f	f	d	d	b	b	0	0
g	g	f	e	d	c	b	a	0

for all $x \in B$, then (G, B) is a soft BCI-algebra over X. since $G(a) = G(b) = G(c) = \{0, a, b, c\}$ is a BCI-subalgebra of X.

Let A be a fuzzy BCK/BCI-subalgebra of X with membership function μ_{A}. Let us consider the family of α-level sets for the function μ_{A} given by

$$F(\alpha) = \{x \in X | \mu_A(x) \geq \alpha\}, \alpha \in [0, 1].$$

Then, $F(\alpha)$ is a BCK/BCI-subalgebra of X. If we know the family F, we can find the functions $\mu_A(x)$ by means of the following formula:

$$\mu_A(x) = \sup\{\alpha \in [0, 1] | x \in F(\alpha)\}.$$

Thus, every fuzzy BCK/BCI-subalgebra A may be considered as the soft BCK/BCI-algebra $(F, [0, 1])$.

Theorem 2.20 *Let (F, A) be a soft BCK/BCI-algebra over X. If B is a subset of A, then $(F|_B, B)$ is a soft BCK/BCI-algebra over X.*

Proof Straightforward. □

The following example shows that there exists a soft set (F, A) over X such that

(i) (F, A) is not a soft BCI-algebra over X.
(ii) there exists a subset B of A such that $(F|_B, B)$ is a soft BCI-algebra over X.

Example 24 Let (F, A) be a soft set over X given in Example 23. Note that (F, A) is not a soft BCI-algebra over X. But if we take $B = \{a, b, c\} \subset A$, then $(F|_B, B)$ is a soft BCI-algebra over X.

Theorem 2.21 *Let (F, A) and (G, B) be two soft BCK/BCI-algebras over X. If $A \cap B \neq \phi$, then the intersection $(F, A)\tilde{\cap}(G, B)$ is a soft BCK/BCI-algebra over X.*

Proof Let $(F, A)\tilde{\cap}(G, B) = (H, C)$, where $C = A \cap B$ and $H(x) = F(x) \cap G(x)$ for all $x \in C$. Note that $H : C \to \mathcal{P}(X)$ is a mapping, and therefore (H, C) is a soft set over X. Since (F, A) and (G, B) are soft BCK/BCI-algebras over X, it follows that $H(x) \cap F(x)$ is a BCK/BCI-subalgebra of X for all $x \in C$. Hence $(H, C) = (F, A)\tilde{\cap}(G, B)$ is a soft BCK/BCI-algebra over X. □

Corollary 2.2 *Let (F, A) and (G, A) be two soft BCK/BCI-algebras over X. Then, their intersection $(F, A)\widetilde{\cap}(G, A)$ is a soft BCK/BCI-algebra over X.*

Proof Straightforward. □

Theorem 2.22 *Let (F, A) and (G, A) be two soft BCK/BCI-algebras over X. If A and B are disjoint, then the union $(F, A)\widetilde{\cup}(G, A)$ is a soft BCK/BCI-algebra over X.*

Proof Let $(F, A)\widetilde{\cup}(G, B) = (H, C)$, where $C = A \cup B$ and for every $e \in C$,

$$H(e) = \begin{cases} F(e) & \text{if } e \in A - B \\ G(e) & \text{if } e \in B - A \\ F(e) \cup G(e) & \text{if } e \in A \cap B \end{cases}$$

Since $A \cap B = \emptyset$, either $x \in A - B$ or $x \in B - A$ for all $x \in C$. If $x \in A \backslash B$, then $H(x) = F(x)$ is a BCK/BCI-subalgebra of X since (F, A) is a soft BCK/BCI-algebra over X. If $x \in B \backslash A$, then $H(x) = G(x)$ is a BCK/BCI-subalgebra of X since (G, B) is a soft BCK/BCI-algebra over X. Hence $(H, C) = (F, A)\widetilde{\cup}(G, A)$ is a soft BCK/BCI-algebra over X. □

Theorem 2.23 *If (F, A) and (G, B) are soft BCK/BCI-algebras over X, then $(F, A) \wedge (G, B)$ is a soft BCK/BCI-algebra over X.*

Proof We have $(F, A) \wedge (G, B) = (H, A \times B)$, where $H(x, y) = F(x) \cap G(y)$ for all $(x, y) \in A \times B$. Since $F(x)$ and $G(y)$ are BCK/BCI-subalgebras of X, the intersection $F(x) \cap G(y)$ is also a BCK/BCI-subalgebra of X. Hence $H(x, y)$ is a BCK/BCI-subalgebra of X for all $(x, y) \in A \times B$, and therefore $(F, A) \wedge (G, B) = (H, A \times B)$ is a soft BCK/BCI-algebra over X. □

Definition 2.17 A soft BCK/BCI-algebra (F, A) over X is said to be trivial (resp., whole) if $F(x) = \{0\}$ (resp., $F(x) = X$) for all $x \in A$.

Let $f : X \to Y$ be a mapping of BCK/BCI-algebras. For a soft set (F, A) over X, $(f(F), A)$ is a soft set over Y where $f(F) : A \to \mathcal{P}(Y)$ is defined by $f(F)(x) = f(F(x))$ for all $x \in A$.

Lemma 2.1 *Let $f : X \to Y$ be a homomorphism of BCK/BCI-algebras. If (F, A) is a soft BCK/BCI-algebra over X, then $(f(F), A)$ is a soft BCK/BCI-algebra over Y.*

Proof For every $x \in A$, we have $f(F)(x) = f(F(x))$ is a BCK/BCI-subalgebra of Y since $F(x)$ is a BCK/BCI-subalgebra of X and its homomorphic image is also a BCK/BCI-subalgebra of Y. Hence $(f(F), A)$ is a soft BCK/BCI-algebra over Y. □

Theorem 2.24 *Let $f : X \to Y$ be a homomorphism of BCK/BCI-algebras and let (F, A) be a soft BCK/BCI-algebra over X.*

(i) If $F(x) = \ker(f)$ for all $x \in A$, then $(f(F), A)$ is the trivial soft BCK/BCI-algebra over Y.

(ii) If f is onto and (F, A) is whole, then $(f(F), A)$ is the whole soft BCK/BCI-algebra over Y.

Proof (i) Assume that $F(x) = \ker(f)$ for all $x \in A$. Then, $f(F)(x) = f(F(x)) = \{0_Y\}$ for all $x \in A$. Hence $(f(F), A)$ is the trivial soft BCK/BCI-algebra over Y by Lemma 2.1 and Definition 2.17.

(ii) Suppose that f is onto and (F, A) is whole. Then, $F(x) = X$ for all $x \in A$, and so $f(F)(x) = f(F(x)) = f(X) = Y$ for all $x \in A$. It follows from Lemma 2.1 and Definition 2.17 that $(f(F), A)$ is the whole soft BCK/BCI-algebra over Y. □

Definition 2.18 Let (F, A) and (G, B) be two soft BCK/BCI-algebras over X. Then (F, A) is called a soft subalgebra of (G, B), if it satisfies:

(i) $A \subset B$,
(ii) $F(x)$ is a BCK/BCI-subalgebra of $G(x)$ for all $x \in A$.

Example 25 Let (F, A) be a soft BCK-algebra over X which is given in Example 22. Let $B = \{a, c, d\}$ be a subset of A and let $G : B \to \mathcal{P}(X)$ be a set-valued function defined by

$$G(x) = \{y \in X | xRy \Leftrightarrow y \in x^{-1}I\}$$

for all $x \in B$, where $I = \{0, a\}$ and $x^{-1}I = \{y \in X | x \wedge y \in I\}$. Then $G(a) = X$, $G(c) = \{0, a, b, d\}$ and $G(d) = \{0, a, b, c\}$ are BCK-subalgebras of $F(a)$, $F(c)$ and $F(d)$, respectively. Hence (G, B) is a soft subalgebra of (F, A).

Theorem 2.25 *Let (F, A) and (G, A) be two soft BCK/BCI-algebras over X.*

(i) If $F(x) \subset G(x)$ for all $x \in A$, then (F, A) is a soft subalgebra of (G, A).
(ii) If $B = \{0\}$ and (H, B), (F, X) are soft BCK/BCI-algebras over X, then $(H, B)$is a soft subalgebra of (F, X).

Proof Straightforward. □

Theorem 2.26 *Let (F, A) be a soft BCK/BCI-algebra over X and let (G_1, B_1) and (G_2, B_2) be soft subalgebras of (F, A). Then*
(i) $(G_1, B_1) \tilde{\cap} (G_2, B_2)$ is a soft subalgebra of (F, A).
(ii) $B_1 \cap B_2 = \emptyset \Rightarrow (G_1, B_1) \tilde{\cup} (G_2, B_2)$ is a soft subalgebra of (F, A).

Proof (i) Follows from definitions directly.
(ii) Assume that $B_1 \cap B_2 = \phi$. We can write $(G_1, B_1) \tilde{\cup} (G_2, B_2) = (G, B)$ where $B = B_1 \cup B_2$ and

$$G(x) = \begin{cases} G_1(x) & \text{if } x \in B_1 - B_2 \\ G_2(x) & \text{if } x \in B_2 - B_1 \\ G_1(x) \cup G_2(x) & \text{if } x \in B_1 \cap B_2 \end{cases}$$

for all $x \in B$. Since (G_i, B_i) is a subalgebra of (F, A) for $i = 1, 2$, $B = B_1 \cup B_2 \subset A$ and $G_i(x)$ is a BCK/BCI-subalgebra of $F(x)$ for all $x \in B_i$, $i = 1, 2$. Since $B_1 \cap B_2 = \phi$, $G(x)$ is a BCK/BCI-subalgebra of $F(x)$ for all $x \in B$. Therefore $(G_1, B_1) \tilde{\cup} (G_2, B_2) = (G, B)$ subalgebra of (F, A). □

Theorem 2.27 *Let $f : X \to Y$ be a homomorphism of BCK/BCI-algebras and let (F, A) and (G, B) be soft BCK/BCI-algebras over X. Then*

$$(F, A) \text{ sub algebra of } (G, B) \Rightarrow (f(F), A) \text{ sub algebra of } (f(G), B).$$

Proof Assume that (F, A) is a sub algebra of (G, B). Let $x \in A$. Then $A \subset B$ and $F(x)$ is a BCK/BCI-subalgebra of $G(x)$. Since f is a homomorphism, $f(F)(x) = f(F(x))$ is a BCK/BCI-subalgebra of $f(G(x)) = f(G)(x)$ and, therefore, $(f(F), A)$ is a BCK/BCI-subalgebra of $(f(G), B)$. □

2.3 Soft Rings and Ideals

The main purpose of this section is to introduce basic notions of soft rings, ideals and related concepts. A soft ring over a ring R is a parametrized family of subrings of R. Results, definitions and examples here are due to [1]. Here R denotes a commutative ring and all soft sets are considered over R.

Definition 2.19 Let (F, A) be a non-null soft set over a ring R. Then (F, A) is called a soft ring over R if $F(x)$ is a subring of R for all $x \in A$.

Example 26 Let $R = A = \mathbb{Z}_6 = \{0, 1, 2, 3, 4, 5\}$. Consider the set-valued function $F : A \longrightarrow \mathcal{P}(R)$ given by $F(x) = \{y \in R \mid x \cdot y = 0\}$. Then $F(0) = R$, $F(1) = \{0\}$, $F(2) = \{0, 3\}$, $F(3) = \{0, 2, 4\}$, $F(4) = \{0, 3\}$, $F(5) = \{0\}$. As we see, all of these sets are subrings of R. Hence, (F, A) is a soft ring over R.

Theorem 2.28 *Let (F, A) and (G, B) be soft rings over R. Then:*

1. *$(F, A) \wedge (G, B)$ is a soft ring over R if it is non-null.*
2. *The intersection $(F, A)\tilde{\cap}(G, B)$ is a soft ring over R if it is non-null.*

Proof (1) Let $(F, A) \wedge (G, B) = (H, C)$, where $C = A \times B$ and $H(a, b) = F(a) \cap G(b)$, for all $(a, b) \in C$. Since (H, C) is non-null, $H(a, b) = F(a) \cap G(b) \neq \emptyset$. Since the intersection of any number of subrings of R is a subring of R, $H(a, b)$ is a subring of R. Hence, (H, C) is a soft ring over R.
(2) Let $(F, A)\tilde{\cap}(G, B) = (H, C)$, where $H(x) = F(x) \cap G(x) \neq \emptyset$, for some $x \in A \cap B$. We observe that $F(x) \cap G(x)$ is a subring of R, since $H(x) \neq \emptyset$ and $F(x)$, $G(x)$ are subring of R. Consequently, $(H, C) = (F, A)\tilde{\cap}(G, B)$ is a soft ring over R if it is non-null. □

Definition 2.20 Let (F, A) and (G, B) be soft rings over R. Then (G, B) is called a soft subring of (F, A) if it satisfies the following:

1. $B \subset A$.
2. $G(x)$ is a subring of $F(x)$, for all$x \in \text{Supp}(G, B)$.

Example 27 Let $R = A = 2\mathbb{Z}$ and $B = 6\mathbb{Z} \subset A$. Consider the set-valued functions $F : A \longrightarrow \mathcal{P}(R)$ and $G : B \longrightarrow \mathcal{P}(R)$ given by $F(x) = \{nx | n \in \mathbb{Z}\}$ and $G(x) = \{5nx | n \in \mathbb{Z}\}$. As we see, for all $x \in B$, $G(x) = 5x\mathbb{Z}$ is a subring of $x\mathbb{Z} = F(x)$. Hence, (G, B) is a soft subring of (F, A).

Theorem 2.29 *Let (F, A) and (G, B) be soft rings over R. Then we have the following:*

1. *If $G(x) \subset F(x)$, for all $x \in B \subset A$, then (G, B) is a soft subring of (F, A).*
2. *$(F, A)\widetilde{\cap}(G, B)$ is a soft subring of both (F, A) and (G, B) if it is non-null.*

Proof (1) Clear.

(2) Let $(F, A)\widetilde{\cap}(G, B) = (H, C)$. Since $A \cap B \subset A$ and $H(x) = F(x) \cap G(x)$ is a subring of $F(x)$, (H, C) is a soft subring of (F, A). Similarly, we see that (H, C) is a soft subring of (G, B).

□

Example 28 Let $R = \mathbb{Z}$, $A = 2\mathbb{Z}$ and $B = 3\mathbb{Z}$. Consider the functions $F : A \longrightarrow \mathcal{P}(R)$ and $G : B \longrightarrow \mathcal{P}(R)$ defined by $F(x) = \{2nx \mid n \in \mathbb{Z}\} = 2x\mathbb{Z}$ and $G(x) = \{3nx \mid n \in \mathbb{Z}\} = 3x\mathbb{Z}$. Let $(F, A)\widetilde{\cap}(G, B) = (H, C)$ where $C = A \cap B = 6\mathbb{Z}$. For every $x \in C$, we have $H(x) = F(x) \cap G(x) = 6x\mathbb{Z}$ which is a subring of both $F(x) = 2x\mathbb{Z}$ and $G(x) = 3x\mathbb{Z}$. Consequently, $(F, A)\widetilde{\cap}(G, B)$ is a soft subring of both (F, A) and (G, B).

Theorem 2.30 *Let $(F_i, A_i)_{i \in I}$ be a nonempty family of soft rings over R. Then, clearly we have*

1. *$\wedge_{i \in I} (F_i, A_i)$ is a soft ring over R if it is non-null.*
2. *$\widetilde{\cap}_i (F_i, A_i)$ is a soft ring over R if it is non-null.*
3. *If $\{A_i | i \in I\}$ are pairwise disjoint, then $\widetilde{\bigcup}_{i \in I} (F_i, A_i)$ is a soft ring over R.*

The notion of ideals plays a vital role in the theory of classical algebra. For this reason, we introduce the soft ideals of a soft ring. Note that, if I is an ideal of a ring R, we write $I \lhd R$.

Definition 2.21 Let (F, A) be a soft ring over R. A non-null soft set (γ, I) over R is called soft ideal of (F, A) which will be denoted by $(\gamma, I)\widetilde{\lhd}(F, A)$, if it satisfies the following conditions:

1. $I \subset A$.
2. $\gamma(x)$ is an ideal of $F(x)$ for all $x \in \text{supp}(\gamma, I)$.

Example 29 Let $R = A = \mathbb{Z}_4 = \{0, 1, 2, 3\}$ and $I = \{0, 1, 2\}$. Let us consider the set-valued function $F : A \longrightarrow \mathcal{P}(R)$ given by $F(x) = \{y \in R \,|\, x \cdot y \in \{0, 2\}\}$. Then $F(0) = R$, $F(1) = \{0\}$, $F(2) = \mathbb{Z}_4$ and $F(3) = \{0, 2\}$. As we see, all these sets are subrings of R. Hence, (F, A) is a soft ring over R. On the other hand, consider the function $\gamma : I \longrightarrow \mathcal{P}(R)$ given by $\gamma(x) = \{y \in R \mid x \cdot y = 0\}$. As we see, $\gamma(0) = R \lhd R$, $\gamma(1) = \{0\} \lhd F(1) = \{0\}$ and $\gamma(2) = \{0, 2\} \lhd F(2) = \mathbb{Z}_4$. Hence, (γ, I) is a soft ideal of (F, A).

Theorem 2.31 *Let (γ_1, I_1) and (γ_2, I_2) be a soft ring (F, A) over R. Then (γ_1, I_1) $\widetilde{\cap}(\gamma_2, I_2)$ is a soft ideal of (F, A) if it is non-null.*

Proof Clear. □

Theorem 2.32 *Let (γ_1, I_1) and (γ_2, I_2) be soft ideals of soft rings (F, A) and (G, B) over R respectively. Then $(\gamma_1, I_1)\widetilde{\cap}(\gamma_2, I_2)$ is a soft ideal of $(F, A)\widetilde{\cap}(G, B)$ if it is non-null.*

Proof Let $(\gamma_1, I_1)\widetilde{\cap}(\gamma_2, I_2) = (\gamma, I)$, where $I = I_1 \cap I_2$ and $\gamma(x) = \gamma_1(x) \cap \gamma_2(x)$ for all $x \in I$. Similarly, we have $(F, A)\widetilde{\cap}(G, B) = (H, C)$ and $C = A \cap B$ where $H(x) = F(x) \cap G(x)$ for all $x \in C$. Since $I_1 \cap I_2$ is non-null, there exists an $x \in \operatorname{Supp}(\gamma, I)$ such that $\gamma(x) = \gamma_1(x) \cap \gamma_2(x) \neq \emptyset$. Since $I_1 \cap I_2 \subset A \cap B$, we need to show that $\gamma(x)$ is an ideal of ring $H(x)$ for all $x \in \operatorname{Supp}(\gamma, I)$. Because of the facts that $\gamma_1(x) \subset F(x)$ and $(\gamma_2(x) \subset G(x)$, we see that $\gamma_1(x) \cap \gamma_2(x) \subset F(x) \cap G(x)$. Hence, $\gamma(x)$ is a subring of R. Finally we shall show that $r \cdot a \in \gamma(x)$ for all $r \in H(x)$ and for all $a \in \gamma(x)$. Since $\gamma_1(x)$ is an ideal of $F(x)$, for $r \in H(x) = F(x) \cap G(x)$ and $a \in \gamma(x) = \gamma_1(x) \cap \gamma_2(x)$, we observe that $r \cdot a \in \gamma_1(x)$ and $r \cdot a \in \gamma_2(x)$. Hence $r \cdot a \in \gamma(x)$. □

Example 30 Let $R = M_2(\mathbb{Z})$, i.e., 2×2 matrices with integer terms, $A = 3\mathbb{Z}$, $B = 5\mathbb{Z}$, $I_1 = 6\mathbb{Z}$ and $I_2 = 10\mathbb{Z}$. Consider the functions $F : A \longrightarrow \mathcal{P}(R)$ and $G : B \longrightarrow \mathcal{P}(R)$ defined by $F(x) = \left\{\begin{bmatrix} nx & 0 \\ 0 & nx \end{bmatrix} | n \in \mathbb{Z}\right\}$ and $G(x) = \left\{\begin{bmatrix} nx & nx \\ 0 & nx \end{bmatrix} | n \in \mathbb{Z}\right\}$ which are subrings of R. Thus, (F, A) and (G, B) are soft rings over R. Consider the set-valued functions $\gamma_1 : I_1 \longrightarrow \mathcal{P}(R)$ and $\gamma_2 : I_2 \longrightarrow \mathcal{P}(R)$ defined by $\gamma_1(x) = \left\{\begin{bmatrix} nx & nx \\ 0 & 0 \end{bmatrix} | n \in \mathbb{Z}\right\}$ and $v_2(x) = \left\{\begin{bmatrix} 0 & nx \\ 0 & nx \end{bmatrix} | n \in \mathbb{Z}\right\}$ which are ideals of $F(x)$ and $G(x)$ respectively. For all $x \in I_1 \cap I_2, \gamma_1(x) \cap \gamma_2(x) = \left\{\begin{bmatrix} 0 & nx \\ 0 & 0 \end{bmatrix} | n \in \mathbb{Z}\right\} \lhd F(x) \cap G(x) = \left\{\begin{bmatrix} nx & 0 \\ 0 & nx \end{bmatrix} | n \in \mathbb{Z}\right\}$. This indicates that $(\gamma_1, I_1)\widetilde{\cap}(\gamma_2, I_2)$ is a soft ideal of $(F, A)\widetilde{\cap}(G, B)$.

Theorem 2.33 *Let (F, A) be a soft ring over R and (γ_1, I_1), (γ_2, I_2) be soft ideals of (F, A) over R. If I_1 and I_2 are disjoint, then $(\gamma_1, I_1) \widetilde{\bigcup} (\gamma_2, I_2)$ is a soft ideal of (F, A).*

Proof Take $(\gamma_1, I_1) \widetilde{\cup} (\gamma_2, I_2) = (\beta, I)$ where $I_1 \cup I_2 = I$ and for all $x \in I$,

$$\beta(x) = \begin{cases} \gamma_1(x) & \text{if } x \in I_1 - I_2, \\ \gamma_2(x) & \text{if } x \in I_2 - I_1, \\ \gamma_1(x) \cup \gamma_2(x) & \text{if } x \in I_1 \cap I_2. \end{cases}$$

Since $(\gamma_1, I_1)\widetilde{\lhd}(F, A)$ and $(\gamma_2, I_2)\widetilde{\lhd}(F, A)$ we see that $I \subset A$. For every $x \in \operatorname{Supp}(\beta, I)$, $x \in I_1 - I_2$ or $x \in I_2 - I_1$ since I_1 and I_2 are disjoint. If $x \in I_1 - I_2$, then $\beta(x) = I_1(x) \neq \emptyset$ is an ideal of $F(x)$ since $(\gamma_1, I_1)\widetilde{\lhd}(F, A)$. Similarly, if $x \in I_2 - I_1$, then $\beta(x) = I_2(x) \neq \emptyset$ is an ideal of $F(x)$ since $(\gamma_2, I_2)\widetilde{\lhd}(F, A)$. Thus, $\beta(x) \lhd F(x)$ for all $x \in \mathrm{Supp}(\beta, I)$. Hence, (β, I) is a soft ideal of (F, A). □

Theorem 2.34 *Let (F, A) be a soft ring over R and $(\gamma_k, I_k)_{k \in K}$ be a nonempty family of soft ideals of (F, A). Then we have the following:*

1. *$\widetilde{\cap}(\gamma_k, I_k)$ is a soft ideal of (F, A) if it is non-null.*
2. *$\wedge_{k \in I}(\gamma_k, I_k)$ is a soft ideal of (F, A) if it is non-null.*
3. *If $\{I_k | k \in K\}$ are pairwise disjoint, then $\widetilde{\bigcup}_{k \in I}(\gamma_k, I_k)$ is a soft ideal of (F, A) if it is non-null.*

Proof (1) It is an obvious result since the intersection of an arbitrary nonempty family of ideals of a ring is an ideal of it. (2) and (3) are similar to (1). □

2.3.1 Idealistic Soft Rings

Definition 2.22 Let (F, A) be a non-null soft set over R. Then (F, A) is called an idealistic soft ring over R if $F(x)$ is an ideal of R for all $x \in \mathrm{Supp}(F, A)$.

Example 31 In Example 29, (F, A) is an idealistic soft ring over R since $F(x)$ is an ideal of R for all $x \in A$.

Proposition 2.11 *Let (F, A) be a soft set over R and $B \subset A$. If (F, A) is an idealistic soft ring over R, then so is (F, B) whenever it is non-null.*

Proof Obvious. □

Theorem 2.35 *Let (F, A) and (G, B) be idealistic soft rings over R. Then $(F, A)\widetilde{\cap}(G, B)$ is an idealistic soft ring over R if it is non-null.*

Proof Take $(F, A)\widetilde{\cap}(G, B) = (H, C)$ where $C = A \cap B$ and $H(x) = F(x) \cap G(x)$, for all $x \in C$. Assume that (H, C) is a non-null soft set over R. So, if $x \in \mathrm{Supp}(H, C)$ then $H(x) = F(x) \cap G(x) \neq \emptyset$ and the nonempty sets $F(x)$ and $G(x)$ are ideals of R. Therefore, since the intersection of any nonempty family of ideals of a ring is an ideal of it, $H(x)$ is an ideal of R for all $x \in \mathrm{Supp}(H, C)$. Consequently, $(H, C) = (F, A)\widetilde{\cap}(G, B)$ is an idealistic soft ring over R. □

Theorem 2.36 *Let (F, A) and $G, B)$ be idealistic soft rings over R. If A and B are disjoint, then $(F, A)\widetilde{\cup}(G, B)$ is an idealistic soft ring over R.*

Proof Take $(F, A)\widetilde{\cup}(G, B) = (H, C)$ where $C = A \cup B$ and for all $x \in C$,

$$H(x) = \begin{cases} F(x) & \text{if } x \in A - B, \\ G(x) & \text{if } x \in B - A, \\ F(x) \cup G(x) & \text{if } x \in A \cap B. \end{cases}$$

Assume that $A \cap B = \emptyset$. Under this assumption, if $x \in \text{Supp}(H, C)$ then $x \in A - B$ or $x \in B - A$. If $x \in A - B$, then $H(x) = F(x)$ is an ideal of R since (F, A) is an idealistic soft ring over R. Similarly, if $x \in B - A$, then $H(x) = G(x)$ is an ideal of R since (G, A) is an idealistic soft ring over R. Hence, for all $x \in \text{Supp}(H, C)$, $H(x)$ is an ideal of R. As a result, $(H, C) = (F, A)\widetilde{\cup}(G, B)$ is an idealistic soft ring over R. □

In Theorem 2.36, if A and B are not disjoint, then the result is not true in general, because the union of two different ideals of a ring R may not be an ideal of R.

Example 32 Let $R = \mathbb{Z}_{10} = \{0, 1, 2, 3, 4, 5, 6, 7, 8, 9\}$, $A = \{0, 4\}$ and $B = \{4\}$. Consider the set-valued function $F : A \longrightarrow \mathcal{P}(R)$ given by $F(x) = \{y \in R \mid x \cdot y = 0\}$. Then $F(0) = R \lhd R$ and $F(4) = \{0, 5\} \lhd R$. Hence, (F, A) is an idealistic soft ring over R. Now, consider the function $G : B \longrightarrow \mathcal{P}(R)$ given by $G(x) = \{0\} \cup \{y \in R \mid x + y \in \{0, 2, 4, 6, 8\}\}$. As we see, $G(4) = \{0, 2, 4, 6, 8\} \lhd R$. Therefore, (G, B) is an idealistic soft ring over R. Since $F(4) \cup G(4) = \{0, 2, 4, 5, 6, 8\}$ is not an ideal of R, $(F, A)\widetilde{\cup}(G, B)$ is not an idealistic soft ring over R.

Theorem 2.37 *(F, A) and (G, B) be idealistic soft rings over R. Then $(F, A) \wedge (G, B)$ is an idealistic soft ring over R if it is non-null.*

Proof We take $(F, A) \wedge (G, B) = (H, C)$ where $C = A \times B$ and $H(a, b) = F(a) \cap G(b)$ for all $(a, b) \in C$. Assume that (H, C) is a non-null soft set over R. If $(x, y) \in \text{Supp}(H, C)$, then $H(x, y) = F(x) \cap G(y) \neq \emptyset$. Since (F, A) and (G, B) are idealistic soft rings over R, the nonempty sets $F(x)$ and $G(x)$ are ideals of R. Therefore, being an intersection of two ideals, $H(x, y)$ is an ideal of R for all $(x, y) \in \text{Supp}(H, C)$. Consequently, $(H, C) = (F, A) \wedge (G, B)$ is an idealistic soft ring over R. □

Example 33 Let $R = \left\{\begin{bmatrix} x & y \\ 0 & z \end{bmatrix} | x, y, z \in \mathbb{Z}\right\}$, $A = 6\mathbb{Z}$ and $B = 10\mathbb{Z}$. Consider the functions $F : A \longrightarrow \mathcal{P}(R)$ and $G : B \longrightarrow \mathcal{P}(R)$ defined by
$F(x) = \left\{\begin{bmatrix} nx & nx \\ 0 & 0 \end{bmatrix} | n \in \mathbb{Z}\right\}$ and $G(x) = \left\{\begin{bmatrix} 0 & nx \\ 0 & nx \end{bmatrix} | n \in \mathbb{Z}\right\}$.
(F, A) and (G, B) are idealistic soft rings over R. Let $(F, A) \wedge (G, B) = (H, C)$ where $C = A \times B$. Then, for all $(x, y) \in C$, we have

$$H(x, y) = F(x) \cap G(y) = \left\{\begin{bmatrix} 0 & tn \\ 0 & 0 \end{bmatrix} \mid n \in \mathbb{Z}\right\} \lhd R$$

where t is equal to the least common multiple of x and y.

Definition 2.23 An idealistic soft ring (F, A) over a ring R is said to be trivial if $F(x) = \{0\}$ for every $x \in A$. An idealistic soft ring (F, A) over R is said to be whole if $F(x) = R$ for all $x \in A$.

Example 34 Let p be a prime integer, $R = \mathbb{Z}_p$ and $A = \mathbb{Z}_p - \{0\}$. Consider the set-valued function $F : A \longrightarrow \mathcal{P}(R)$ given by $F(x) = \{y \in R|(x \cdot y)^{p-1} = 1\} \cup \{0\}$.

Then for all $x \in A$, we have $F(x) = R \lhd R$. Hence, (F, A) is a whole idealistic soft ring over R. Now, consider the function $G : A \longrightarrow \mathcal{P}(R)$ given by $G(x) = \{y \in R | \ xy = 0\}$. As we see, for all $x \in A$ we have $G(x) = \{0\} \lhd R$. Hence, (G, A) is a trivial idealistic soft ring over R.

Let (F, A) be a soft set over R and $f : R \longrightarrow R'$ be a mapping of rings. Then we can define a soft set $(f(F), A)$ over R' where $f(F) : A \longrightarrow P(R')$; is defined as $f(F)(x) = f(F(x))$ for all $x \in A$. Here, by definition, we see that $\text{Supp}(f(F), A) = \text{Supp}(F, A)$.

Proposition 2.12 *Let $f : R \longrightarrow R'$ be a ring epimorphism. If (F, A) is an idealistic soft ring over R, then $(f(F), A)$ is an idealistic soft ring over R'.*

Proof Since (F, A) is a non-null soft set and (F, A) is an idealistic soft ring over R, we observe that $(f(F), A)$ is a non-null soft set over R'. We see that, for all $x \in \text{Supp}(f(F), A)$, $f(F)(x) = f(F(x)) \neq \emptyset$. Since the nonempty set $F(x)$ is an ideal of R and f is an epimorphism, $f(F(x))$ is an ideal of R'. Therefore, $f(F(x))$ is an ideal of R' for all $x \in \text{Supp}(f(F), A)$. Consequently, $(f(F), A)$ is an idealistic soft ring over R'. □

Theorem 2.38 *Let (F, A) be an idealistic soft ring over R and $f : R \longrightarrow R'$ be a ring epimorphism.*

1. *If $F(x) = ker(f)$ for all $x \in A$, then $(f(F), A)$ is the trivial idealistic soft ring over R'.*
2. *If (F, A) is whole, then $(f(F), A)$ is the whole idealistic soft ring over R'.*

Proof (1) Suppose that $F(x) = ker(f)$ for all $x \in A$. Then $f(F)(x) = f(F(x)) = \{0_{R'}\}$ for all $x \in A$. So, $(f(F), A)$ is the trivial idealistic soft ring over R' by Proposition 2.12 and Definition 2.23.

(2) Assume that (F, A) is whole. Then $F(x) = R$ for all $x \in A$. Hence, $f(F)(x) = f(F(x)) = f(R) = R'$ for all $x \in A$. As a result, by Proposition 2.12 and Definition 2.23, $(f(F), A)$ is the whole idealistic soft ring over R'.

□

Definition 2.24 Let (F, A) and (G, B) be soft rings over the rings R and R' respectively. Let $f : R \longrightarrow R'$ and $g : A \longrightarrow B$ be two mappings. The pair (f, g) is called a soft ring homomorphism if the following conditions are satisfied:

1. f is a ring epimorphism,
2. g is surjective,
3. $f(F(x)) = G(g(x)) \ \forall x \in A$.

If we have a soft ring homomorphism between (F, A) and (G, B), (F, A) is said to be soft homomorphic to (G, B),which is denoted by $(F, A) \sim (G, B)$. In addition, if f is a ring isomorphism and g is bijective mapping, then (f, g) is called a soft ring isomorphism. In this case, we say that (F, A) is softly isomorphic to (G, B), which is denoted by $(F, A) \simeq (G, B)$.

Example 35 Consider the rings $R = \mathbb{Z}$ and $R' = \{0\} \times \mathbb{Z}$. Let $A = 2\mathbb{Z}$ and $B = \{0\} \times 6\mathbb{Z}$. We see that (F, A) is a soft ring over R and (G, B) is a soft ring over R'. Consider the set-valued functions $F : A \longrightarrow \mathcal{P}(R)$ and $G : B \longrightarrow P(R')$ which are given by $F(x) = x18\mathbb{Z}$ and $G((0, y)) = \{0\} \times 6y\mathbb{Z}$. Then the function $f : R \longrightarrow R'$ which is given by $f(x) = (0, x)$ is a ring isomorphism. Moreover, the function $g : A \longrightarrow B$ which is defined by $g(y) = (0, 3y)$ is a surjective map. As we see, for all $x \in A$, we have $f(F(x)) = f(18x\mathbb{Z}) = \{0\} \times 18x\mathbb{Z}$ and $G(g(x)) = G(\{0\} \times 6x\mathbb{Z}) = \{0\} \times 18x\mathbb{Z}$. Consequently, (f, g) is a soft ring isomorphism and $(F, A) \simeq (G, B)$.

2.4 Soft Modules

Soft module theory, which extends the notion of module by including some algebraic structures in soft sets, will be introduced in this section. A soft module defined is a parameterized family of submodules, and has many properties similar to those of modules. Definitions, examples and results given here are due to Qiu-Mei Sun et al. [133].

First we recall some basic definitions and results related to module theory [9, 60].

Definition 2.25 Let R be a ring with identity. M is said to be a left R -module if left scalar multiplication $\lambda : R \times M \rightarrow M$ via $(a, x) \mapsto$ ax satisfying the axioms $\forall r, r_1, r_2, 1 \in R; m, m_1, m_2 \in M$:

(1) M is an abelian group (which we shall write additively);
(2) $r\,(m_1 + m_2) = rm_1 + rm_2$, $(r_1 + r_2)\, m = r_1 m + r_2 m$
(3) $(r_1 r_2)\, m = r_1\,(r_2 m)$
(4) $1m = m$

Left R-module is denoted by ${}_R M$ or M for short. Similarly we can define right R-module and denote it by M_R.

For ${}_R M$ and M_S, M is said to be S-R-bimodule if the following conditions are satisfied:
$s(mr) = (sm)r \quad \forall s \in S, m \in M, r \in R$
Let M be a left R-module, then an abelian subgroup N of M is a left R - submodule of M in case it is closed under scalar multiplication by R. This relationship will be denoted by $N < M$.

Proposition 2.13 *A is a subset of M, then the following statements are equivalent*

(1) $N < M$;
(2) A is an abelian subgroup of M and $ar \in A$ for all $a \in A, r \in R$;
(3) $a_1 + a_2 \in A$, $ar \in A$ for all $a, a_1, a_2 \in A, r \in R$.

Definition 2.26 N is a non-trivial submodule of M

(1) if there is no submodule of M which contain N, we call N the maximal submodules.
(2) if there is no non-zero submodule of M which contained in N, we call N the minimal submodules.

Definition 2.27 Let $\{M_i | i \in I\}$ be a nonempty family of submodules
(1) if $\{M_i | i \in I\}$ is a family of maximal submodules, then $\bigcap_{i \in I} M_i$ is *a* submodule of M called Jacobson radical of module. This is denoted by radM.
(2) if $\{M_i | i \in I\}$ is a family of minimal submodules, then $\sum_{i \in I} M_i$ is *a* submodule of M called socle of module. This is denoted by socM.

Definition 2.28 Let $\{M_i | i \in I\}$ be a nonempty family of R -modules, $P = \prod_{i \in I} M_i = \{(x_i) \, | x_i \in M_i\}$ is a direct product set, if the operations on the product are given by

$$(x_i) + (y_i) = (x_i + y_i) \quad r\,(x_i) = (r x_i)$$

then P do induce a left R -module structure called direct product of $\{M_i | i \in I\}$, which will be denoted by $\prod_{i \in I} M_i$.

Proposition 2.14 *Let $\{M_i | i \in I\}$ be a family of submodules of M, then $\bigcap_{i \in I} M_i$ and $\sum_{i \in I} M_i$ are all submodules of M.*

Definition 2.29 All the elements (x_i) in the direct product $\prod_{i \in I} M_i$, where x_i is zero for almost all $i \in I$ except finite one, establish a submodule of $\prod_{i \in I} M_i$ which called direct sum of $\{M_i | i \in I\}$, will be denoted by $\bigoplus_{i \in I} M_i$.

Definition 2.30 The homomorphism sequence of R-modules $\cdots \to M_{n-1} \xrightarrow{f_{n-1}} M_n \xrightarrow{f_n} M_{n+1} \to \cdots$ is called exact sequence of modules if $\operatorname{Im} f_{n-1} = Ker f_n$ for all $n \in N$, and we call the exact sequence of modules form as $0 \to M' \xrightarrow{f} M \xrightarrow{g} M'' \to 0$ the short exact sequence of modules.

Now we define soft modules. Throughout this section, let M be a left R-module, A be any nonempty set. $F : A \to P(M)$ refer to a set-valued function and the pair (F, A) is a soft set over M.

Definition 2.31 Let (F, A) be a soft set over M. (F, A) is said to be a soft module over M if and only if $F(x) < M$ for all $x \in A$.

Definition 2.32 Let (F, A) and (G, B) be two soft modules over M. Then (G, B) is soft submodule of (F, A) if

(1) $B \subset A$ and
(2) $G(x) < F(x), \forall x \in B$

This is denoted by $(G, B) \tilde{<} (F, A)$.

Definition 2.33 Let (F, A) be a soft module over M, then

(1) (F, A) is said to be a null soft module over M if $F(x) = 0$ for all $x \in A$, where 0 is zero element of M.
(2) (F, A) is said to be an absolute soft module over M if $F(x) = M$ for all $x \in A$.

Definition 2.34 Let $E = \{e\}$, where e is unit of A. Then every soft module (F, A) over M at least have two soft submodules (F, A) and (F, E) called trivial soft submodules.

Proposition 2.15 *Let (F, A) and (G, B) be two soft modules over M then*

(1) $(F, A)\widetilde{\cap}(G, B)$ is a soft module over M.
(2) $(F, A)\widetilde{\cup}(G, B)$ is a soft module over M if $A \cap B = \emptyset$

Proof (1) From Definition we know that $(F, A)\widetilde{\cap}(G, B) = (H, C)$ is a soft set over M, where $C = A \cap B$ and $H(x) = F(x) \cap G(x) < M$ for all $x \in C$ since (F, A) and (G, B) are soft module over M, so $(F, A)\widetilde{\cap}(G, B)$ is a soft module over M.

(2) We know $(F, A)\widetilde{\cup}(G, B) = (H, C)$ is a soft set, where $C = A \cup B$ and

$$H(x) = \begin{cases} F(x) & x \in A - B \\ G(x) & x \in B - A \\ F(x) \cup G(x) & x \in A \cap B \end{cases}$$

Now $x \in A - B$ or $x \in B - A$ as $A \cap B = \emptyset$, thus (H, C) is a soft module over M since (F, A) and (G, B) are soft modules.

The following results are the trivial consequences of the definitions and propositions already given in this section.

Proposition 2.16 *Let (F, A) and (G, B) be two soft modules over M. Then $(F, A)+(G, B)$ denoted as $(H, A \times B)$, where $H(x, y) = F(x) + G(y) \forall (x, y) \in A \times B$, is is soft module over M.*

Proposition 2.17 *Let (F, A) and (G, B) be two soft modules over M and N respectively. Then $(F, A) \times (G, B) = (H, A \times B)$ denoted as $H(x, y) = F(x) \times G(y)$ for all $(x, y) \in A \times B$, is a soft module over $M \times N$.*

Proposition 2.18 *Let (F, A) and (G, B) be two soft modules over M. (G, B) is soft submodule of (F, A) if $G(x) \subseteq F(x), \forall x \in A$.*

Proposition 2.19 *Let (F, A) be a soft module over M, and $\{(G_i, B_i) \,|i \in I\}$ be a nonempty family of soft submodules of (F, A). Then*

(1) $\widetilde{\sum}_{i\in I} (G_i, B_i)$ is soft submodule of (F, A).
(2) $\widetilde{\bigcap}_{i\in I} (G_i, B_i)$ is soft submodule of (F, A).
(3) $\widetilde{\bigcup}_{i\in I} (G_i, B_i)$ is soft submodule of (F, A), if $B_i \cap B_j = \emptyset$ for all $i, j \in I$

Proposition 2.20 *Let (F, A) and (G, B) be two soft modules over M, and (G, B) be soft submodule of (F, A). If $f : M \to N$ is homomorphism of module, then $(f(F), A)$ and $(f(G), B)$ are all soft modules over N, and $(f(G), B)$ is soft sub module of $(f(F), A)$.*

Proof It is easy to prove that the image of a homomorphism of submodule is a submodule. □

Definition 2.35 Let (F, A) and (G, B) be two soft modules over M and N respectively, $f : M \to N, g : A \to B$ be two functions. Then we say that (f, g) is a soft homomorphism if the following conditions are satisfied: (1) $f : M \to N$ is homomorphism of module; (2) $f(F(x)) = G(g(x)), \forall x \in A$. At the same time, we say (F, A) is soft homomorphic to (G, B), which denoted by $(F, A) \simeq (G, B)$.

In this definition, if f is an isomorphism from M to N and g is a one-to-one mapping from A onto B, then we say that (F, A) is a soft isomorphism and that (F, A) is a soft isomorphic to (G, B), this is denoted by $(F, A) \cong (G, B)$.

Definition 2.36 Let (F, A) and (G, B) be two soft modules over M, and (G, B) be soft submodule of (F, A). We say (G, B) is maximal soft submodule of (F, A) if $G(x)$ is maximal submodule of $F(x)$ for all $x \in B$. We say (G, B) is minimal soft submodule of (F, A) if $G(x)$ is minimal submodule of $F(x)$ for all $x \in B$.

Now the following proposition follows clearly

Proposition 2.21 *Let (F, A) be a soft module over M.*

(1) if $\{(G_i, B_i) \mid i \in I\}$ is a nonempty family of maximal soft submodules of (F, A), then $\bigcap_{i\in I} (G_i, B_i)$ is maximal soft submodule of (F, A).
(2) if $\{(G_i, B_i) \mid i \in I\}$ is a nonempty family of minimal soft submodules of (F, A), then $\sum_{i\in I} (G_i, B_i)$ is minimal soft submodule of (F, A).

Proposition 2.22 *(1) Let (F, A) be a soft module over M and $f : M \to N$ be a homomorphism. If $F(x)$ =Kerf for all $x \in A$, then $(f(F), A)$ is the null soft module over N.*
(2) Let (F, A) be an absolute soft module over M and let $f : M \to N$ be an epimorphism, then $(f(F), A)$ is an absolute soft module over N.

Proof Proof is straight forward. □

Proposition 2.23 *Let (F, A) be a null soft module over module P and (G, B) be an absolute soft module over module Q. if $0 \to P \xrightarrow{f} M \xrightarrow{g} Q \to 0$ is a short exact sequence, then $0 \to F(x) \xrightarrow{\tilde{f}} M \xrightarrow{\tilde{g}} G(y) \to 0$ is a short exact sequence for all $x \in A, y \in B$.*

Proof $F(x) = 0, \forall x \in A$ since (F, A) is a null soft module over P, so $\tilde{f}$ is a monomorphism. $G(y) = Q, \forall y \in B$ since (G, B) is an absolute soft module over $Q \cdot g : M \to Q$ is an epimorphism as $0 \to P \xrightarrow{f} M \xrightarrow{g} Q \to 0$ is a short exact sequence, so $\widetilde{g}$ is an epimorphism. □

Proposition 2.24 *Let (F, A) be a null soft module over module P and (G, B) be an absolute soft module over module M. if $0 \to P \xrightarrow{f} M \xrightarrow{g} Q \to 0$ is a short exact sequence, then $0 \to f(F)(x) \xrightarrow{\tilde{f}} M \xrightarrow{\tilde{g}} g(G)(y) \to 0$ is a short exact sequence for all $x \in A$, $y \in B$.*

Proof $F(x) = 0, \forall x \in A$ since (F, A) is a null soft module over P. Kerf $= 0$. So Kerf $= F(x), \forall x \in A$, consequently $(f(F), A)$ is null soft module over M. (G, B) is an absolute soft module over M and $g : M \to Q$ is an epimorphism, so $(g(G), B)$ is an absolute soft module over Q, thus $0 \to f(F)(x) \xrightarrow{\tilde{f}} M \xrightarrow{\tilde{g}} g(G)(y) \to 0$ is a short exact sequence for all $x \in A$, $y \in B$ by Proposition 2.23. □

2.5 Soft Lattice Structures

The concept of soft lattices has been evolved from soft sets. Initial works related to this can be seen in F. Li [72] where the basic operations on soft lattices are defined. Further, a different approach with an application of soft sets to lattices has been done by E. Kuppusamy [68]. In this section, we give operations on soft lattices with their properties. Results given in this section is mainly taken from Jobish et al. [57]. For basic definitions and results related to lattice theory see [21, 85].

Definition 2.37 A partially ordered set (or a poset) is an ordered pair $(S, \leq)$ where S is a set and '$\leq$' is a partial order on S. We say that S is partially ordered by $\leq$.

A relation which is reflexive, transitive and anti-symmetric is called a partial order or ordering.

Definition 2.38 A partially ordered set $(S, \leq)$ is called a lattice if for every two-element subset of S(i.e. every subset of the form $\{a, b\}$ where a, b are distinct elements of S) has both least upper bound (denoted by $a \vee b$) and the greatest lower bound (denoted by $a \wedge b$) exist in S. It is called a complete lattice if the least upper bound and the greatest lower bound exist for every subset of S.

Let L be a complete lattice. Its universal bounds are denoted by 0_L and 1_L. We presume that L is consistent. i.e., 0_L is distinct from 1_L.Thus $0_L \leq \alpha \leq 1_L$ for all $\alpha \in L$. We note $\vee\emptyset = 0_L$ and $\wedge\emptyset = 1_L$. If it will not cause any confusion, we denote 0_L and 1_L by 0 and 1 for short respectively.

Definition 2.39 A quasi complementation is a mapping $' : L \to L$ which is involutive (i.e., $(\alpha')' = \alpha$ for all $\alpha \in L$) and order reversing. (i.e., $\alpha \leq \beta$ implies $\beta' \leq \alpha'$).

In $(L, ')$ the DeMorgan's laws hold. i.e.,

$$(\vee A)' = \wedge\{\alpha' : \alpha \in A\} \text{ and } (\wedge A)' = \vee\{\alpha' : \alpha \in A\}, \forall A \subset L.$$

In particular, $0' = 1$ and $1' = 0$. An element $b \in L$ is called a complement of an element $a \in L$ if $b \wedge a = 0, b \vee a = 1$.

More over if every element in L is having a unique complement, and then L will be called a uniquely complemented lattice. A non empty subset R of L is said to be a sub lattice of L if $a, b \in R$ implies $a \vee b, a \wedge b \in R$.

If R is a sub lattice of a uniquely complemented lattice L, then lattice complement of R in L is defined as $R^c = \{a' : a \in R\}$. From uniqueness of complementation and DeMorgan's laws it clearly follows that R^c is also a sub-lattice.

Definition 2.40 A lattice L is called distributive, if L satisfies either of the logically equivalent FD1 or FD2 below:

FD1: $\forall a, b, c \in L, a \wedge (b \vee c) = (a \wedge b) \vee (a \wedge c)$
FD2: $\forall a, b, c \in L, a \vee (b \wedge c) = (a \vee b) \wedge (a \vee c)$.

Definition 2.41 A complete lattice L is called completely distributive if it satisfies either of the logically equivalent CD1 or CD2 below:

CD1: $\wedge_{i \in I}(\vee_{j \in J_i} a_{i,j}) = \vee_{\phi \in \prod_{i \in I} J_i}(\wedge_{i \in I} a_{i,\phi(i)})$
CD2: $\vee_{i \in I}(\wedge_{j \in J_i} a_{i,j}) = \wedge_{\phi \in \prod_{i \in I} J_i}(\vee_{i \in I} a_{i,\phi(i)})$,

for all $\{\{a_{i,j} : j \in J_i\} : i \in I\} \subset P(L) \backslash \{\emptyset\}, I \neq \emptyset$.

Definition 2.42 Given two non empty sub lattices A, B of L, the lattice meet $A \wedge B$ and the lattice join $A \vee B$ are defined as $A \wedge B = \{a \wedge b : a \in A, b \in B\}$ and $A \vee B = \{a \vee b : a \in A, b \in B\}$ respectively.

Clearly $A \vee B$ and $A \wedge B$ are sub lattices of L.

Definition 2.43 Let L be any complete lattice and E be a set of parameters then a soft lattice over L is a triplet of the form (f, E, L),where $f : E \rightarrow P(L)$ and $f(e)$ is a sub lattice of L for every $e \in E$. Alternatively we represent (f, E, L) as f_E^L.

Definition 2.44 For two soft lattices $(f, A, L), (g, B, L)$ over L, we say (f, A, L) is a sub soft lattice of (g, B, L) if

1. $A \subset B$
2. $\forall \varepsilon \in A, g(\varepsilon)$ *and* $f(\varepsilon)$ are identical

It is denoted by $(f, A, L) \subset (g, B, L)$

Definition 2.45 A soft lattice (f, A, L) is said to be null soft lattice, if $\forall \delta \in E, f(\delta) = \{0_L\}$. A soft lattice (f, E, L) is said to be absolute soft lattice, if $\forall \delta \in X, f(\delta) = \{1_L\}$.

The null soft lattice will be represented by Φ and the absolute soft lattice by $\mathcal{L}$. Where E is the attribute set and $A \subset E$.

If $A \subset E$, we use notations Φ_A and $\mathcal{L}_A$ to soft lattices with respect to the parameter subset A, where for each $\forall a \in A, f(a) = \{0_L\}$ and $f(a) = \{1_L\}$ respectively.

Definition 2.46 Two soft lattice (f, A, L), (g, B, L) are said to be soft equal if $(f, A, L) \subset (g, B, L)$, and $(g, B, L) \subset (f, A, L)$.

Definition 2.47 The lattice complement of a soft lattice (f, A, L) is denoted by $(f, A, L)^c$ is defined as $(f, A, L)^c = (f^c, A, L)$ where $f^c(x) = [f(x)]^c$ where $[f(x)]^c$ is the lattice complement of $f(x)$ in L.

Definition 2.48 The set relative complement of a soft lattice (f, A, L) is denoted by $(f, A, L)^r$ and is defined by $(f, A, L)^r = (f^r, A, L)$, where $f^r : A \to P(L)$ is a mapping given by $f^r(\alpha) = L \backslash f(\alpha), \forall \alpha \in A$

Definition 2.49 (AND and OR operations in Soft Lattices :) Let (f, X, L), (g, Y, L) be two soft Lattices.
'(f, X, L) AND (g, Y, L)'is defined by:

$$(f, X, L) \wedge (g, Y, L) = (o, X \times Y, L)$$
$$\text{where } \forall(\alpha, \beta) \in X \times Y, o(\alpha, \beta) = f(\alpha) \wedge g(\beta).$$

'(f, X, L) OR (g, Y, L)'is defined by:

$$(f, X, L) \vee (g, Y, L) = (h, X \times Y, L)$$
$$\text{where } \forall(\alpha, \beta) \in X \times Y, h(\alpha, \beta) = f(\alpha) \vee g(\beta);$$

where $\wedge$, $\vee$ are the lattice meet and lattice join respectively.

Definition 2.50 Let (f, X, L), (g, Y, L) be two soft lattices then,

1. Intersection of (f, X, L) and (g, Y, L) is the soft lattice $(h, X \cap Y, L)$, in which, $\forall \delta \in X \cap Y, h(\delta) = f(\delta) \wedge g(\delta)$, denoted by $(f, X, L) \sqcap (g, Y, L)$
2. Extended intersection of (f, X, L) and (g, Y, L) is the soft lattice $(h, X \cup Y, L)$ where

$$h(\delta) = \begin{cases} f(\delta), & \text{if } \delta \in X \backslash Y, \\ g(\delta), & \text{if } \delta \in Y \backslash X, \\ f(\delta) \wedge g(\delta), & \text{if } \delta \in X \cap Y. \end{cases}$$

   and is denoted by $(f, X, L) \sqcap_E (g, Y, L)$
3. Union of (f, X, L) and (g, Y, L) is the soft lattice, $(h, X \cup Y, L)$ where

$$h(\delta) = \begin{cases} f(\delta), & \text{if } \delta \in X \backslash Y, \\ g(\delta), & \text{if } \delta \in Y \backslash X, \\ f(\delta) \vee g(\delta), & \text{if } \delta \in X \cap Y. \end{cases}$$

   and is denoted by $(f, X, L) \sqcup (g, Y, L)$

4. Restricted union of (f, X, L) and (g, Y, L) with $X \cap Y \neq \emptyset$ is the soft lattice-$(h, X \cap Y, L)$, in which $\forall \delta \in X \cap Y, h(\delta) = f(\delta) \vee g(\delta)$.
   We denote as $(f, X, L) \sqcup_R (g, Y, L)$.
5. Let (f, X, L) and (g, Y, L) be two soft lattices over L such that $X \cap Y \neq \phi$. The restricted difference of (f, X, L) and (g, Y, L) is denoted by $(f, X, L) \sim_R (g, Y, L)$ and is defined as $(f, X, L) \sim_R (g, Y, L) = (h, X \cap Y, L)$.
   Where $h(\delta) = f(\delta) \backslash g(\delta), \forall \delta \in X \cap Y \neq \phi$.

2.5.1 DeMorgan's Laws in Soft Lattices

Theorem 2.39 *Let (f, X, L), (g, Y, L) be two soft lattices, then the soft lattice versions of DeMorgan's Laws are,*

1. $((f, X, L) \vee (g, Y, L))^c = (f, X, L)^c \wedge (g, Y, L)^c$
2. $((f, X, L) \wedge (g, Y, L))^c = (f, X, L)^c \vee (g, Y, L)^c$
3. $((f, X, L) \sqcup_R (g, Y, L))^c = (f, X, L)^c \sqcap (g, Y, L)^c$
4. $((f, X, L) \sqcap (g, Y, L))^c = (f, X, L)^c \sqcup_R (g, Y, L)^c$

Proof 1. $((f, X, L) \vee (g, Y, L))^c = (f, X, L)^c \wedge (g, Y, L)^c$
Consider soft lattices (f, X, L) and (g, Y, L) then

$$\begin{aligned}
(f, X, L) \vee (g, Y, L) &= (h, X \times Y, L), \\
&\quad \text{where} \forall (\alpha, \beta) \in X \times Y, h(\alpha, \beta) \\
&= f(\alpha) \vee g(\beta) \\
\text{Therefore}((f, X, L) \vee (g, Y, L))^c &= (h, X \times Y, L)^c = (h^c, (X \times Y), L) \\
\text{again, } (f, X, L)^c \wedge (g, Y, L)^c &= (f^c, X, L) \wedge (g^c, Y, L) \\
&= (o, X \times Y, L) \\
&\quad \text{where } \forall (\alpha, \beta) \in X \times Y, o(\alpha, \beta) \\
&= f^c(\alpha) \wedge g^c(\beta) \\
\text{Now,} h^c(\alpha, \beta) &= [f(\alpha) \vee g(\beta)]^c \\
&= [f(\alpha)]^c \wedge [g(\beta)]^c \\
&= f^c(\alpha) \wedge g^c(\beta) \\
&= o(\alpha, \beta).
\end{aligned}$$

Which proves, $((f, X, L) \vee (g, Y, L))^c = (f, X, L)^c \wedge (g, Y, L)^c$.

2. $((f, X, L) \wedge (g, Y, L))^c = (f, X, L)^c \vee (g, Y, L)^c$

$$\begin{aligned}
\text{We have,}(f, X, L) \wedge (g, Y, L) &= (h, X \times Y, L). \\
&\quad \text{where } \forall (\alpha, \beta) \in X \times Y, h(\alpha, \beta) \\
&= f(\alpha) \wedge g(\beta)
\end{aligned}$$

$$\text{Therefore } ((f, X, L) \wedge (g, Y, L))^c = (h, X \times Y, L)^c = (h^c, X \times Y, L)$$

$$\text{again, } (f, X, L)^c \vee (g, Y, L)^c = (f^c, X, L) \vee (g^c, Y, L)$$

$$= (o, X \times Y, L)$$

$$\text{where } \forall(\alpha, \beta) \in X \times Y, o(\alpha, \beta) = f^c(\alpha) \vee g^c(\beta)$$

$$\text{Now, } h^c(\alpha, \beta) = [f(\alpha) \wedge g(\beta)]^c = [f(\alpha)]^c \vee [g(\beta)]^c = f^c(\alpha) \vee g^c(\beta) = o(\alpha, \beta)$$

Which proves the conclusion : $((f, X, L) \wedge (g, Y, L))^c = (f, X, L)^c \vee (g, Y, L)^c$

3. $((f, X, L) \sqcup_R (g, Y, L))^c = (f, X, L)^c \sqcap (g, Y, L)^c$
Consider soft lattices (f, X, L) and (g, Y, L) over L. Then, restricted union is defined as,

$$(f, X, L) \sqcup_R (g, Y, L) = (h, X \cap Y, L)$$

$$\text{where } \forall \alpha \in X \cap Y \neq \phi, h(\alpha) = f(\alpha) \vee g(\alpha).$$

$$\text{Now, } h^c(\alpha) = [f(\alpha) \vee g(\alpha)]^c, \forall \alpha \in X = [f(\alpha)]^c \wedge [g(\alpha)]^c = f^c(\alpha) \wedge g^c(\alpha) \text{ accordingly.}$$

$$\text{Thus we have, } (h, X \cap Y, L)^c = (h^c, X \cap Y, L) = (f, X, L)^c \sqcap (g, Y, L)^c.$$

$$\text{i.e.,} ((f, X, L) \sqcup_R (g, Y, L))^c = (f, X, L)^c \sqcap (g, Y, L)^c$$

proves the conclusion.

4. $((f, X, L) \sqcap (g, X, L))^c = (f, X, L)^c \sqcup_R (g, X, L)^c$
Consider soft lattices (f, X, L) and (g, Y, L) over L.

$$\text{Then, } (f, X, L) \sqcap (g, Y, L) = (h, X \cap Y, L)$$

$$\text{where } \forall \alpha \in X, h(\alpha) = f(\alpha) \wedge g(\alpha)$$

$$\text{Now, } h^c(\alpha) = [f(\alpha) \wedge g(\alpha)]^c, \forall \alpha \in X = [f(\alpha)]^c \vee [g(\alpha)]^c = f^c(\alpha) \vee g^c(\alpha)$$

$$\text{But we have } (f, X, L)^c \sqcup_R (g, Y, L)^c = (k, X \cap Y, L)$$

$$\text{where } k(\alpha) = f^c(\alpha) \vee g^c(\alpha)$$

$$\text{Thus, } h^c(\alpha) = k(\alpha)$$

$$\text{i.e.,} ((f, X, L) \sqcap (g, X, L))^c = (f, X, L)^c \sqcup_R (g, X, L)^c.$$

□

The following theorem shows that results similar to Theorem 2.39 will be holding for set relative complement also.

Theorem 2.40 *Let (f, X, L), (g, Y, L) be two soft lattices, then the set relative complement versions of DeMorgan's Laws are:*

1. $((f, X, L) \vee (g, Y, L))^r = (f, X, L)^r \wedge (g, Y, L)^r$
2. $((f, X, L) \wedge (g, Y, L))^r = (f, X, L)^r \vee (g, Y, L)^r$
3. $((f, X, L) \sqcup_R (g, X, L))^r = (f, X, L)^r \sqcap (g, X, L)^r$
4. $((f, X, L) \sqcap (g, X, L))^r = (f, X, L)^r \sqcup_R (g, X, L)^r$

Proof Proof is similar to that of theorem 2.39 and we omit it. □

2.5.2 Properties of Soft Lattice Operations

Here, we discuss the basic properties of various operations on soft lattices defined earlier.

Theorem 2.41 *Let (f, A, L), (g, B, L), (h, C, L) be three soft lattice over L. Then we have,*

1. $(f, A, L) \sqcup ((g, B, L) \sqcup (h, C, L)) = ((f, A, L) \sqcup (g, B, L)) \sqcup (h, C, L)$.
2. $(f, A, L) \sqcup \mathcal{L}_A = \mathcal{L}_A$, $(f, A, L) \sqcup \mathcal{L} = \mathcal{L}$, $(f, A, L) \sqcup \Phi = (f, A, L)$, $(f, A, L) \sqcap \Phi_A = \Phi_A$
3. $(f, A, L) \sqcup (g, A, L) = \Phi_A \iff (f, A, L) = \Phi_A$ and $(g, A, L) = \Phi_A$.
4. $(f, A, L) \sqcup ((g, B, L) \sqcap (h, C, L)) = ((f, A, L) \sqcup (g, B, L)) \sqcap ((f, A, L) \sqcup (h, C, L))$.
5. $((f, A, L) \sqcap (g, B, L)) \sqcup (h, C, L) = ((f, A, L) \sqcup (h, C, L)) \sqcap ((g, B, L) \sqcup (h, C, L))$.

Proof The proofs of (1) and (2) follows directly from definitions.
(3) Suppose that $(f, A, L) \sqcup (g, A, L) = (h, A, L)$, $h(a) = f(a) \vee g(a)$, $\forall a \in A$. Since $(h, A, L) = \Phi_A$ from the assumption,

$$\begin{aligned} h(a) &= f(a) \vee g(a) = \{0_L\}, \forall a \in A \\ &\iff f(x) = \{0_L\} \text{ and } g(x) = \{0_L\} \\ &\iff (f, A, L) = \Phi_A \text{ and } (g, A, L) = \Phi_A. \end{aligned}$$

(4) and (5) can be proved easily so we omit here. □

Remark 2.2 (f, A, L) needs not be a soft sub-lattice of $(f, A, L) \sqcup (g, B, L)$. But if $(f, A, L) \subset (g, B, L)$, then $(f, A, L) \subset (f, A, L) \sqcup (g, B, L)$, and in this case

$$(f, A, L) = (f, A, L) \sqcup (g, B, L).$$

Remark 2.3 Let E be parameter set. Then the set of all soft lattices over a lattice L denoted by $S_L(E)$ forms a bounded distributive lattice $(S_L(E), \mathcal{L}_A, \Phi_A, \sqcup, \sqcap)$.

Theorem 2.42 *Let* $(f, A, L), (g, B, L), (h, C, L)$ *be three soft lattices over* L. *Then we have:*

1. $(f, A, L) \sqcup_R [(g, B, L) \sqcup_R (h, C, L)] = [(f, A, L) \sqcup_R (g, B, L)] \sqcup_R (h, C, L)$.
2. $(f, A, L) \sqcup_R \mathcal{L}_A = \mathcal{L}_A, (f, A, L) \sqcup_R \Phi_A = (f, A, L)$.
3. (f, A, L) need not be a soft sub-lattice of $(f, A, L) \sqcup_R (g, B, L)$.
   But if $(f, A, L) \subset (g, B, L)$, then $(f, A, L) \subset (f, A, L) \sqcup_R (g, B, L)$, and-$(f, A, L) = (f, A, L) \sqcup_R (g, B, L)$.
4. $(f, A, L) \sqcup_R (g, A, L) = \Phi_A \Longleftrightarrow (f, A, L) = \Phi_A$ and $(g, A, L) = \Phi_A$.
5. $(f, A, L) \sqcup_R [(g, B, L) \sqcap (h, C, L)] = [(f, A, L) \sqcup_R (g, B, L)] \sqcap [(f, A, L) \sqcup_R (h, C, L)]$.
6. $[(f, A, L) \sqcap (g, B, L)] \sqcup_R (h, C, L) = [(f, A, L) \sqcup_R (h, C, L)] \sqcap [(g, B, L) \sqcup_R (h, C, L)]$.
7. $(f, A, L) \sqcup_R [(g, B, L) \sqcap_E (h, C, L)] = [(f, A, L) \sqcup_R (g, B, L)] \sqcap_E [(f, A, L) \sqcup_R (h, C, L)]$.
8. $[(f, A, L) \sqcap_E (g, B, L)] \sqcup_R (h, C, L) = [(f, A, L) \sqcup_R (h, C, L)] \sqcap_E [(g, B, L) \sqcup_R (h, C, L)]$.

Proof Proof follow from the definitions and are similar to that of Theorem 2.41. As an illustration we give proof of (7).

Suppose that $(g, B, L) \sqcap_E (h, C, L) = (t, B \cup C, L)$, where,$\forall \delta \in B \cup C$

$$t(\delta) = \begin{cases} g(\delta), & \text{if } \delta \in B \backslash C, \\ h(\delta), & \text{if } \delta \in C \backslash B, \\ g(\delta) \wedge h(\delta), & \text{if } \delta \in B \cap C. \end{cases}$$

Assume that $(f, A, L) \sqcup_R (t, B \cup C, L) = (m, A \cap (B \cup C), L)$

where $m(x) = f(x) \vee t(x), \forall x \in A \cap (B \cup C)$.

Rewriting 'm'by replacing 't', then-

$$m(\delta) = \begin{cases} f(\delta) \vee g(\delta), & \text{if } \delta \in A \cap (B \setminus C) = (A \cap B) \setminus (A \cap C), \\ f(\delta) \vee h(\delta), & \text{if } \delta \in A \cap (C \setminus B) = (A \cap C) \setminus (A \cap B), \\ f(\delta) \vee (g(\delta) \wedge h(\delta)), & \text{if } \delta \in A \cap (B \cap C), \forall \delta \in A \cap (B \cup C). \end{cases}$$

Now consider RHS.

Suppose that$(f, A, L) \sqcup_R (g, B, L) = (q, A \cap B, L)$

where $q(x) = f(x) \vee g(x) \forall x \in A \cap B \neq \emptyset$

Assume$(f, A, L) \sqcup_R (h, C, L) = (w, A \cap C, L)$

where $w(x) = f(x) \vee h(x) \forall x \in A \cap C \neq \emptyset$

Let $(q, A \cap B, L) \sqcap_E (w, A \cap C, L) = (n, (A \cap B) \cup (A \cap C), L)$, where

$$n(\delta) = \begin{cases} q(\delta), & \text{if } \delta \in (A \cap B) \setminus (A \cap C), \\ w(\delta), & \text{if } \delta \in (A \cap C) \setminus (A \cap B), \\ q(\delta) \wedge w(\delta), & \text{if } \delta \in (A \cap B) \cap (A \cap C) = A \cap (B \cap C). \end{cases}$$

for all $\delta \in (A \cap B) \cup (A \cap C)$. Again, we can rewrite 'n'as,

$$n(\delta) = \begin{cases} f(\delta) \vee g(\delta), & \text{if } \delta \in (A \cap B) \setminus (A \cap C), \\ f(\delta) \vee h(\delta), & \text{if } \delta \in (A \cap C) \setminus (A \cap B), \\ (f(\delta) \vee g(\delta)) \wedge (f(\delta) \vee h(\delta)), & \text{if } \delta \in A \cap (B \cap C). \end{cases}$$

as 'm'and 'n'same gives the proof. □

Results similar to Theorem 2.41 and 2.42 will follow for other versions of union and intersections also.

Theorem 2.43 *Let* (f, A, L), (g, B, L), (h, C, L) *be three soft lattices over L. Then we have:*

1. $(f, A, L) \sqcap_E ((g, B, L) \sqcap_E (h, C, L)) = ((f, A, L) \sqcap_E (g, B, L)) \sqcap_E (h, C, L)$.
2. $(f, A, L) \sqcap_E \mathcal{L} = (f, A, L), (f, A, L) \sqcap_E \Phi = \Phi$.
3. $(f, A, L) \sqcap_E (g, B, L) \not\subseteq (g, B, L)$, in general. But if$(f, A, L) \subset (g, B, L)$, then-$(f, A, L) \sqcap_E (g, B, L) \subset (g, B, L)$. More over$(f, A, L) \sqcap_E (g, B, L) = (g, B, L)$
4. $(f, A, L) \sqcap_E ((g, B, L) \sqcup_R (h, C, L)) = ((f, A, L) \sqcap_E (g, B, L)) \sqcup_R ((f, A, L) \sqcap_E (h, C, L))$.
5. $((f, A, L) \sqcup_R (g, B, L)) \sqcap_E (h, C, L) = ((f, A, L) \sqcap_E (h, C, L)) \sqcup_R ((g, B, L) \sqcap_E (h, C, L))$.

Proof (1) and (2) clear.
(3) Since $A \cup B \not\subseteq A$, $(f, A, L) \sqcap_E (g, B, L) \not\subseteq (f, A, L)$, in general.
Suppose that $(f, A, L) \subset (g, B, L)$ and $(f, A, L) \sqcap_E (g, B, L) = (h, C, L)$, where,

$$h(\delta) = \begin{cases} f(\delta), & \text{if } \delta \in A \setminus B, \\ g(\delta), & \text{if } \delta \in B \setminus A, \\ f(\delta) \wedge g(\delta), & \text{if } \delta \in A \cap B, \forall \delta \in A \cup B. \end{cases}$$

Since $A \subset B$, then it is obvious that $A \cup B = B$.

For having, $h(x)$ and $g(x)$ the same approximations for all $x \in B$, Let $x \in B$, then either $x \in B - A$ or $x \in A \cap B = A$. If $x \in B \setminus A$, then $h(x) = g(x)$, and if $x \in A \cap B = A$, then $h(x) = f(x) \wedge g(x) = g(x) \wedge g(x) = g(x) \forall x \in A$, since $f(x)$ and $g(x)$ are the same approximations for all $x \in A$. Thus $g(x)$ and $h(x)$ are the identical approximations for all $x \in B$.

(4) Suppose that $(g, B, L) \sqcup_R (h, C, L) = (m, B \cap C, L)$

where $m(x) = g(x) \vee h(x)$,

$\forall x \in B \cap C \neq \emptyset$.

Assume that $(f, A, L) \sqcap_E (m, (B \cap C), L) = (n, A \cup (B \cap C), L)$, where

$$n(\delta) = \begin{cases} f(\delta), & \text{if } \delta \in A \backslash (B \cap C), \\ m(\delta), & \text{if } \delta \in (B \cap C) \backslash A, \\ f(\delta) \wedge m(\delta), & \text{if } \delta \in A \cap (B \cap C), \forall \delta \in C = A \cup (B \cap C). \end{cases}$$

By taking properties of operations in set theory,

$$n(\delta) = \begin{cases} f(\delta), & \text{if } \delta \in (A \setminus B) \cup (A \setminus C), \\ g(\delta) \vee h(\delta), & \text{if } \delta \in (B \setminus A) \cap (C \setminus A), \\ f(\delta) \wedge (g(\delta) \vee h(\delta)), & \text{if } \delta \in A \cap (B \cap C). \end{cases}$$

Now consider the right-hand side of the equality.
Suppose that $(f, A, L) \sqcap_E (g, B, L) = (t, A \cup B, L)$, where

$$t(\delta) = \begin{cases} f(\delta), & \text{if } \delta \in A \backslash B, \\ g(\delta), & \text{if } \delta \in B \backslash A, \\ f(\delta) \wedge g(\delta), & \text{if } \delta \in A \cap B, \forall \delta \in C = A \cup B. \end{cases}$$

And suppose $(f, A, L) \sqcap_E (h, C, L) = (w, A \cup C, L)$, where

$$w(\delta) = \begin{cases} f(\delta), & \text{if } \delta \in A \backslash C, \\ h(\delta), & \text{if } \delta \in C \backslash A, \\ f(\delta) \wedge h(\delta), & \text{if } \delta \in A \cap C, \forall \delta \in A \cup C. \end{cases}$$

Let $(t, A \cup B, L) \sqcup_R (w, A \cup C, L) = (p, (A \cup B) \cap (A \cup C), L)$
where $p(x) = t(x) \vee w(x)$,
$\forall x \in (A \cup B) \cap (A \cup C)$

By considering the definitions of 't'and 'w'along with 'p', we can write below the equalities:

$$p(\delta) = \begin{cases} f(\delta), & \text{if } \delta \in (A \backslash B) \cup (A \backslash C), \\ g(\delta) \vee h(\delta), & \text{if } \delta \in (B \setminus A) \cap (C \setminus A), \\ (f(\delta) \wedge g(\delta)) \vee (f(\delta) \wedge h(\delta)), & \text{if } \delta \in (A \cap B) \cap (A \cap C). \end{cases}$$

$\forall \delta \in (A \cup B) \cap (A \cup C)$.

This follows that 'n'and 'p'are the same set-valued mapping, gives proof.
By using similar techniques, (5) too can be proved. Therefore we skip the proof. □

Theorem 2.44 *Properties of the soft lattice intersection ($\sqcap$) operation. Let (f, A, L), (g, B, L) and (h, C, L) be any three soft lattices over L. Then we have as,*

1. $(f, A, L) \sqcap ((g, B, L) \sqcap (h, C, L)) = ((f, A, L) \sqcap (g, B, L)) \sqcap (h, C, L)$.
2. $(f, A, L) \sqcap \mathcal{L} = (f, A, L), (f, A, L) \sqcap \Phi = \Phi$.
3. $(f, A, L) \sqcap (g, B, L)) \not\subseteq (f, A, L)$, *in general. But if* $(f, A, L) \subset (g, B, L)$, *then-*
   $(f, A, L) \sqcap (g, B, L) \subset (f, A, L)$,*moreover* $(f, A, L) \sqcap (g, B, L) = (f, A, L)$.
4. $(f, A, L) \sqcap ((g, B, L) \sqcup_R (h, C, L)) = ((f, A, L) \sqcap (g, B, L)) \sqcup_R ((f, A, L) \sqcap (h, C, L))$.
5. $((f, A, L) \sqcup_R (g, B, L)) \sqcap (h, C, L) = ((f, A, L) \sqcap (h, C, L)) \sqcup_R ((g, B, L) \sqcap (h, C, L))$.
6. $(f, A, L) \sqcap ((g, B, L) \vee (h, C, L)) = ((f, A, L) \sqcap (g, B, L)) \vee ((f, A, L) \sqcap (h, C, L))$.
7. $((f, A, L) \vee (g, B, L)) \sqcap (h, C, L) = ((f, A, L) \sqcap (h, C, L)) \vee ((g, B, L) \sqcap (h, C, L))$.
8. $(f, A, L) \sqcap ((g, B, L) \sim_R (h, C, L)) = ((f, A, L) \sqcap (g, B, L)) \sim_R ((f, A, L) \sqcap (h, C, L))$.
9. $((f, A, L) \sim_R (g, B, L)) \sqcap (h, C, L) = ((f, A, L) \sqcap (h, C, L)) \sim_R ((g, B, L) \sqcap (h, C, L))$.

Proof The proof of 1, 2, 4, 5, 6, 7 and 9 are obvious.

$$\text{(3) Let } (f, A, L) \sqcap (g, B, L) = (h, C, L), \text{ where } C = A \cap B, \text{ and-}$$
$$h(x) = f(x) \wedge g(x), \forall x \in C.$$

Since 'h'and 'f'do not need to be the same set-valued mapping for all $x \in A \cap B$, $(f, A, L) \sqcap (g, B, L) \not\subseteq (f, A, L)$, in general.

Now assume that $(f, A, L) \subseteq (g, B, L)$ then it is obvious that $A \cap B = A \subset A$. Now we need to show that $h(e)$ and $f(e)$ are the same approximations for all $e \in A \cap B = A$. Since $(f, A, L) \subseteq (g, B, L)$ and $f(e)$ and $g(e)$ are the same approximation for all $e \in A$, it follows that $h(e) = f(e) \wedge g(e) = f(e) \wedge f(e) = f(e), \forall e \in A$.

(8) For the LHS of the equality, suppose that $(g, B, L) \sim_R (h, C, L) = (t, B \cap C, L)$, where $t(x) = g(x) \backslash h(x), \forall x \in B \cap C \neq \emptyset$.

$$\begin{aligned}
\text{Assume}(f, A, L) \sqcap (t, B \cap C, L) &= (w, A \cap (B \cap C), L), \\
\text{where } w(x) &= f(x) \wedge t(x) \\
&= f(x) \wedge (g(x) \setminus h(x)) \\
&= (f(x) \wedge g(x)) \setminus (f(x) \wedge h(x)), \\
&\quad \forall x \in A \cap (B \cap C) \neq \emptyset.
\end{aligned}$$

Now, for RHS of the equality.

$$\text{Assume } (f, A, L) \sqcap (g, B, L) = (m, A \cap B, L)$$
$$\text{where } m(x) = f(x) \wedge g(x), \forall x \in A \cap B \neq \emptyset.$$

$$\text{Let } (f, A.L) \sqcap (h, C, L) = (n, A \cap C, L).$$
$$\text{where } n(x) = f(x) \wedge h(x), \forall x \in A \cap C.$$

$$\begin{aligned} \text{Suppose that}(m, A \cap B, L) \sim_R (n, A \cap C, L) &= (k, (A \cap B) \cap (A \cap C), L) \\ &= (k, A \cap (B \cap C), L). \\ &\quad \text{where } k(x) = m(x) \setminus n(x) \\ &= (f(x) \wedge g(x)) \setminus (f(x) \wedge h(x)). \\ &\quad \forall x \in (A \cap B) \cap (A \cap C). \end{aligned}$$

Since 'w'and 'k'are the same set-valued mapping for all $x \in (A \cap B) \cap (A \cap C) = A \cap (B \cap C)$, completes the proof. □

Theorem 2.45 *Properties of restricted difference ($\sim_R$) operation in soft lattice case. Let (f, A, L), (g, B, L), and (h, C, L) be three soft lattices over L. Then we have,*

1. $(f, A, L) \sim_R \Phi = (f, A, L)$.
2. $(f, A, L) \sim_R (f, A, L) = \Phi$.
3. $\mathcal{L} \sim_R (f, A, L) = (f, A, L)^r$.
4. *Restricted difference holds a right distribution law over intersection, restricted union, extended intersection and union.*
5. $(f, A, L) \sim_R ((g, B, L) \wedge (h, C, L)) = ((f, A, L) \sim_R (g, B, L)) \sqcup_R ((f, A, L) \sim_R (h, C, L))$.
6. $(f, A, L) \sim_R ((g, B, L) \sqcup_R (h, C, L)) = ((f, A, L) \sim_R (g, B, L)) \wedge ((f, A, L) \sim_R (h, C, L))$.
7. $(f, A, L) \sim_R ((g, B, L) \sqcap_E (h, C, L)) = ((f, A, L) \sim_R (g, B, L)) \vee ((f, A, L) \sim_R (h, C, L))$.
8. $(f, A, L) \sim_R ((g, B, L) \vee (h, C, L)) = ((f, A, L) \sim_R (g, B, L)) \sqcap_E ((f, A, L) \sim_R (h, C, L))$.

Proof (1) Let $\Phi_A = (m, A, L)$.

$$\begin{aligned} \text{Then } (f, A, L) \sim_R \Phi_A &= (f, A, L) \sim_R (m, A, L) \\ &= (h, A, L). \\ &\quad \text{where } h(e) = f(e) \setminus m(e), \forall e \in A. \end{aligned}$$

Since $m(e) = \emptyset, \forall e \in A$, it follows that $h(e) = f(e) \setminus \emptyset = f(e)$. This means that '$f$'and '$h$'are the same set-valued mapping. Thus the required proof.

(4) We show that restricted difference holds a right distribution law over restricted union and extended intersection, respectively. The others can be shown similarly. First we handle the left-hand side of the equality of,

$$((f, A, L) \sqcup_R (g, B, L)) \sim_R (h, C, L) = ((f, A, L) \sim_R (h, C, L)) \sqcup_R ((g, B, L) \sim_R (h, C, L)).$$

Suppose that $(f, A, L) \sqcup_R (g, B, L) = (t, A \cap B, L)$

where $t(x) = f(x) \vee g(x), \forall x \in A \cap B \neq \emptyset.$

Assume $(t, A \cap B, L) \sim_R (h, C, L) = (p, (A \cap B) \cap C, L),$

$$\text{where } p(x) = t(x) \setminus h(x) = (f(x) \vee g(x)) \setminus h(x) = (f(x) \setminus h(x)) \vee (g(x) \setminus h(x)), \quad \forall x \in (A \cap B) \cap C \neq \emptyset.$$

Now we handle the right-hand side of the equality.

$$\begin{aligned}
\text{Assume that,} (f, A, L) \sim_R (h, C, L) &= (m, A \cap C, L). \\
\text{where } m(x) &= f(x) \setminus h(x), \forall x \in A \cap C \neq \emptyset. \\
\text{Let } (g, B, L) \sim_R (h, C, L) &= (n, B \cap C, L). \\
\text{where } n(x) &= g(x) \setminus h(x), \forall x \in B \cap C \neq \emptyset. \\
\text{Suppose that} (m, A \cap C, L) \sqcup_R (n, B \cap C, L) &= (q, (A \cap C) \cap (B \cap C), L) \\
&= (q, (A \cap B) \cap C, L). \\
\text{where } q(x) &= m(x) \vee n(x) \\
&= (f(x) \setminus h(x)) \vee (g(x) \setminus h(x)), \\
&\quad \forall x \in (A \cap C) \cap (B \cap C).
\end{aligned}$$

Since 'p'and 'q'are the same set-valued mapping for all $x \in (A \cap C) \cap (B \cap C) = (A \cap B) \cap C$, the proof is for this part.

Now we show that,

$$((f, A, L) \sqcap_E (g, B, L)) \sim_R (h, C, L) = ((f, A, L) \sim_R (h, C, L)) \sqcap_E ((g, B, L) \sim_R (h, C, L)).$$

First we investigate the left-hand side of the equality.

Suppose that $(f, A, L) \sqcap_E (g, B, L) = (x, A \cup B, L)$, where

$$x(\delta) = \begin{cases} f(\delta), & \text{if } \delta \in A \backslash B, \\ g(\delta), & \text{if } \delta \in B \backslash A, \\ f(\delta) \wedge g(\delta), & \text{if } \forall \delta \in A \cap B. \end{cases}$$

Assume that $(x, A \cup B, L) \sim_R (h, C, L) = (y, (A \cup B) \cap C, L)$,
where $y(e) = x(e) \setminus h(e), \forall e \in (A \cup B) \cap C$.

By taking the properties of operations in set theory, we can write below the equalities for y:

$$y(\delta) = \begin{cases} f(\delta) \setminus h(\delta), & \text{if } \delta \in (A \setminus B) \cap C = (A \cap C) \setminus (B \cap C), \\ g(\delta) \setminus h(\delta), & \text{if } \delta \in (B \setminus A) \cap C = (B \cap C) \setminus (A \cap C), \\ (f(\delta) \wedge g(\delta)) \setminus h(\delta), & \text{if } \delta \in (A \cap B) \cap C, \forall \delta \in (A \cup B) \cap C. \end{cases}$$

Now we investigate the RHS of the equality.

Assume that $(f, A, L) \sim_R (h, C, L) = (k, A \cap C, L)$,
where $k(x) = f(x) \setminus h(x), \forall x \in A \cap C$.
Assume $(g, B, L) \sim_R (h, C, L) = (j, B \cap C, L)$,
where $j(x) = g(x) \setminus h(x), \forall x \in B \cap C$.

Let $(k, A \cap C, L) \sqcap_E (j, B \cup C, L) = (v, (A \cap C) \cup (B \cap C), L)$, where,

$$v(\delta) = \begin{cases} k(\delta), & \text{if } \delta \in (A \cap B) \setminus (B \cap C), \\ j(\delta), & \text{if } \delta \in (B \cap C) \setminus (A \cap C), \\ k(\delta) \wedge j(\delta), & \text{if } \delta \in (A \cap C) \cap (B \cap C), \forall \delta \in (A \cap C) \cup (B \cap C). \end{cases}$$

By taking the definitions of 'k'and 'j', we can rewrite 'v'as below:

$$v(\delta) = \begin{cases} f(\delta) \setminus h(\delta), & \text{if } \delta \in (A \cap C) \setminus (B \cap C), \\ g(\delta) \setminus h(\delta), & \text{if } \delta \in (B \cap C) \setminus (A \cap C), \\ (f(\delta) \setminus h(\delta)) \wedge (g(\delta) \setminus h(\delta)), & \text{if } \delta \in (A \cap C) \cup (B \cap C) \end{cases}$$

This follows that 'y' and 'v' are the same set-valued mapping. Therefore we complete the proof.

The remaining proofs can be obtained directly from the relevant definitions. □

Chapter 3
Topological Structures of Soft Sets

Topology is a major branch of mathematics with many applications in the fields of physical and computer sciences. Topology can be defined as the study of qualitative properties of certain objects, called topological spaces that are invariant under certain kinds of transformations, called continuous maps. These properties are generally described in terms of open sets. The concept of topological space is often generalized by replacing open sets by more general ones. Fuzzy topology introduced by Chang [29] and modified later by Lowen [74] is a classical example of this type of generalization. In a similar manner, topological structures on soft sets are more generalized methods and they can be useful for measuring the similarities and dissimilarities between the objects in a universe which are soft sets.

There are two versions of soft topology defined on soft sets, one by Shabir and Naz [125] and other by Cagman et al. [28]. The basic difference in these approaches is that the first one considers a subcollection of set of all soft sets in an initial universe with a fixed set of parameters and the second one considers a subcollection from the set of all soft subsets of a given soft set in a universe. In this chapter we consider both approaches with some typical topological notions. Obviously all standard notions in topology can be studied using both approaches. Sections 3.1–3.4 use Shabir and Naz [125] approach and the remaining sections use Cagman et al. [28] approach. Basic definitions and results of topological spaces are given in [38, 62, 89, 129, 149].

3.1 Soft Topological Spaces

Shabir and Naz [125] introduced the concept of soft topological spaces which are defined over an initial universe with a fixed set of parameters. Some concepts in soft topological spaces such as soft interior, soft closure, soft continuity, soft separation

S. J. John, *Soft Sets*, Studies in Fuzziness and Soft Computing 400,
https://doi.org/10.1007/978-3-030-57654-7_3

axioms etc. which are based on the definition of soft topology by Shabir and Naz were introduced and studied by many authors (see for example [3, 14, 49, 86, 105, 118, 168]). The discussions, results and examples included in Sects. 3.1–3.4 are taken from [118, 125].

Let X be an initial universe set and E be the non-empty set of parameters. Throughout this chapter union of soft sets ($\cup$) stands for $\tilde{\cup}$ and intersection ($\cap$) stands for $\tilde{\cap}$. Also we use $\tilde{\Phi}$, $\tilde{X}$ respectively for $\tilde{\Phi}_E$, $\tilde{X}_E$ if the parameter set E is understood.

Definition 3.1 The difference (H, E) of two soft sets (F, E) and (G, E) over X, denoted by $(F, E)\backslash(G, E)$, is defined as $H(e) = F(e)\backslash G(e)$ for all $e \in E$.

Definition 3.2 Let (F, E) be a soft set over X and $x \in X$. We say that $x \in (F, E)$ read as x belongs to the soft set (F, E) whenever $x \in F(\alpha)$ for all $\alpha \in E$.

Note that for any $x \in X$, $x \notin (F, E)$, if $x \notin F(\alpha)$ for some $\alpha \in E$.

Definition 3.3 Let Y be a non-empty subset of X, then $\tilde{Y}$ denotes the soft set (Y, E) over X for which $Y(\alpha) = Y$, for all $\alpha \in E$. In particular, (X, E) will be denoted by $\tilde{X}$.

Definition 3.4 Let $x \in X$, then (x, E) denotes the soft set over X for which $x(\alpha) = \{x\}$, for all $\alpha \in E$.

Definition 3.5 Let (F, E) be a soft set over X and Y be a non-empty subset of X. Then the sub soft set of (F, E) over Y denoted by $({}^Y F, E)$, is defined as follows

$$ {}^Y F(\alpha) = Y \cap F(\alpha), \text{ for all } \alpha \in E. $$

In other words $({}^Y F, E) = \tilde{Y} \cap (F, E)$.

Definition 3.6 Let $\mathcal{T}$ be the collection of soft sets over X with parameter set E, then $\mathcal{T}$ is said to be a soft topology on X if

1. $\tilde{\Phi}_E, \tilde{X}_E$ belong to $\mathcal{T}$
2. the union of any number of soft sets in $\mathcal{T}$ belongs to $\mathcal{T}$
3. the intersection of any two soft sets in $\mathcal{T}$ belongs to $\mathcal{T}$.

The triplet $(X, \mathcal{T}, E)$ is called a soft topological space over X.

Definition 3.7 Let $(X, \mathcal{T}, E)$ be a soft space over X, then the members of $\mathcal{T}$ are said to be soft open sets in X.

Definition 3.8 Let $(X, \mathcal{T}, E)$ be a soft space over X. A soft set (F, E) over X is said to be a soft closed set in X, if its relative complement $(F, E)^r$ or equivalently denoted as $(F, E)'$ belongs to $\mathcal{T}$.

Proposition 3.1 *Let $(X, \mathcal{T}, E)$ be a soft space over X. Then*

1. *$\tilde{\Phi}$, $\tilde{X}$ are closed soft sets over X.*
2. *the intersection of any number of soft closed sets is a soft closed set over X.*

3. *the union of any two soft closed sets is a soft closed set over X.*

Proof Follows from the definition of soft topological spaces and De Morgan's laws. □

Definition 3.9 Let X be an initial universe set, E be the set of parameters and $\mathcal{T} = \{\widetilde{\Phi}, \widetilde{X}\}$. Then $\mathcal{T}$ is called the soft indiscrete topology on X and $(X, \mathcal{T}, E)$ is said to be a soft indiscrete space over X.

Definition 3.10 Let X be an initial universe set, E be the set of parameters and let $\mathcal{T}$ be the collection of all soft sets which can be defined over X. Then $\mathcal{T}$ is called the soft discrete topology on X and $(X, \mathcal{T}, E)$ is said to be a soft discrete space over X.

Proposition 3.2 *Let $(X, \mathcal{T}, E)$ be a soft space over X. Then the collection $\mathcal{T}_\alpha = \{F(\alpha)|(F, E) \in \mathcal{T}\}$ for each $\alpha \in E$, defines a topology on X.*

Proof By definition, for any $\alpha \in E$, we have $\mathcal{T}_\alpha = \{F(\alpha)|(F, E) \in \mathcal{T}\}$. Now

1. $\widetilde{\Phi}, \widetilde{X} \in \mathcal{T}$ implies that $\emptyset, X \in \mathcal{T}_\alpha$.
2. Let $\{F_i(\alpha)|i \in I\}$ be a collection of sets in $\mathcal{T}_\alpha$ since $(F_i, E) \in \mathcal{T}$, for all $i \in I$ so that $\cup_{i\in I} (F_i, E) \in \mathcal{T}$ thus $\cup_{i\in F} F_i(\alpha) \in \mathcal{T}_\alpha$.
3. Let $F(\alpha), G(\alpha) \in \mathcal{T}_\alpha$ for some $(F, E), (G, E) \in \mathcal{T}$ since $(F, E) \cap (G, E) \in \mathcal{T}$ so $F(\alpha) \cap G(\alpha) \in \mathcal{T}_\alpha$.

Thus $\mathcal{T}_\alpha$ defines a topology on X for each $\alpha \in E$. □

Proposition 3.2 shows that corresponding to each parameter $\alpha \in E$, we have a topology $\mathcal{T}_\alpha$ on X. Thus a soft topology on X gives a parameterized family of topologies on X.

Example 36 Let $X = \{h_1, h_2, h_3\}, E = \{e_1, e_2\}$ and $\mathcal{T} = \{\widetilde{\Phi}, \widetilde{X}, (F_1, E), (F_2, E), (F_3, E), (F_4, E)\}$ where $(F_1, E), (F_2, E), (F_3, E), (F_4, E)$ are soft sets over X, defined as follows

$F_1(e_1) = \{h_2\}, \quad F_1(e_2) = \{h_1\}$
$F_2(e_1) = \{h_2, h_3\}, \quad F_2(e_2) = \{h_1, h_2\}$
$F_3(e_1) = \{h_1, h_2\}, \quad F_3(e_2) = X$
$F_4(e_1) = \{h_1, h_2\}, \quad F_4(e_2) = \{h_1, h_3\}.$

Then $\mathcal{T}$ defines a soft topology on X and hence $(X, \mathcal{T}, E)$ is a soft topological space over X. It can be easily seen that

$$\mathcal{T}_{e_1} = \{\emptyset, X, \{h_2\}, \{h_2, h_3\}, \{h_1, h_2\}\}$$

and

$$\mathcal{T}_{e_2} = \{\emptyset, X, \{h_1\}, \{h_1, h_3\}, \{h_1, h_2\}\}$$

are topologies on X.

Proposition 3.3 *Let $(X, \mathcal{T}_1, E)$ and $(X, \mathcal{T}_2, E)$ be two soft topological spaces over the same universe X, then $(X, \mathcal{T}_1 \cap \mathcal{T}_2, E)$ is a soft topological space over X.*

Proof 1. $\widetilde{\Phi}, \widetilde{X}$ belong to $\mathcal{T}_1 \cap \mathcal{T}_2$.
2. Let $\{(F_i, E) \,|i \in I\}$ be a family of soft sets in $\tau_1 \cap \mathcal{T}_2$ Then $(F_i, E) \in \mathcal{T}_1$ and $(F_i, E) \in \mathcal{T}_2$, for all $i \in I$, so $\cup_{i\in I} (F_i, E) \in \mathcal{T}_1$ and $\cup_{\text{iel}} (F_i, E) \in \mathcal{T}_2$. Thus $\cup_{i\in I} (F_i, E) \in \mathcal{T}_1 \cap \mathcal{T}_2$.
3. Let $(F, E), (G, E) \in \mathcal{T}_1 \cap \mathcal{T}_2$. Then $(F, E), (G, E) \in \mathcal{T}_1$ and $(F, E), (G, E) \in \mathcal{T}_2$ since $(F, E) \cap (G, E) \in \mathcal{T}_1$ and $(F, E) \cap (G, E) \in \mathcal{T}_2$, so $(F, E) \cap (G, E) \in \mathcal{T}_1 \cap \mathcal{T}_2$.

Thus $\mathcal{T}_1 \cap \mathcal{T}_2$ defines a soft topology on X and $(X, \mathcal{T}_1 \cap \mathcal{T}_2, E)$ is a soft topological space over X. □

Remark 3.1 The union of two soft topologies on X may not be a soft topology on X even if the collection corresponding to each parameter defines a topology on X.

Example 37 Let $X = \{h_1, h_2, h_3\}$, $E = \{e_1, e_2\}$ and
$\mathcal{T}_1 = \{\widetilde{\Phi}, \widetilde{X}, (F_1, E), (F_2, E), (F_3, E), (F_4, E)\}$ $\mathcal{T}_2 = \{\Phi, \widetilde{X}, (G_1, E), (G_2, E), (G_3, E), (G_4, E)\}$ be two soft topologies defined on X where $(F_1, E), (F_2, E), (F_3, E), (F_4, E), (G_1, E), (G_2, E), (G_3, E)$ and (G_4, E) are soft sets over X, defined as follows
$F_1(e_1) = \{h_2\}, \quad F_1(e_2) = \{h_1\}$
$F_2(e_1) = \{h_2, h_3\}, \quad F_2(e_2) = \{h_1, h_2\}$
$F_3(e_1) = \{h_1, h_2\}, \quad F_3(e_2) = X$
$F_4(e_1) = \{h_1, h_2\}, \quad F_4(e_2) = \{h_1, h_3\}$
and
$G_1(e_1) = \{h_2\}, \quad G_1(e_2) = \{h_1\}$
$G_2(e_1) = \{h_2, h_3\}, G_2(e_2) = \{h_1, h_2\}$
$G_3(e_1) = \{h_1, h_2\}, G_3(e_2) = \{h_1, h_2\}$
$G_4(e_1) = \{h_2\}, \quad G_4(e_2) = \{h_1, h_3\}$.
Now, we define

$$\begin{aligned}\mathcal{T} &= \mathcal{T}_1 \cup \mathcal{T}_2 \\ &= \left\{\widetilde{\Phi}, \widetilde{X}, (F_1, E), (F_2, E), (F_3, E), (F_4, E), (G_3, E), (G_4, E)\right\}\end{aligned}$$

If we take

$$(F_2, E) \cup (G_3, E) = (H, E)$$

then

$$\begin{aligned}H(e_1) &= F_2(e_1) \cup G_3(e_1) \\ &= \{h_2, h_3\} \cup \{h_1, h_2\} \\ &= X\end{aligned}$$

and

$$\begin{aligned}H(e_2) &= F_2(e_2) \cup G_3(e_2) \\ &= \{h_1, h_2\} \cup \{h_1, h_2\} \\ &= \{h_1, h_2\}\end{aligned}$$

but $(H, E) \notin \mathcal{T}$. Thus $\mathcal{T}$ is not a soft topology on X.

Definition 3.11 Let $(X, \mathcal{T}, E)$ be a soft topological space over X, (G, E) be a soft set over X and $x \in X$. Then (G, E) is said to be a soft neighborhood of x if there exists a soft open set (F, E) such that $x \in (F, E) \tilde{\subset} (G, E)$.

Proposition 3.4 *Let $(X, \mathcal{T}, E)$ be a soft topological space over X, then*

1. *each $x \in X$ has a soft neighborhood.*
2. *if (F, E) and (G, E) are soft neighborhoods of some $x \in X$, then $(F, E) \cap (G, E)$ is also a soft neighborhood of x.*
3. *if (F, E) is a soft neighborhood of $x \in X$ and $(F, E) \tilde{\subset} (G, E)$, then (G, E) is also a soft neighborhood of $x \in X$.*

Proof 1. For any $x \in X$, $x \in \tilde{X}$ and since $\tilde{X} \in \mathcal{T}$, so $x \in \tilde{X} \tilde{\subset} \tilde{X}$. Thus $\tilde{X}$ is a soft neighborhood of x.
2. Let (F, E) and (G, E) be the soft neighborhoods of $x \in X$, then there exist $(F_1, E), (F_2, E) \in \mathcal{T}$ such that
$$x \in (F_1, E) \tilde{\subset} (F, E).$$
Thus
$$x \in (F_1, E) \tilde{\subset} (G, E).$$
Now $x \in (F_1, E)$ and $x \in (F_2, E)$ implies that $x \in (F_1, E) \cap (F_2, E)$ and $(F_1, E) \cap (F_2, E) \in \mathcal{T}$. So we have
$$x \in (F_1, E) \cap (F_2, E) \tilde{\subset} (F, E) \cap (G, E).$$
Thus $(F, E) \cap (G, E)$ is a soft neighborhood of x.
3. Let (F, E) be a soft neighborhood of $x \in X$ and $(F, E) \tilde{\subset} (G, E)$. By definition there exists a soft open set (F_1, E) such that
$$x \in (F_1, E) \tilde{\subset} (F, E) \tilde{\subset} (G, E).$$
Thus
$$x \in (F_1, E) \tilde{\subset} (G, E).$$
Hence (G, E) is a soft neighborhood of x. □

Proposition 3.5 *Let (X, T, E) be a soft topological space over X. For any soft open set (F, E) over X, (F, E) is a soft neighborhood of each point of $\bigcap_{\alpha \in E} F(\alpha)$.*

Proof Let $(F, E) \in \mathcal{T}$. For any $x \in \bigcap_{\alpha \in E} F(\alpha)$, we have $x \in F(\alpha)$ for each $\alpha \in E$. Thus
$$x \in (F, E) \tilde{\subset} (F, E)$$
and so (F, E) is a soft neighborhood of x. □

Definition 3.12 Let $(X, \mathcal{T}, E)$ be a soft topological space over X and Y be a non-empty subset of X. Then

$$\mathcal{T}_Y = \{({}^Y F, E) \,|(F, E) \in \mathcal{T}\}$$

is said to be the soft relative topology on Y and $(Y, \mathcal{T}_Y, E)$ is called a soft subspace of $(X, \mathcal{T}, E)$.

We can easily verify that $\mathcal{T}_Y$ is, in fact, a soft topology on Y.

Remark 3.2 (1) Any soft subspace of a soft discrete topological space is a soft discrete topological space.
(2) Any soft subspace of a soft indiscrete topological space is a soft indiscrete topological space.

Proposition 3.6 *Let $(X, \mathcal{T}, E)$ be a soft topological space over X and Y be a non-empty subset of X. Then $(Y, \mathcal{T}_{\alpha Y})$ is a subspace of $(X, \mathcal{T}_\alpha)$ for each $\alpha \in E$.*

Proof Since $(Y, \mathcal{T}_Y, E)$ is a soft topological space over Y so $(Y, \mathcal{T}_{\alpha Y})$ is a topological space for each $\alpha \in E$

$$\begin{aligned}\mathcal{T}_{\alpha Y} &= \{{}^Y F(\alpha)|(F, E) \in \mathcal{T}\}\\ &= \{Y \cap F(\alpha), |(F, E) \in \mathcal{T}\}\\ &= \{Y \cap F(\alpha), |F(\alpha) \in \mathcal{T}_\alpha\}.\end{aligned}$$

Thus $(Y, \mathcal{T}_{\alpha Y})$ is a subspace of $(X, \mathcal{T}_\alpha)$. □

Proposition 3.7 *Let $(Y, \mathcal{T}_Y, E)$ be a soft subspace of a soft topological space $(X, \mathcal{T}, E)$ and (F, E) be a soft open set in Y If $\widetilde{Y} \in \mathcal{T}$ then $(F, E) \in \mathcal{T}$.*

Proof Let (F, E) be a soft open set in Y, then there exists a soft open set (G, E) in X such that $(F, E) = \widetilde{Y} \cap (G, E)$. Now, if $\widetilde{Y} \in \mathcal{T}$ then $\widetilde{Y} \cap (G, E) \in \mathcal{T}$ by the third axiom of the definition of a soft topological space and hence $(F, E) \in \mathcal{T}$. □

Theorem 3.1 *Let $(Y, \mathcal{F}_Y, E)$ be a soft subspace of soft topological space $(X, \mathcal{T}, E)$ and (F, E) be a soft set over X, then*

1. *(F, E) is soft open in Y if and only if $(F, E) = \widetilde{Y} \cap (G, E)$ for some $(G, E) \in \mathcal{T}$.*
2. *(F, E) is soft closed in Y if and only if $(F, E) = \widetilde{Y} \cap (G, E)$ for some soft closed set (G, E) in X.*

Proof (1) Follows from the definition of soft subspace.
(2) If (F, E) is soft closed in Y then we have If (F, E) is soft closed in Y then we have $(F, E) = \widetilde{Y} \backslash (G, E)$, for some $(G, E) \in \mathcal{T}_Y$.
'Now, $(G, E) = \widetilde{Y} \cap (H, E)$, for some $(H, E) \in \mathcal{T}$

For any $\alpha \in E$,

$$\begin{aligned} F(\alpha) &= Y(\alpha)\backslash G(\alpha) \\ &= Y\backslash G(\alpha) \\ &= Y\backslash(Y(\alpha) \cap H(\alpha)) \\ &= Y\backslash(Y \cap H(\alpha)) \\ &= Y\backslash H(\alpha) \\ &= Y \cap (X\backslash H(\alpha)) \\ &= Y \cap (H(\alpha))^c \\ &= Y(\alpha) \cap (H(\alpha))^c. \end{aligned}$$

Thus $(F, E) = \widetilde{Y} \cap (H, E)'$ where $(H, E)'$ is soft closed in X as $(H, E) \in \mathcal{T}$.

Conversely, assume that $(F, E) = Y \cap (G, E)$ for some soft closed set (G, E) in X. This means that $(G, E)' \in \mathcal{T}$.

Now, if $(G, E) = (X, E)\backslash(H, E)$ where $(H, E) \in \mathcal{T}$ then for any $\alpha \in E$

$$\begin{aligned} F(\alpha) &= Y(\alpha) \cap G(\alpha) \\ &= Y \cap G(\alpha) \\ &= Y \cap (X(\alpha)\backslash H(\alpha)) \\ &= Y \cap (X\backslash H(\alpha)) \\ &= Y\backslash H(\alpha) \\ &= Y\backslash(Y \cap H(\alpha)) \\ &= Y(\alpha)\backslash(Y(\alpha) \cap H(\alpha)). \end{aligned}$$

Thus $(F, E) = \widetilde{Y}\backslash(\widetilde{Y} \cap (H, E))$. Since $(H, E) \in \mathcal{T}$, $(Y \cap (H, E)) \in \mathcal{T}_Y$ and hence (F, E) is soft closed in Y. □

3.2 Closure, Interior, Boundary and Limit Points

Definition 3.13 Let $(X, \mathcal{T}, E)$ be a soft topological space over X and (F, E) be a soft set over X. Then the soft closure of (F, E), denoted by $\overline{(F, E)}$ is the intersection of all soft closed super sets of (F, E).

Clearly $\overline{(F, E)}$ is the smallest soft closed set over X which contains (F, E).

Theorem 3.2 *Let $(X, \mathcal{T}, E)$ be a soft topological space over X, (F, E) and (G, E) are soft sets over X. Then*

1. $\overline{\widetilde{\Phi}} = \widetilde{\Phi}$ *and* $\overline{\widetilde{X}} = \widetilde{X}$.
2. $(F, E)\widetilde{\subseteq}\overline{(F, E)}$.
3. (F, E) *is a soft closed set if and only if* $(F, E) = \overline{(F, E)}$.

4. $\overline{\overline{(F, E)}} = \overline{(F, E)}$.
5. $(F, E)\tilde{\subseteq}(G, E)$ *implies* $\overline{(F, E)}\tilde{\subseteq}\overline{(G, E)}$.
6. $\overline{(F, E) \cup (G, E)} = \overline{(F, E)} \cup \overline{(G, E)}$.
7. $\overline{(F, E) \cap (G, E)}\tilde{\subseteq}\overline{(F, E)} \cap \overline{(G, E)}$.

Proof (1) and (2) are obvious.

(3) If (F, E) is a soft closed set over X then (F, E) is itself a soft closed set over X which contains (F, E). So (F, E) is the smallest soft closed set containing (F, E) and $(F, E) = \overline{(F, E)}$. Conversely suppose that $(F, E) = \overline{(F, E)}$. Since $\overline{(F, E)}$ is a soft closed set, so $\overline{(F, E)}$ is a soft closed set over X.

(4) Since $\overline{(F, E)}$ is a soft closed set therefore by part (3) we have $\overline{\overline{(F, E)}} = \overline{(F, E)}$.

(5) Suppose that $(F, E)\tilde{\subseteq}(G, E)$. Then every soft closed super set of (G, E) will also contain (F, E). This means every soft closed super set of (G, E) is also a soft closed super set of (F, E). Hence the intersection of soft closed super sets of (F, E) is contained in the soft intersection of soft closed super sets of (G, E). Thus $\overline{(F, E)}\tilde{\subseteq}\overline{(G, E)}$.

(6) Since $(F, E)\tilde{\subseteq}(F, E) \cup (G, E)$ and $(G, E)\tilde{\subseteq}(F, E) \cup (G, E)$. So by part (5), $\overline{(F, E)}\tilde{\subseteq}\overline{(F, E) \cup (G, E)}$ and $\overline{(G, E)}\tilde{\subseteq}\overline{(F, E) \cup (G, E)}$. Thus $\overline{(F, E)} \cup \overline{(G, E)}\tilde{\subseteq}\overline{(F, E) \cup (G, E)}$. Conversely suppose that $(F, E)\tilde{\subseteq}\overline{(F, E)}$ and $(G, E)\tilde{\subseteq}\overline{(G, E)}$. So $(F, E) \cup (G, E)\tilde{\subseteq}\overline{(F, E)} \cup \overline{(G, E)}$. Now $\overline{(F, E)} \cup \overline{(G, E)}$ is a soft closed set over X being the union of two soft closed sets. Then $\overline{(F, E) \cup (G, E)}\tilde{\subseteq}\overline{(F, E)} \cup \overline{(G, E)}$. Thus $\overline{(F, E) \cup (G, E)} = \overline{(F, E)} \cup \overline{(G, E)}$.

(7) Since $(F, E) \cap (G, E)\tilde{\subseteq}(F, E)$ and $(F, E) \cap (G, E)\tilde{\subseteq}(G, E)$. So by part (5) $\overline{(F, E) \cap (G, E)}\tilde{\subseteq}\overline{(F, E)}$ and $\overline{(F, E) \cap (G, E)}\tilde{\subseteq}\overline{(G, E)}$. Thus $\overline{(F, E) \cap (G, E)}\tilde{\subseteq}\overline{(F, E)} \cap \overline{(G, E)}$. □

Definition 3.14 Let $(X, \mathcal{T}, E)$ be a soft topological space over X and (F, E) be a soft set over X. Then we associate with (F, E) a soft set over X, denoted by $(\overline{F}, E)$ and defined as $\overline{F}(\alpha) = \overline{F(\alpha)}$, where $\overline{F}(\alpha)$ is the closure of $F(\alpha)$ in $\mathcal{T}_\alpha$ for each $\alpha \in E$.

Proposition 3.8 *Let $(X, \mathcal{T}, E)$ be a soft topological space over X and (F, E) be a soft set over X. Then $(\overline{F}, E)\tilde{\subseteq}\overline{(F, E)}$.*

Proof For any $\alpha \in E$, $\overline{F(\alpha)}$ is the smallest closed set in $(X, \mathcal{T}_\alpha)$ which contains $F(\alpha)$. Moreover if $\overline{(F, E)} = (H, E)$ then $H(\alpha)$ is also a closed set in $(X, \mathcal{T}_\alpha)$ containing $F(\alpha)$. This implies that $\overline{F}(\alpha) = \overline{F(\alpha)} \subseteq H(\alpha)$. Thus $(\overline{F}, E)\tilde{\subseteq}\overline{(F, E)}$. □

Corollary 3.1 *Let $(X, \mathcal{T}, E)$ be a soft topological space over X and (F, E) be a soft set over X. Then $(\overline{F}, E) = \overline{(F, E)}$ if and only if $(\overline{F}, E)' \in \mathcal{T}$.*

Proof If $(\overline{F}, E) = \overline{(F, E)}$ then $(\overline{F}, E)$ is a soft closed set and so $(\overline{F}, E)' \in \mathcal{T}$. Conversely if $(\overline{F}, E)' \in \mathcal{T}$ then $(\overline{F}, E)$ is a soft closed set containing (F, E). By

Proposition 3.8, $(\overline{F}, E)\widetilde{\subset}\overline{(F, E)}$ and by the definition of soft closure of (F, E), any soft closed set over X which contains (F, E) will contain (F, E). Therefore $\overline{(F, E)}\widetilde{\subset}(\overline{F}, E)$. Thus $(\overline{F}, E) = \overline{(F, E)}$. □

Example 38 Let $X = \{h_1, h_2, h_3\}, E = \{e_1, e_2\}$ and $\mathcal{T} = \{\widetilde{\Phi}, \widetilde{X}, (F_1, E), (F_2, E), (F_3, E), \ldots, (F_7, E)\}$ where

$$F_1(e_1) = \{h_1, h_2\}, \quad F_1(e_2) = \{h_1, h_2\}$$
$$F_2(e_1) = \{h_2\}, \quad F_2(e_2) = \{h_1, h_3\}$$
$$F_3(e_1) = \{h_2, h_3\}, \quad F_3(e_2) = \{h_1\}$$
$$F_4(e_1) = \{h_2\}, \quad F_4(e_2) = \{h_1\}$$
$$F_5(e_1) = \{h_1, h_2\}, \quad F_5(e_2) = X$$
$$F_6(e_1) = X \quad F_6(e_2) = \{h_1, h_2\}$$
$$F_7(e_1) = \{h_2, h_3\}, \quad F_7(e_2) = \{h_1, h_3\}.$$

Then $(X, \mathcal{T}, E)$ is a soft topological space over X.

Let (F, E) and (G, E) are defined as follows:

$$F(e_1) = \{h_1, h_3\}, \quad F(e_2) = \emptyset$$
$$G(e_1) = \{h_2, h_3\}, \quad G(e_2) = \{h_1, h_2\}$$

Then $(F, E) \cap (G, E) = ((F \cap G), E)$ is given by

$$(F \cap G)(e_1) = \{h_3\}, \quad (F \cap G)(e_2) = \emptyset$$

Now,

$$\overline{(F, E)} = \widetilde{X} \cap (F_2, E)' \cap (F_4, E)' = (F_2, E)'.$$

and

$$\overline{(G, E)} = \widetilde{X}.$$

Therefore

$$\overline{(F, E)} \cap \overline{(G, E)} = \overline{(F, E)}.$$

Also

$$\overline{(F, E) \cap (G, E)} = \cap\left\{\widetilde{X}, (F_1, E)', (F_2, E)', (F_4, E)', (F_5, E)'\right\}$$
$$= (F_5, E)'$$

So

$$\overline{(F, E) \cap (G, E)}\widetilde{\subset}\overline{(F, E)} \cap \overline{(G, E)}.$$

but

$$\overline{(F, E) \cap (G, E)} \neq \overline{(F, E)} \cap \overline{(G, E)}.$$

Next we see that

$$\mathcal{T}_{e_1} = \{\emptyset, X, \{h_2\}, \{h_2, h_3\}, \{h_1, h_2\}\}.$$

and

$$\mathcal{T}_{e_1} = \{\emptyset, X, \{h_1\}, \{h_1, h_3\}, \{h_1, h_2\}\}.$$

Here $(\overline{F}, E)$ is given by

$$\overline{F}(e_1) = \{h_1, h_3\}, \ \overline{F}(e_2) = \emptyset.$$

Clearly

$$(\overline{F, E})\widetilde{\subset}(\overline{F}, E) \text{ but } \overline{(F, E)} \neq (\overline{F}, E).$$

Definition 3.15 Let $(X, \mathcal{T}, E)$ be a soft topological space over X, (G, E) be a soft set over X and $x \in X$. Then x is said to be a soft interior point of (G, E) if there exists a soft open set (F, E) such that $x \in (F, E)\widetilde{\subset}(G, E)$.

Proposition 3.9 *Let $(X, \mathcal{T}, E)$ be a soft topological space over X, (G, E) be a soft set over X and $x \in X$. If x is a soft interior point of (G, E) then x is an interior point of $G(\alpha)$ in $(X, \mathcal{T}_\alpha)$, for each $\alpha \in E$.*

Proof For any $\alpha \in E$, $G(\alpha) \subseteq X$. If $x \in X$ is a soft interior point of (G, E) then there exists $(F, E) \in \mathcal{T}$ such that $x \in (F, E) \tilde{\subseteq} (G, E)$. This means that, $x \in F(\alpha) \subseteq G(\alpha)$. As $F(\alpha) \in \mathcal{T}_\alpha$, $F(\alpha)$ is an open set in $\mathcal{T}_\alpha$ and $x \in F(\alpha)$. This implies that x is an interior point of $G(\alpha)$ in $\mathcal{T}_\alpha$. □

Definition 3.16 Let $(X, \mathcal{T}, E)$ be a soft topological space over X then soft interior of soft set (F, E) over X is denoted by $(F, E)^\circ$ and is defined as the union of all soft open sets contained in (F, E).

Thus $(F, E)^\circ$ is the largest soft open set contained in (F, E).

Now consider the soft topological space $(X, \mathcal{T}, E)$ over X in Example 36 and $(F, E) = \{\{h_2\}, \{h_1, h_3\}\}$ be soft set of soft topological space X. Then $(F, E)^\circ = \{\{h_2\}, \{h_1\}\}$.

Theorem 3.3 *Let $(X, \mathcal{T}, E)$ be a soft topological space over X and (F, E) and (G, E) are soft sets over X. Then*

1. $\Phi^\circ = \Phi$ *and* $\tilde{X}^\circ = \tilde{X}$.
2. $(F, E)^\circ \tilde{\subseteq} (F, E)$.
3. $((F, E)^\circ)^\circ = (F, E)^\circ$.
4. (F, E) *is a soft open set if and only if* $(F, E)^\circ = (F, E)$.
5. $(F, E) \tilde{\subseteq} (G, E)$ *implies* $(F, E)^\circ \tilde{\subseteq} (G, E)^\circ$.
6. $(F, E)^\circ \cap (G, E)^\circ = ((F, E) \cap (G, E))^\circ$.
7. $(F, E)^\circ \cup (G, E)^\circ \tilde{\subseteq} ((F, E) \cup (G, E))^\circ$.

Proof (1) and (2) are obvious.

(3) Since $((F, E)^\circ)^\circ$ is soft open and $((F, E)^\circ)^\circ$ is the union of all soft open subsets in X contained in $(F, E)^\circ$, $(F, E)^\circ \tilde{\subseteq} ((F, E)^\circ)^\circ$. But $((F, E)^\circ)^\circ \tilde{\subseteq} (F, E)^\circ$ by (2). Hence $((F, E)^\circ)^\circ = (F, E)^\circ$.

(4) If (F, E) is a soft open sets over X then (F, E) is itself a soft open set over X which contains (F, E). So $(F, E)^\circ$ is the largest soft open set contained in (F, E) and $(F, E) = (F, E)^\circ$.
Conversely, suppose that $(F, E) = (F, E)^\circ$. Since $(F, E)^\circ$ is a soft open set, (F, E) is a soft open set over X.

(5) Suppose that $(F, E) \tilde{\subseteq} (G, E)$. Since $(F, E)^\circ \tilde{\subseteq} (F, E) \tilde{\subseteq} (G, E)$. $(F, E)^\circ$ is a soft open subset of (G, E), so by definition of $(G, E)^\circ$, $(F, E)^\circ \tilde{\subseteq} (G, E)^\circ$.

(6) From (5), we have $((F, E) \cap (G, E)) \tilde{\subseteq} (F, E)$, $((F, E) \cap (G, E)) \tilde{\subseteq} (G, E)$ implies $((F, E) \cap (G, E))^\circ \tilde{\subseteq} (F, E)^\circ$, $((F, E) \cap (G, E))^\circ \tilde{\subseteq} (G, E)^\circ$ so that $(F, E)^\circ \cap (G, E)^\circ$ is a soft open subset of $((F, E) \cap (G, E))^\circ$. Hence $(F, E)^\circ \cap (G, E)^\circ \tilde{\subseteq} ((F, E) \cap (G, E))^\circ$. Thus $(F, E)^\circ \cap (G, E)^\circ = ((F, E) \cap (G, E))^\circ$.

(7) Since $(F, E)\tilde{\subseteq}((F, E) \cup (G, E))$ and $(G, E)\tilde{\subseteq}((F, E) \cup (G, E))$. By (5), $(F, E)^\circ\tilde{\subseteq}((F, E) \cup (G, E))^\circ$ and $(G, E)^\circ\tilde{\subseteq}((F, E) \cup (G, E))^\circ$. So that $(F, E)^\circ \cup (G, E)^\circ\tilde{\subseteq}((F, E) \cup (G, E))^\circ$ since $(F, E)^\circ \cup (G, E)^\circ$ is a soft open set. □

The following example shows that the equality does not hold in Theorem 3.3(7).

Example 39 Let us consider the soft topological space $(X, \mathcal{T}, E)$ over X in Example 36 and $(F, E) = \{\{h_2\}, \{h_1, h_3\}\}, (G, E) = \{\{h_1, h_3\}, \{h_1, h_2, h_3\}\}$ are soft sets of soft topological space X. Then $(F, E)^\circ = \{\{h_2\}, \{h_1\}\}$ and $(G, E)^\circ = \Phi \cdot (F, E) \cup (G, E) = \{\{h_2\}, \{h_1, h_3\}\} \cup \{\{h_1, h_3\}, \{h_1, h_2, h_3\}\} = \{\{h_1, h_2, h_3\}, \{h_1, h_2, h_3\}\} = \widetilde{X}$. Now $((F, E) \cup (G, E))^\circ = (\widetilde{X})^\circ = \widetilde{X}$ and $(F, E)^\circ \cup (G, E)^\circ = \{\{h_2\}, \{h_1\}\} \cup \Phi = \{\{h_2\}\}$. So that $(F, E)^\circ \cup (G, E)^\circ\tilde{\subseteq}((F, E) \cup (G, E))^\circ$ but $((F, E) \cup (G, E))^\circ \tilde{\subsetneq} (F, E)^\circ \cup (G, E)^\circ$.

Theorem 3.4 *Let (F, E) be a soft set of soft topological space over X. Then*

1. $((F, E)')^\circ = \left(\overline{(F, E)}\right)'$.
2. $\overline{(F, E)'} = ((F, E)^\circ)'$.
3. $(F, E)^\circ = (\overline{(F, E)'})'$.
4. $\overline{(F, E)'} = ((F, E)^\circ)'$.

Definition 3.17 Let $(X, \mathcal{T}, E)$ be a soft topological space over X then the soft exterior of soft set (F, E) over X is denoted by $(F, E)_\circ$ and is defined as $(F, E)_\circ = ((F, E)')^\circ$.

Thus x is called a soft exterior point of (F, E) if there exists a soft open set (G, E) such that $x \in (G, E)\tilde{\subseteq}(F, E)'$. We observe that $(F, E)_\circ$ is the largest soft open set contained in $(F, E)'$.

Example 40 Let us consider the soft topological space $(X, \mathcal{T}, E)$ over X in Example 36 and $(F, E) = \{\{h_1\}, \{h_2\}\}$ be the soft set of soft topological space X. Then $(F, E)' = \{(h_2, h_3), \{h_1, h_3\}\}$ and so, $(F, E)_\circ = \left((F, E)'\right)^\circ = \{\{h_2\}, \{h_1\}\}$. Clearly we have

Theorem 3.5 *Let (F, E) be a soft set of soft topological space over$X$. Then*

1. $((F, E) \cup (G, E))_\circ = (F, E)_\circ \cap (G, E)_\circ$.
2. $(F, E)_\circ \cup (G, E)_\circ\tilde{\subseteq}((F, E) \cap (G, E))_\circ$.

The following example shows that the equality does not hold in Theorem 3.5(2).

Example 41 Let $X = \{h_1, h_2, h_3\}, E = \{e_1, e_2\}$ and $\mathcal{T} = \{\widetilde{\Phi}, \widetilde{X}, (G_1, E), (G_2, E), (G_3, E), (G_4, E), (G_5, E)\}$ be two soft topologies defined on X where $(G_1, E), (G_2, E), (G_3, E), (G_4, E)$ and (G_5, E) are soft sets over X, defined as follows

$G_1(e_1) = \{h_2\}, \quad G_1(e_2) = \{h_1\}$
$G_2(e_1) = \{h_2, h_3\}, \quad G_2(e_2) = \{h_1, h_2\}$
$G_3(e_1) = \{h_1, h_2\}, \quad G_3(e_2) = \{h_1, h_2\}$
$G_4(e_1) = \{h_2\}, \quad G_4(e_2) = \{h_1, h_3\},$
$G_5(e_1) = \{h_1, h_2\}, \quad G_5(e_2) = X.$

Then $\mathcal{T}$ defines a soft topology on X and hence $(X, \mathcal{T}, E)$ is a soft topological space over X.

Let us take $(F, E) = \{\{h_1, h_2\}, \{h_2\}\}, (G, E) = \{\{h_3\}, \phi\}$.

Now $(F, E)_\circ = \left((F, E)'\right)^\circ = \{\{h_3\}, \{h_1, h_2\}\}^\circ = \phi$ and
$(G, E)_\circ = \left((G, E)'\right)^\circ = \{\{h_1, h_2\}, X\}^\circ = \{\{h_1, h_2\}, \{h_1, h_2, h_3\}\}$.
$(F, E)_\circ \cup (G, E)_\circ = \{\{h_1, h_2\}, \{h_1, h_2, h_3\}\}$.

Also
$(((F, E) \cap (G, E))_\circ = \left(((F, E) \cap (G, E))'\right)^\circ = \left((\{\{h_1, h_2\}, \{h_2\}\} \cap \{\{h_3\}, \phi\}'\right)^\circ$
$= \left((\phi)'\right)^\circ = (\widetilde{X})^\circ = \widetilde{X}$.

Definition 3.18 Let $(X, \mathcal{T}, E)$ be a soft topological space over X then the soft boundary of soft set (F, E) over X is denoted by $\underline{(F, E)}$ and is defined as $\underline{(F, E)} = \overline{(F, E)} \cap \overline{((F, E)')}$. Obviously $\underline{(F, E)}$ is a smallest soft closed set over X containing (F, E).

Remark 3.3 From the above definition it follows directly that the soft sets (F, E) and $(F, E)'$ have same soft boundary.

In the above Example 41, let us take $(F, E) = \{\{h_2\}, \{h_1, h_2\}\}$ then $\overline{(F, E)} = \widetilde{X}$, $(F, E)' = \{\{h_1, h_3\}, \{h_3\}\}$ and $\overline{(F, E)'} = \{\{h_1, h_3\}, \{h_2, h_3\}\}$. Thus the soft boundary of (F, E) is $\underline{(F, E)} = \overline{(F, E)} \cap \overline{((F, E)')} = \widetilde{X} \cap \{\{h_1, h_3\}, \{h_2, h_3\}\} = \{\{h_1, h_3\}, \{h_2, h_3\}\}$.

Theorem 3.6 *Let (F, E) be a soft set of soft topological space over X. Then the following hold:*

1. $\underline{((F, E))'} = (F, E)^\circ \cup \left((F, E)'\right)^\circ = (F, E)^\circ \cup (F, E)_\circ$.
2. $\overline{(F, E)} = (F, E)^\circ \cup \underline{(F, E)}$.
3. $\underline{(F, E)} = \overline{(F, E)} \cap \overline{(F, E)'} = \overline{(F, E)} - (F, E)^\circ$.
4. $(F, E)^\circ = (F, E) \backslash \underline{(F, E)}$.

Proof

$$\begin{aligned}(1)\ (F, E)^\circ \cup \left((F, E)'\right)^\circ &= \left(((F, E)^\circ)'\right)' \cup \left(\left(((F, E)')^\circ\right)'\right)' \\ &= \left[((F, E)^\circ)' \cap \left((F, E)'\right)^\circ\right]' \\ &= [\overline{(F, E)'} \cap \overline{(F, E)}]' \\ &= \underline{((F, E))'}.\end{aligned}$$

$$\begin{aligned}(2)\ (F, E)^\circ \cup \underline{(F, E)} &= (F, E)^\circ \cup (\overline{(F, E)} \cap \overline{(F, E)'}) \\ &= \left[(F, E)^\circ \cup \overline{(F, E)}\right] \cap \left[(F, E)^\circ \cup \overline{(F, E)'}\right] \\ &= \overline{(F, E)} \cap \left[(F, E)^\circ \cup ((F, E)^\circ)'\right] \\ &= \overline{(F, E)} \cap \widetilde{X} \\ &= \overline{(F, E)}.\end{aligned}$$

$$
\begin{aligned}
(3)\ \underline{(F,E)} &= \overline{(F,E)} - (F,E)^\circ \\
&= \overline{(F,E)} \cap ((F,E)^\circ)' \\
&= \overline{(F,E)} \cap \overline{(F,E)'}
\end{aligned}
$$

$$
\begin{aligned}
(4)\ (F,E)\setminus\underline{(F,E)} &= (F,E) \cap (F,E)' \\
&= (F,E) \cap \big((F,E)^\circ \cup ((F,E)')^\circ\big) \\
&= \big[(F,E) \cap (F,E)^\circ\big] \cup \big[(F,E) \cap \big((F,E)'\big)^\circ\big] \\
&= (F,E)^\circ \cup \tilde{\Phi} \\
&= (F,E)^\circ.
\end{aligned}
$$

Remark 3.4 By (3) of above Theorem 3.6, it is clear that $\underline{(F,E)} = \underline{(F,E)'}$.

Theorem 3.7 *Let (F,E) be a soft set of soft topological space over X. Then:*

1. *(F,E) is a soft open set over X if and only if $(F,E) \cap \underline{(F,E)} = \tilde{\Phi}$.*
2. *(F,E) is a soft closed set over X if and only if $(F,E)\tilde{\subseteq}\underline{(F,E)}$.*

Proof (1) Let (F,E) be a soft open set over X. Then $(F,E)^\circ = (F,E)$ implies

$$
\begin{aligned}
(F,E) \cap \underline{(F,E)} &= (F,E)^\circ \cap \underline{(F,E)} \\
&= \tilde{\Phi}.
\end{aligned}
$$

Conversely, let $(F,E) \cap \underline{(F,E)} = \tilde{\Phi}$. Then $(F,E) \cap \overline{(F,E)} \cap \overline{(F,E)'} = \tilde{\Phi}$ or $(F,E) \cap \overline{(F,E)'} = \tilde{\Phi}$ or $\overline{(F,E)'}\tilde{\subseteq}(F,E)'$, which implies $(F,E)'$ is a soft closed set and hence (F,E) is a soft open set.

(2) Let (F,E) be a soft closed set over X. Then $\overline{(F,E)} = (F,E)$. Now $\underline{(F,E)} = \overline{(F,E)} \cap \overline{(F,E)'}\tilde{\subseteq}\overline{(F,E)}$. That is, $\underline{(F,E)}\tilde{\subseteq}(F,E)$.

Conversely, $\underline{(F,E)}\tilde{\subseteq}(F,E)$. Then $\underline{(F,E)} \cap (F,E)' = \tilde{\Phi}$. Since $\underline{(F,E))} = \underline{(F,E)'} = \tilde{\Phi}$, we have $\underline{(F,E)'} \cap (F,E)' = \tilde{\Phi}$. By(1), $(F,E)'$ is soft open and hence (F,E) is soft closed. □

Theorem 3.8 *Let (F,A) and (G,B) be soft sets of soft topological space over X. Then the following hold:*

1. $\underline{((F,A) \cup (G,B))}\tilde{\subseteq}\big[\underline{(F,A)} \cap \underline{(G,B)}\big] \cup [\underline{(G,B)} \cap \overline{((F,A)')}]$.
2. $\underline{[(F,A) \cap (G,B)]}\tilde{\subseteq}[\underline{(F,A)} \cap \overline{(G,B)}] \cup [\underline{(G,B)} \cap \overline{(F,A)}]$.

Proof

$$
\begin{aligned}
(1)\ \underline{((F,A) \cup (G,B))} &= \overline{((F,A) \cup (G,B))} \cap \overline{(((F,A) \cup (G,B))')} \\
&= (\overline{(F,A)} \cup \overline{(G,B)}) \cap \overline{((F,A)' \cap (G,B)')} \\
&\tilde{\subseteq} (\overline{(F,A)} \cup \overline{(G,B)}) \cap \left[\overline{(F,A)}' \cap \overline{(G,B)'}\right] \\
&= (\overline{(F,A)} \cap \overline{(F,A)'}) \cap (\overline{(G,B)'} \cup \overline{(G,B)}) \cap [\overline{(F,A)'} \cap \overline{(G,B)'}] \\
&= (\underline{(F,A)} \cap \overline{(G,B)'}) \cup (\underline{(G,B)} \cap \overline{(F,A)'}) \\
&\tilde{\subseteq} \big[\underline{(F,A)} \cap \underline{(G,B)}\big] \cup [\underline{(G,B)} \cap \overline{((F,A)')}].
\end{aligned}
$$

$$
\begin{aligned}
(2)\ \underline{((F,A)\cap(G,B))} &= \overline{((F,A)\cap(G,B))}\cap\overline{(((F,A)\cap(G,B))')}\\
&\widetilde{\subseteq}[\overline{(F,A)}\cap\overline{(G,B)}]\cap[\overline{((F,A)'\cup(G,B)')}]\\
&=[\overline{(F,A)}\cap\overline{(G,B)}]\cap[\overline{(F,A)'}\cup\overline{(G,B)'}]\\
&=[(\overline{(F,A)}\cap\overline{(G,B)})\cap\overline{(F,A)'}]\cup[(\overline{(F,A)}\cap\overline{(G,B)})\cap\overline{(G,B)'}]\\
&=(\underline{(F,A)}\cap\overline{(G,B)})\cup(\overline{(F,A)}\cap\underline{(G,B)}).
\end{aligned}
$$

Theorem 3.9 *Let (F, E) be a soft set of soft topological space over X. Then the following holds:*

$$\underline{(\underline{(\underline{(F,E)})})} = \underline{(\underline{(F,E)})}.$$

Proof

$$
\begin{aligned}
\underline{(\underline{(\underline{(F,E)})})} &= \overline{(\underline{(\underline{(F,E)})})}\cap\overline{((\underline{(\underline{(F,E)})})')}\\
&= \overline{(\underline{(\underline{(F,E)})})}\cap\overline{((\underline{(\underline{(F,E)})})')}.
\end{aligned}
$$

Now consider

$$
\begin{aligned}
\overline{((\underline{(\underline{(F,E)})})')} &= [\overline{(\underline{(F,E)})}\cap((\underline{(F,E)})')]'\\
&= [\underline{(F,E)}\cap\overline{(\underline{(F,E)})'}]'\\
&= (\underline{(F,E)})'\cup(\overline{(\underline{(F,E)})'})'.
\end{aligned}
$$

Therefore

$$
\begin{aligned}
\overline{((\underline{(\underline{(F,E)})})')} &= \overline{[(\overline{(\underline{(F,E)})})'\cup((\overline{(\underline{(F,E)})'}))']}\\
&= (\overline{(\overline{(\underline{(F,E)})})'})\cup(\overline{((\overline{(\underline{(F,E)})'}))'})\\
&= (G,E)\cup((\overline{(\underline{(G,E)})'}))') = \widetilde{X}.
\end{aligned}
$$

where $(G,E) = \overline{\overline{((\underline{(\underline{(F,E)})})')}}$. Now we have

$$\underline{(\underline{(\underline{(F,E)})})} = \underline{(\underline{(F,E)})}\cap\widetilde{X} = \underline{(\underline{(F,E)})}. \qquad \square$$

Theorem 3.10 *Let (F, E) and (G, E) be soft open sets of soft topological space over X. Then the following hold:*

1. $((F,E)\backslash(G,E))^{\circ}\widetilde{\subseteq}(F,E)^{\circ}\backslash(G,E)^{\circ}$.
2. $\underline{(F,E)^{\circ}}\widetilde{\subseteq}\underline{(F,E)}$

Proof

$$
\begin{aligned}
(1)\ ((F,E)\setminus(G,E))^\circ &= \left((F,E)\cap(G,E)'\right)^\circ \\
&= (F,E)^\circ \cap \left((G,E)'\right)^\circ \\
&= (F,E)^\circ \cap \left(\overline{(G,E)}\right)' \quad \text{(by Theorem 3.4(1))} \\
&= (F,E)^\circ \setminus \overline{(G,E)} \\
&\tilde{\subseteq} (F,E)^\circ \setminus (G,E)^\circ .
\end{aligned}
$$

$$
\begin{aligned}
(2)\ \underline{(F,E)^\circ} &= \overline{(F,E)^\circ} \cap \left(\overline{((F,E)^\circ)'}\right) \\
&\tilde{\subseteq} \overline{(F,E)^\circ} \cap (\overline{((F,E)')}) \quad \text{(by Theorem 3.4(4))} \\
&\tilde{\subseteq} \overline{(F,E)} \cap \overline{((F,E)')} = \underline{(F,E)}.
\end{aligned}
$$

Theorem 3.11 *Let (F, E) be a soft set of soft topological space over X. Then $\underline{(F, E)} = \tilde{\Phi}$ if and only if (F, E) is a soft closed set and a soft open set.*

Proof Suppose that $\underline{(F, E)} = \tilde{\Phi}$. First we prove that (F, E) is a soft closed set. Consider

$$
\begin{aligned}
\underline{(F,E)} = \tilde{\Phi} &\Rightarrow \overline{(F,E)} \cap \overline{((F,E)')} = \tilde{\Phi} \\
&\Rightarrow \overline{(F,E)} \tilde{\subseteq} (\overline{(F,E)'})' \\
&= (F,E)^\circ \quad \text{(by Theorem 3.4(3))} \\
&\Rightarrow \overline{(F,E)} \tilde{\subseteq} (F,E) \Rightarrow \overline{(F,E)} = (F,E).
\end{aligned}
$$

This implies that (F, E) is a soft closed set. We now prove that (F, E) is a soft open set. $(F, E) = \tilde{\Phi} \Rightarrow \overline{(F, E)} \cap \overline{(F, E)'}$ or $(F, E) \cap ((F, E)^\circ)' = \tilde{\Phi} \Rightarrow (F, E) \tilde{\subseteq} (F, E)^\circ \Rightarrow (F, E)^\circ = (F, E)$. This implies that (F, E) is a soft open set.

Conversely, suppose that (F, E) is soft open and soft closed set. Then

$$
\begin{aligned}
\underline{(F,E)} &= \overline{(F,E)} \cap \overline{(F,E)'} \\
&= \overline{(F,E)} \cap ((F,E)^\circ)' \quad \text{(by Theorem 3.4(4))} \\
&= (F,E) \cap (F,E)' = \tilde{\Phi}.
\end{aligned}
$$

Definition 3.19 Let $(X, \mathcal{T}, E)$ be a soft topological space, (F, E) be soft subset of X, and $x \in X$. If every soft neighborhood of x soft intersects (F, E) in some point other than x itself, then x is called a soft limit point of (F, E). The set of all soft limit points of (F, E) is denoted by $(F, E)^d$.

In other words, if $(X, \mathcal{T}, E)$ is a soft topological space, (F, E) is a soft subset of X, and $x \in X$, then $x \in (F, E)^d$ if and only if $(G, E) \tilde{\cap} ((F, E) \setminus \{x\}) \neq \tilde{\Phi}$, for all soft open neighborhoods (G, E) of x.

Remark 3.5 Form Definition 3.19, it follows that x is a soft limit point of (F, E) if and only if $x \in \overline{((F, E)\backslash\{x\})}$.

Theorem 3.12 *Let $(X, \mathcal{T}, E)$ be a soft topological space and (F, E) be soft subset of X. Then, $(F, E)\tilde{\cup}(F, E)^d = \overline{(F, E)}$.*

Proof If $x \in (F, E)\tilde{\cup}(F, E)^d$, then $x \in (F, E)$ or $x \in (F, E)^d$. If $x \in (F, E)$, then $x \in \overline{(F, E)}$. If $x \in (F, E)^d$, then $(G, E)\tilde{\cap}((F, E)\backslash\{x\}) \neq \tilde{\Phi}$, for all soft open neighborhoods (G, E) of x, and so $(G, E)\tilde{\cap}(F, E) \neq \Phi$, for all soft open neighborhoods (G, E) of x; hence, $x \in \overline{(F, E)}$.

Conversely, if $x \in \overline{(F, E)}$, then $x \in (F, E)$ or $x \notin (F, E)$. If $x \in (F, E)$, it is trivial that $x \in (F, E)\tilde{\cup}(F, E)^d$. If $x \notin (F, E)$, then $(G, E)\tilde{\cap}((F, E)\backslash\{x\}) \neq \tilde{\Phi}$, for all soft open neighborhoods (G, E) of x. Therefore, $x \in (F, E)^d$ implies $x \in (F, E)\tilde{\cup}(F, E)^d$. So, $(F, E)\tilde{\cup}(F, E)^d = \overline{(F, E)}$. Hence the proof. □

Theorem 3.13 *Let $(X, \mathcal{T}, E)$ be a soft topological space, and (F, E) be soft subset of X. Then (F, E) is soft closed if and only if $(F, E)^d\tilde{\subseteq}(F, E)$.*

Proof (F, E) is soft closed if and only if $(F, E) = \overline{(F, E)}$ if and only if $(F, E) = (F, E)\tilde{\cup}(F, E)^d$ if and only if $(F, E)^d\tilde{\subseteq}(F, E)$. This completes the proof. □

Theorem 3.14 *Let $(X, \mathcal{T}, E)$ be a soft topological space, and (F, E), (G, E) are soft subsets of X. Then,*

(1) $(F, E)\tilde{\subseteq}(G, E) \Rightarrow (F, E)^d\tilde{\subseteq}(G, E)^d$.
(2) $((F, E)\tilde{\cap}(G, E))^d\tilde{\subseteq}(F, E)^d\tilde{\cap}(G, E)^d$.
(3) $((F, E)\tilde{\cup}(G, E))^d = (F, E)^d\tilde{\cup}(G, E)^d$.
(4) $\left((F, E)^d\right)^d \tilde{\subseteq}(F, E)^d$.
(5) $\overline{(F, E)^d} = (F, E)^d$.

Proof (1) Let $(F, E)\tilde{\subseteq}(G, E)$. Since $(F, E)\backslash\{x\}\tilde{\subseteq}(G, E)\backslash\{x\}$, $\overline{(F, E)\backslash\{x\}}\tilde{\subseteq}\overline{(G, E)\backslash\{x\}}$, and we obtain $(F, E)^d\tilde{\subseteq}(G, E)^d$.

(2) $((F, E)\tilde{\cap}(G, E))\tilde{\subseteq}(F, E)$ and $((F, E)\tilde{\cap}(G, E))\tilde{\subseteq}(G, E)$. Then by (1), $((F, E)\tilde{\cap}(G, E))^d \tilde{\subseteq}(F, E)^d$ and $((F, E)\tilde{\cap}(G, E))^d\tilde{\subseteq}(G, E)^d$. Therefore $((F, E)\tilde{\cap}(G, E))^d \tilde{\subseteq} (F, E)^d \tilde{\cap}(G, E)^d$.

(3) For all $x \in ((F, E)\tilde{\cup}(G, E))^d$ implies $x \in \overline{((F, E)\tilde{\cup}(G, E))\backslash\{x\}^c}$. Therefore

$$\begin{aligned}\overline{((F, E)\tilde{\cup}(G, E))\backslash\{x\}} &= \overline{((F, E) \cup (G, E)) \cap \{x\}^c}\\ &= \overline{((F, E)\tilde{\cap}\{x\}^c)\tilde{\cup}((G, E)\tilde{\cap}\{x\}^c)}\\ &= \overline{((F, E)\tilde{\cap}\{x\}^c)}\tilde{\cup}\overline{(G, E)\tilde{\cap}\{x\}^c)}\\ &= \overline{((F, E)\backslash\{x\})}\tilde{\cup}\overline{((G, E)\backslash\{x\})}\end{aligned}$$

if and only $x \in (F, E)^d\tilde{\cup}(G, E)^d$. Hence $((F, E)\tilde{\cup}(G, E))^d = (F, E)^d \tilde{\cup}(G, E)^d$.

(4) Suppose that $x \notin (F, E)^d$. Then $x \notin \overline{(F, E)\backslash\{x\}}$. This implies that there is a soft open set (G, E) such that $x \in (G, E)$ and $(G, E)\tilde{\cap}((F, E)\backslash\{x\}) = \tilde{\Phi}$. We prove that $x \notin \left((F, E)^d\right)^d$. Suppose on the contrary that $x \in \left((F, E)^d\right)^d$. Then $x \in \overline{(F, E)^d\backslash\{x\}}$. Since $x \in (G, E)$, we have $(G, E)\tilde{\cap}\left((F, E)^d\backslash\{x\}\right) \neq$

$\widetilde{\Phi}$. Therefore there is $y \neq x$ such that $y \in (G, E)\widetilde{\cap}(F, E)^d$. It follows that $y \in ((G, E)\backslash\{x\})\widetilde{\cap}((F, E)\backslash\{y\})$. Hence $((G, E)\backslash\{x\})\widetilde{\cap}((F, E)\backslash\{y\}) \neq \widetilde{\Phi}$, a contradiction to the fact that $(G, E)\widetilde{\cap}((F, E)\backslash\{x\}) = \widetilde{\Phi}$. This implies that $x \in \left((F, E)^d\right)^d$ and so $\left((F, E)^d\right)^d \tilde{\subseteq}(F, E)^d$.

(5) This is a consequence of (2), (3), (5) and Theorem 3.12.
This completes the proof. □

Now we prove the following theorem:

Theorem 3.15 *If* $(X, \mathcal{T}, E)$ *be a soft topological space such that* $\forall x, y(\neq x)$ *in* X, *there exist soft nbds of* x *and* y *whose intersection is empty and* Y *be a nonempty subset of* X *containing finite number of points, then* Y *is soft closed.*

Proof Let us take $Y = \{x\}$. Now we show that Y is soft closed. If y is a point of X different from x, then x and y have disjoint soft neighborhoods (F, E) and (G, E), respectively. Since (F, E) does not soft intersect $\{y\}$, point x cannot belong to the soft closure of the set $\{y\}$. As a result, the soft closure of the set $\{x\}$ is $\{x\}$ itself, so it is soft closed. Since Y is arbitrary, this is true for all subsets of X containing finite number of points. Hence the proof. □

The condition on X imposed in theorem above is specifically known as soft Hausdorff separation axiom. A detailed discussion on soft separation axioms is provided in the next section.

3.3 Separation Axioms

Definition 3.20 Let $(X, \mathcal{T}, E)$ be a soft topological space over X and $x, y \in X$ such that $x \neq y$. If there exist soft open sets (F, E) and (G, E) such that
$x \in (F, E)$ and $y \notin (F, E)$ or
$y \in (G, E)$ and $x \notin (G, E)$, then $(X, \mathcal{T}, E)$ is called a soft T_0-space.

Definition 3.21 Let $(X, \mathcal{T}, E)$ be a soft topological space over X and $x, y \in X$ such that $x \neq y$. If there exist soft open sets (F, E) and (G, E) such that
$x \in (F, E)$ and $y \notin (F, E)$ and
$y \in (G, E)$ and $x \notin (G, E)$, then $(X, \mathcal{T}, E)$ is called a soft T_1-space.

Theorem 3.16 *Let* $(X, \mathcal{T}, E)$ *be a soft topological space over* X. *If* (x, E) *is a soft closed set in* $\mathcal{T}$ *for each* $x \in X$ *then* $(X, \mathcal{T}, E)$ *is a soft* T_1*-space.*

Proof Suppose that for each $x \in X$, (x, E) is a soft closed set in $\mathcal{T}$. Then $(x, E)'$ is a soft open set in $\mathcal{T}$. Let $x, y \in X$ such that $x \neq y$. For $x \in X$, $(x, E)'$ is a soft open set such that $y \in (x, E)'$ and $x \notin (x, E)'$. Similarly $(y, E)' \in \mathcal{T}$ is such that $x \in (y, E)'$ and $y \notin (y, E)'$. Thus $(X, \mathcal{T}, E)$ is a soft T_1 -space over X. □

Remark 3.6 The converse of Theorem 3.16 does not hold in general.

Example 42 Let $X = \{h_1, h_2\}$, $E = \{e_1, e_2\}$ and $\mathcal{T} = \{\widetilde{\Phi}, \widetilde{X}, (F_1, E), (F_2, E), (F_3, E)\}$ where

$$F_1(e_1) = X, \quad F_1(e_2) = \{h_2\},$$
$$F_2(e_1) = \{h_1\}, \quad F_2(e_2) = X,$$
$$F_3(e_1) = \{h_1\}, \quad F_3(e_2) = \{h_2\}.$$

Then $(X, \mathcal{T}, E)$ is a soft topological space over X. We have

$$\mathcal{T}_{e_1} = \{\emptyset, X, \{h_1\}\}$$

and

$$\mathcal{T}_{e_2} = \{\emptyset, X, \{h_2\}\}$$

Neither $(X, \mathcal{T}_{e_1})$ nor $(X, \mathcal{T}_{e_2})$ is a T_1-space but $h_1, h_2 \in X$ and $h_1 \neq h_2$, also

$$h_2 \in (F_1, E) \quad \text{but} \quad h_1 \notin (F_1, E)$$

and

$$h_1 \in (F_2, E) \quad \text{but} \quad h_2 \notin (F_2, E)$$

Thus $(X, \mathcal{T}, E)$ is a soft T_1-space over X.

We note that for $(h_1, E), (h_2, E)$ over X, where

$$h_1(e_1) = \{h_1\}, \quad h_1(e_2) = \{h_1\},$$
$$h_2(e_1) = \{h_2\}, \quad h_2(e_2) = \{h_2\}.$$

The relative complement sets $(h_1, E)', (h_2, E)'$ over X are defined by

$$h'(e_1) = \{h_2\}, \quad h'(e_2) = \{h_2\},$$
$$h'(e_1) = \{h_1\}, \quad h'(e_2) = \{h_1\}.$$

Neither $(h_1, E)'$ nor $(h_2, E)'$ belong to $\mathcal{T}$. This shows that the converse of Theorem 3.16 does not hold.

We also note that, if $(X, \mathcal{T}, E)$ is a discrete soft topological space,then it is a soft T_1-space but $(X, \mathcal{T}_\alpha)$ is not a T_1-space for every parameter $\alpha \in E$. The following two propositions give us the conditions to fix this problem.

Proposition 3.10 *Let $(X, \mathcal{T}, E)$ be a soft topological space over X and $x, y \in X$ such that $x \neq y$. If there exist soft open sets (F, E) and (G, E) such that $x \in (F, E)$ and $y \in (F, E)'$ or $y \in (G, E)$ and $x \in (G, E)'$, then $(X, \mathcal{T}, E)$ is a soft T_0-space and $(X, \mathcal{T}_\alpha)$ is a T_0-space, for each $\alpha \in E$.*

Proof Let $x, y \in X$ such that $x \neq y$ and (F, E) and (G, E) are soft open sets over X such that $x \in (F, E)$ and $y \in (F, E)'$ or $y \in (G, E)$ and $x \in (G, E)'$. If $y \in (F, E)'$ then $y \in (F(\alpha))'$ for each $\alpha \in E$. This implies that $y \notin F(\alpha)$ for each $\alpha \in E$. Therefore $y \notin (F, E)$. Similarly we can show that if $x \in (G, E)'$ then $x \notin (G, E)$. Hence $(X, \mathcal{T}, E)$ is a soft T_0-space. Now for any $\alpha \in E$, $(X, \mathcal{T}_\alpha)$ is a topological space and $x \in (F, E)$ and $y \in (F, E)'$ or $y \in (G, E)$ and $x \in (G, E)'$. So that $x \in F(\alpha)$ and $y \notin F(\alpha)$, or $y \in G(\alpha)$ and $x \notin G(\alpha)$. Thus $(X, \mathcal{T}_\alpha)$ is a T_0-space. □

Proposition 3.11 *Let $(X, \mathcal{T}, E)$ be a soft topological space over X and $x, y \in X$ such that $x \neq y$. If there exist soft open sets (F, E) and (G, E) such that $x \in (F, E)$ and $y \in (F, E)'$. And $y \in (G, E)$ and $x \in (G, E)'$, then $(X, \mathcal{T}, E)$ is a* soft *T_1-space and $(X, \mathcal{T}_\alpha)$ is a T_1-space, for each $\alpha \in E$.*

Proof The proof is similar to the proof of Proposition 3.10. □

Proposition 3.12 *Let $(X, \mathcal{T}, E)$ be a soft topological space over X and Y be a non-empty subset of X. If $(X, \mathcal{T}, E)$ is a soft T_0-space then $(Y, \mathcal{T}_Y, E)$ is a soft T_0-space.*

Proof Let $x, y \in Y$ such that $x \neq y$. Then there exist soft open sets (F, E) and (G, E) in X such that $x \in (F, E)$ and $y \notin (F, E)$ Or $y \in (G, E)$ and $x \notin (G, E)$. Now $x \in Y$ implies that $x \in \widetilde{Y}$. So $x \in \widetilde{Y}$ and $x \in (F, E)$.

Hence $x \in \widetilde{Y} \cap (F, E) = \left({}^Y F, E\right)$ where $(F, E) \in \mathcal{T}$.

Consider $y \notin (F, E)$, this means that $y \notin F(\alpha)$ for some $\alpha \in E$. Then $y \notin Y \cap F(\alpha) = Y(\alpha) \cap F(\alpha)$. Therefore $y \notin \widetilde{Y} \cap (F, E) = \left({}^Y F, E\right)$. Similarly it can be proved that if $y \in (G, E)$ and $x \notin (G, E)$ then $y \in \left({}^Y G, E\right)$ and $x \notin \left({}^Y G, E\right)$. Thus $(Y, \mathcal{T}_Y, E)$ is a soft T_0-space. □

Proposition 3.13 *Let $(X, \mathcal{T}, E)$ be a soft topological space over X and Y be a non-empty subset of X. If $(X, \mathcal{T}, E)$ is a soft T_1-space then $(Y, \mathcal{T}_Y, E)$ is a soft T_1-space.*

Proof The proof is similar to the proof of Proposition 3.12. □

Definition 3.22 Let $(X, \mathcal{T}, E)$ be a soft topological space over X and $x, y \in X$ such that $x \neq y$. If there exist soft open sets (F, E) and (G, E) such that $x \in (F, E)$, $y \in (G, E)$ and $(F, E) \cap (G, E) = \widetilde{\Phi}$, then $(X, \mathcal{T}, E)$ is called a soft T_2-space.

Proposition 3.14 *Let $(X, \mathcal{T}, E)$ be a soft topological space over X. If $(X, \mathcal{T}, E)$ is a soft T_2-space over X then $(X, \mathcal{T}_\alpha)$ is a T_2 space for each $\alpha \in E$.*

Proof Suppose that $(X, \mathcal{T}, E)$ is a soft T_2-space over X. For any $\alpha \in E$

$$\mathcal{T}_\alpha = \{F(\alpha) | (F, E) \in \mathcal{T}\}.$$

Let $x, y \in X$ such that $x \neq y$, then there exist $(F, E), (G, E) \in \mathcal{T}$ such that

$$x \in (F, E), y \in (G, E) \quad \text{and} \quad (F, E) \cap (G, E) = \widetilde{\Phi}.$$

This implies that

$$x \in F(\alpha), y \in G(\alpha) \text{ and } F(\alpha) \cap G(\alpha) = \emptyset.$$

Thus $(X, \mathcal{T}_\alpha)$ is a T_2-space, for each $\alpha \in E$. □

Remark 3.7 (1) Every soft T_1-space is a soft T_0-space.
(2) Every soft T_2-space is a soft T_1-space.

For,

Let $(X, \mathcal{T}, E)$ be a soft topological space over X and $x, y \in X$ such that $x \neq y$.

(1) If $(X, \mathcal{T}, E)$ is a soft T_1-space then there exist soft open sets (F, E) and (G, E) such that $x \in (F, E)$ and $y \notin (F, E)$, and $y \in (G, E)$ and $x \notin (G, E)$. Obviously then we have $x \in (F, E)$ and $y \notin (F, E)$, or $y \in (G, E)$ and $x \notin (G, E)$. Thus $(X, \mathcal{T}, E)$ is a soft T_0-space.
(2) If $(X, \mathcal{T}, E)$ is a soft T_2-space then there exist soft open sets (F, E) and (G, E) such that

$$x \in (F, E), y \in (G, E) \quad \text{and} \quad (F, E) \cap (G, E) = \widetilde{\Phi}.$$

Since $(F, E) \cap (G, E) = \widetilde{\Phi}$, so $x \notin (G, E)$ and $y \notin (F, E)$. Hence $(X, \mathcal{T}, E)$ is a soft T_1-space.

Example 43 Let $X = \{h_1, h_2\}$, $E = \{e_1, e_2\}$ and $\mathcal{T} = \{\widetilde{\Phi}, \widetilde{X}, (F_1, E), (F_2, E), (F_3, E)\}$ where

$F_1(e_1) = X, \quad F_1(e_2) = \{h_2\},$
$F_2(e_1) = \{h_1\}, \quad F_2(e_2) = X,$
$F_3(e_1) = \{h_1\}, \quad F_3(e_2) = \{h_2\}.$

Then $(X, \mathcal{T}, E)$ is a soft topological space over X. Also $(X, \mathcal{T}, E)$ is a soft T_1-space over X but not a soft T_2-space because $h_1, h_2 \in X$ and there do not exist any soft open sets (F, E) and (G, E) in X such that

$h_1 \in (F, E), h_2 \in (G, E)$ and $(F, E) \cap (G, E) = \widetilde{\Phi}$.

Now consider the following soft topology on X,

$\mathcal{T} = \{\widetilde{\Phi}, \widetilde{X}, (F_1, E)\}$ where

$F_1(e_1) = X, \quad F_1(e_2) = \{h_2\}.$

Then $(X, \mathcal{T}, E)$ is a soft topological space over X. Also $(X, \mathcal{T}, E)$ is a soft T_0-space over X which is not a soft T_1-space because $h_1, h_2 \in X$ but there do not exist soft open sets (F, E) and (G, E), such that

$h_1 \in (F, E)$ and $h_2 \notin (F, E)$. And $h_2 \in (G, E)$ and $h_1 \notin (G, E)$.

Proposition 3.15 *Let $(X, \mathcal{T}, E)$ be a soft topological space over X and Y be a non-empty subset of X. If $(X, \mathcal{T}, E)$ is a soft T_2-space then $(Y, \mathcal{T}_Y, E)$ is a soft T_2-space.*

Proof Let $x, y \in Y$ such that $x \neq y$. Then there exist soft open sets (F, E) and (G, E) in X such that $x \in (F, E)$, $y \in (G, E)$ and $(F, E) \cap (G, E) = \Phi$.

So for each $\alpha \in E$, $x \in F(\alpha)$, $y \in G(\alpha)$ and $F(\alpha) \cap G(\alpha) = \emptyset$. This implies that $x \in Y \cap F(\alpha)$, $y \in Y \cap F(\alpha)$ and $F(\alpha) \cap G(\alpha) = \emptyset$. Hence $x \in ({}^Y F, E)$, $y \in ({}^Y G, E)$ and $({}^Y F, E) \cap ({}^Y G, E) = \Phi$ where $({}^Y F, E), ({}^Y G, E) \in \mathcal{T}_Y$. Thus $(Y, \mathcal{T}_Y, E)$ is a soft T_2-space. □

Definition 3.23 Let $(X, \mathcal{T}, E)$ be a soft topological space over X, (G, E) be a soft closed set in X and $x \in X$ such that $x \notin (G, E)$. If there exist soft open sets (F_1, E) and (F_2, E) such that $x \in (F_1, E)$, $(G, E) \widetilde{\subset} (F_2, E)$ and $(F_1, E) \cap (F_2, E) = \Phi$, then $(X, \mathcal{T}, E)$ is called a soft regular space.

Definition 3.24 Let $(X, \mathcal{T}, E)$ be a soft topological space over X. Then $(X, \mathcal{T}, E)$ is said to be a soft T_3-space if it is soft regular and soft T_1-space.

Remark 3.8 (1) A soft T_3-space may not be a soft T_2-space.
(2) If $(X, \mathcal{T}, E)$ is a soft T_3-space, then $(X, \mathcal{T}_\alpha)$ may not be a T_3-space for each parameter $\alpha \in E$.

Example 44 Let $X = \{h_1, h_2, h_3\}$, $E = \{e_1, e_2\}$ and
$\mathcal{T} = \{\widetilde{\Phi}i, \widetilde{X}, (F_1, E), (F_2, E), (F_3, E), \ldots, (F_{30}, E)\}$ where

$F_1(e_1) = X \quad F_1(e_2) = \emptyset$
$F_2(e_1) = \{h_1\} \quad F_2(e_2) = \emptyset$
$F_3(e_1) = \{h_2\} \quad F_3(e_2) = \emptyset$
$F_4(e_1) = \{h_3\} \quad F_4(e_2) = \emptyset$
$F_5(e_1) = \{h_1, h_2\} \quad F_5(e_2) = \emptyset$
$F_6(e_1) = \{h_1, h_3\} \quad F_6(e_2) = \emptyset$

$F_7(e_1) = \{h_2, h_3\}$ $F_7(e_2) = \emptyset$
$F_8(e_1) = X$ $F_8(e_2) = \{h_1\}$
$F_9(e_1) = \{h_1\}$ $F_9(e_2) = \{h_1\}$
$F_{10}(e_1) = \{h_2\}$ $F_{10}(e_2) = \{h_1\}$
$F_{11}(e_1) = \{h_3\}$ $F_{11}(e_2) = \{h_1\}$
$F_{12}(e_1) = \{h_1, h_2\}$ $F_{12}(e_2) = \{h_1\}$
$F_{13}(e_1) = \{h_1, h_3\}$ $F_{13}(e_2) = \{h_1\}$
$F_{14}(e_1) = \{h_2, h_3\}$ $F_{14}(e_2) = \{h_1\}$
$F_{15}(e_1) = \emptyset$ $F_{15}(e_2) = \{h_1\}$
$F_{16}(e_1) = X$ $F_{16}(e_2) = \{h_2, h_3\}$
$F_{17}(e_1) = \{h_1\}$ $F_{17}(e_2) = \{h_2, h_3\}$
$F_{18}(e_1) = \{h_2\}$ $F_{18}(e_2) = \{h_2, h_3\}$
$F_{19}(e_1) = \{h_3\}$ $F_{19}(e_2) = \{h_2, h_3\}$
$F_{20}(e_1) = \{h_1, h_2\}$ $F_{20}(e_2) = \{h_2, h_3\}$
$F_{21}(e_1) = \{h_1, h_3\}$ $F_{21}(e_2) = \{h_2, h_3\}$
$F_{22}(e_1) = \{h_2, h_3\}$ $F_{22}(e_2) = \{h_2, h_3\}$
$F_{23}(e_1) = \emptyset$ $F_{23}(e_2) = \{h_2, h_3\}$
$F_{24}(e_1) = \{h_1\}$ $F_{24}(e_2) = X$
$F_{25}(e_1) = \{h_2\}$ $F_{25}(e_2) = X$
$F_{26}(e_1) = \{h_3\}$ $F_{26}(e_2) = X$
$F_{27}(e_1) = \{h_1, h_2\}$ $F_{27}(e_2) = X$
$F_{28}(e_1) = \{h_1, h_3\}$ $F_{28}(e_2) = X$
$F_{29}(e_1) = \{h_2, h_3\}$ $F_{29}(e_2) = X$
$F_{30}(e_1) = \emptyset$ $F_{30}(e_2) = X$.

Then $(X, \mathcal{T}, E)$ is a soft topological space over X. We note that $(X, \mathcal{T}, E)$ is a soft T_3-space but it is not a soft T_2-space because $h_2, h_3 \in X$ but there do not exist soft open sets (F, E) and (G, E) such that

$h_2 \in (F, E)$, $h_3 \in (G, E)$ and $(F, E) \cap (G, E) = \Phi$. Thus every soft T_3-space is not necessarily a soft T_2-space.

Now we have,

$$\mathcal{T}_{e_1} = \{\emptyset, X, \{h_1\}, \{h_2\}, \{h_3\}, \{h_1, h_2\}, \{h_1, h_3\}, \{h_2, h_3\}\}$$

and

$$\mathcal{T}_{e_2} = \{\emptyset, X, \{h_1\}, \{h_2, h_3\}\}.$$

Here $(X, \mathcal{T}_{e_2})$ is not a T_3-space. This shows that if $(X, \mathcal{T}, E)$ is a soft T_3-space then $(X, \mathcal{T}_\alpha)$ need not to be a T_3-space for each parameter $\alpha \in E$.

Proposition 3.16 *Let $(X, \mathcal{T}, E)$ be a soft topological space over X and Y be a non-empty subset of X. If $(X, \mathcal{T}, E)$ is a soft T_3-space then $(Y, \mathcal{T}_Y, E)$ is a soft T_3-space.*

Proof By Proposition 3.13, $(Y, \mathcal{T}_Y, E)$ is a soft T_1-space. Let $y \in Y$ and (F, E) be a soft closed set in Y such that $y \notin (F, E)$. Then $y \notin ((Y, E) \cap (G, E))$ where $(F, E) = ((Y, E) \cap (G, E))$ for some soft closed set in X, by Theorem 3.1. But $y \in (Y, E)$, so $y \notin (G, E)$. As $(X, \mathcal{T}, E)$ is a soft T_3-space, so there exist soft open sets (G_1, E) and (G_2, E) in X such that

$y \in (G_1, E), (G, E) \tilde{\subset} (G_2, E)$ and $(G_1, E) \cap (G_2, E) = \tilde{\Phi}$. Now if we take

$$(F_1, E) = (Y, E) \cap (G_1, E) \quad \text{and} \quad (F_2, E) = (Y, E) \cap (G_2, E),$$

then (F_1, E) and $(F_2, E) \in \mathcal{T}_Y$ such that $y \in (F_1, E)$ and $(F, E) \tilde{\subset} (Y, E) \cap (G_2, E) = (F_2, E)$ and $(F_1, E) \cap (F_2, E) \tilde{\subset} (G_1, E) \cap (G_2, E) = \tilde{\Phi}$, i.e., $(F_1, E) \cap (F_2, E) = \tilde{\Phi}$. Thus $(Y, \mathcal{T}_Y, E)$ is a soft T_3-space. □

Definition 3.25 Let $(X, \mathcal{T}, E)$ be a soft topological space over X, (F, E) and (G, E) soft closed sets over X such that $(F, E) \cap (G, E) = \tilde{\Phi}$. If there exist soft open sets (F_1, E) and (F_2, E) such that

$$(F, E) \tilde{\subset} (F_1, E), (G, E) \tilde{\subset} (F_2, E) \text{ and } (F_1, E) \cap (F_2, E) = \tilde{\Phi},$$

then $(X, \mathcal{T}, E)$ is called a soft normal space.

Definition 3.26 Let $(X, \mathcal{T}, E)$ be a soft topological space over X. Then $(X, \mathcal{T}, E)$ is said to be a soft T_4-space if it is soft normal and soft T_1-Space.

Remark 3.9 (1) A soft T_4-space need not be a soft T_3-space.
(2) If $(X, \mathcal{T}, E)$ is a soft T_4-space then $(X, \mathcal{T}_\alpha)$ need not be a T_4-space for each parameter $\alpha \in E$.
(3) If $(X, \mathcal{T}, E)$ is a soft T_4-space and Y is a non-empty subset of X then $(Y, \mathcal{T}_Y, E)$ need not be a soft T_4-space.

Example 45 Let $X = \{h_1, h_2, h_3, h_4\}$, $E = \{e_1, e_2\}$ and $\mathcal{T} = \{\tilde{\Phi}, \tilde{X}, (F_1, E), (F_2, E), (F_3, E), \ldots, (F_8, E)\}$ where

$$F_1(e_1) = \{h_1, h_2, h_4\}, \quad F_1(e_2) = \{h_1, h_2, h_3\},$$
$$F_2(e_1) = \{h_1, h_3, h_4\}, \quad F_2(e_2) = \{h_1, h_2, h_3\},$$
$$F_3(e_1) = \{h_1, h_4\}, \quad F_3(e_2) = \{h_1, h_2, h_3\},$$
$$F_4(e_1) = \{h_2, h_3\}, \quad F_3(e_2) = \{h_1, h_2, h_3\},$$
$$F_5(e_1) = \{h_2\}, \quad F_5(e_2) = \{h_1, h_2, h_3\},$$
$$F_6(e_1) = \{h_3\}, \quad F_6(e_2) = \{h_1, h_2, h_3\},$$
$$F_7(e_1) = \emptyset, \quad F_7(e_2) = \{h_1, h_2, h_3\},$$
$$F_8(e_1) = X, \quad F_8(e_2) = \{h_1, h_2, h_3\}.$$

Then $(X, \mathcal{T}, E)$ is a soft topological space over X. We note that $(X, \mathcal{T}, E)$ is a soft T_4-space but it is not a soft T_3-space because $h_1 \in X$ and $(F_3, E)'$ is a soft closed set in X such that $h_1 \notin (F_3, E)'$ but there do not exist soft open sets (F, E) and (G, E) such that $h_1 \in (F, E)$, $(F_3, E)' \tilde{\subset} (G, E)$ and $(F, E) \cap (G, E) = \tilde{\Phi}$. Thus every soft T_4-space is not necessarily a soft T_3-space.

Now we have,

$$\mathcal{T}_{e_1} = \{\emptyset, X, \{h_1, h_2, h_4\}, \{h_1, h_3, h_4\}, \{h_1, h_4\}, \{h_2, h_3\}, \{h_2\}, \{h_3\}\}$$

and

$$\mathcal{T}_{e_2} = \{\emptyset, X, \{h_1, h_2, h_3\}\}.$$

Here $(X, \mathcal{T}_{e_1})$ and $(X, \mathcal{T}_{e_2})$ are not T_3-spaces. This shows that if $(X, \mathcal{T}, E)$ is a soft T_4-space then $(X, \mathcal{T}_\alpha)$ need not be a T_4-space for each parameter $\alpha \in E$.

We take

$$Y = \{h_1, h_2, h_3\}.$$

Then

$$\mathcal{T}_Y = \{\widetilde{\Phi}, \widetilde{Y}, ({}^Y F_1, E), ({}^Y F_2, E), ({}^Y F_3, E), \ldots, ({}^Y F_8, E)\}$$

where

${}^Y F_1(e_1) = \{h_1, h_2\}, \quad {}^Y F_1(e_2) = Y,$
${}^Y F_2(e_1) = \{h_1, h_3\}, \quad {}^Y F_2(e_2) = Y,$
${}^Y F_3(e_1) = \{h_1\}, \quad {}^Y F_3(e_2) = Y,$
${}^Y F_4(e_1) = \{h_2, h_3\}, \quad {}^Y F_4(e_2) = Y,$
${}^Y F_5(e_1) = \{h_2\}, \quad {}^Y F5(e_2) = Y,$
${}^Y F_6(e_1) = \{h_3\}, \quad {}^Y F_6(e_2) = Y,$
${}^Y F_7(e_1) = \emptyset, \quad {}^Y F7(e_2) = Y,$
${}^Y F_8(e_1) = Y, \quad {}^Y F_8(e_2) = Y.$

We note that $(Y, \mathcal{T}_Y, E)$ is not a soft T_4-space because $({}^Y F_3, E)'$ and $({}^Y F_4, E)'$ are soft closed sets in Y such that $({}^Y F_3, E)' \cap ({}^Y F_4, E)' = \widetilde{\Phi}$ but there do not exist any soft open sets (F, E) and (G, E) in Y such that $({}^Y F_3, E)' \widetilde{\subset} (F, E)$, $(F_4, E)' \widetilde{\subset} (G, E)$ and $(F, E) \cap (G, E) = \widetilde{\Phi}$. Thus a soft subspace of a soft T_4-space may not be a soft T_4-space.

3.4 Continuous Mappings and Connectedness

The concept of soft continuity and connectedness will be discussed in this section. Results and examples provided are taken from Zorlutuna et al. [168] and Sabir Hussain [119].

Definition 3.27 Two soft sets (F, E) and (G, E) over a common universe X are soft disjoint, if $(F, E) \widetilde{\cap} (G, E) = \widetilde{\Phi}$. That is, $\phi = F(e) \cap G(e)$, for all $e \in E$.

Definition 3.28 The soft set $(F, E) \in S(X)_E$ is called a soft point in $\widetilde{X}$, denoted by e_F, if for the element $e \in E$, $F(e) \neq \phi$ and $F(e^c) = \phi$ for all $e^c \in E \backslash \{e\}$.

Definition 3.29 The soft point e_F is said to be in the soft set (G, E), denoted by $e_F \tilde{\in} (G, E)$, if for the element $e \in E$ and $F(e) \widetilde{\subseteq} G(e)$.

Definition 3.30 A soft set (G, E) in a soft topological space (X, τ, E) is called a soft neighborhood (briefly: soft nbd) of the soft point $e_F \in X$, if there exists a soft open set (H, E) such that $e_F \in (H, E) \widetilde{\subseteq} (G, E)$.

The soft neighborhood system of a soft point e_F, denoted by $N_\tau(e_F)$, is the family of all its soft neighborhoods.

Definition 3.31 Let (X, τ, E) and (Y, τ^*, E') be soft topological spaces. Let $u : X \to Y$ and $p : E \to E'$ be mappings. Let $u_p : S(X)_E \to S(Y)_{E'}$ be a soft mapping and $e_F \in \widetilde{X}$.

(a) u_p is soft pu-continuous at $e_F \in \widetilde{X}$, if for each is $(G, E') \in N_{\tau^*}(u_p(e_F))$, there exists an $(H, E) \in N_\tau(e_F)$ such that $u_p(H, E) \widetilde{\subseteq} (G, E')$.

(b) u_p is soft pu-continuous on $\tilde{X}$, if u_p is soft pu-continuous at each soft point in X.

Theorem 3.17 *Let (X, τ, E) and (Y, τ^*, E') be soft topological spaces. Let $u_p : S(X)_E \to S(Y)_{E'}$ be a soft mapping. Then the following statements are equivalent.*

(a) u_p is soft pu-continuous,
(b) For each $(G, E') \in \tau^*, u_p^{-1}((G, E')) \in \tau$.
(c) For (G, E') soft closed in $(Y, \tau^*, E'), u_p^{-1}(G, E')$ is soft closed in (X, τ, E).

Definition 3.32 Let $(X, \mathcal{T}, E)$ be a soft topological space over X. Then $(X, \mathcal{T}, E)$ is said to be soft connected, if there does not exist a pair (F, E) and (G, E) of nonempty soft disjoint soft open subsets of $(X, \mathcal{T}, E)$ such that $\widetilde{X} = (F, E)\widetilde{\cup}(G, E)$, otherwise $(X, \mathcal{T}, E)$ is said to be soft disconnected. In this case, the pair (F, E) and (G, E) is called the soft disconnection of X.

Example 46 Let $X = \{h_1, h_2, h_3\}$-the houses under consideration, $E = \{e_1 =$ beautiful , $e_2 =$ cheap$\}$ and
$\tau = \{\widetilde{\Phi}, \widetilde{X}, (F_1, E), (F_2, E), (F_3, E), (F_4, E), (F_5, E)\}$
where $(F_1, E), (F_2, E), (F_3, E), (F_4, E)$ and (F_5, E) are soft sets over X which gives a collection of approximate description of an object, defined as follows:
$F_1(e_1) = \{h_2\}, F_1(e_2) = \{h_1\}$,
$F_2(e_1) = \{h_2, h_3\}, F_2(e_2) = \{h_1, h_2\}$,
$F_3(e_1) = \{h_1, h_2\}, F_3(e_2) = X$,
$F_4(e_1) = \{h_1, h_2\}, F_4(e_2) = \{h_1, h_3\}$,
$F_5(e_1) = \{h_2\}, F_5(e_2) = \{h_1, h_2\}$.

Then $\mathcal{T}$ defines a soft topology on X and hence $(X, \mathcal{T}, E)$ is a soft topological space over X. Clearly X is soft connected.

Definition 3.33 Let $(X, \mathcal{T}, E)$ be a soft topological space over X. A soft subset (F, E) of a soft topological space $(X, \mathcal{T}, E)$ is soft connected, if it is soft connected as a soft subspace.

Theorem 3.18 *A soft topological space $(X, \mathcal{T}, E)$ is soft disconnected (respt. soft connected) if and only if there exists (respt. does not exist) nonempty soft subset (F, E) of $(X, \mathcal{T}, E)$ which is both soft open and soft closed in $(X, \mathcal{T}, E)$.*

Theorem 3.19 *Let $(X, \mathcal{T}, E)$ and $(Y, \mathcal{T}^*, E')$ be two soft topological spaces and $u : X \to Y$ and $p : E \to E'$ be mappings. Also a soft mapping $u_p : S(X)_E \to S(Y)_{E'}$ is soft pu-continuous and soft onto. If $(X, \mathcal{T}, E)$ is soft connected, then the soft image of $(X, \mathcal{T}, E)$ is also soft connected.*

Proof Let a soft mapping $u_p : S(X)_E \to S(Y)_{E'}$ be soft pu-continuous and soft onto. Contrarily, suppose that $(Y, \mathcal{T}^*, E')$ is soft disconnected and pair (G_1, E') and (G_2, E') is a soft disconnections of $(Y, \mathcal{T}^*, E')$. Since $u_p : S(X)_E \to S(Y)_{E'}$ is soft pu-continuous, $u_p^{-1}(G_1, E')$ and $u_p^{-1}(G_2, E')$ are both soft open in $(X, \mathcal{T}, E)$. Clearly the pair $u_p^{-1}(G_1, E')$ and $u_p^{-1}(G_2, E')$ is a soft disconnection of $(X, \mathcal{T}, E)$, a contradiction. Hence $(Y, \mathcal{T}^*, E')$ is soft connected. This completes the proof. □

Now we characterize soft connectedness in terms of soft boundary.

Theorem 3.20 *A soft topological space $(X, \mathcal{T}, E)$ is soft connected if and only if every nonempty proper soft subspace has a non-empty soft boundary.*

Proof Contrarily suppose that a nonempty proper soft subspace (F, E) of a soft connected space $(X, \mathcal{T}, E)$ has empty soft boundary. Then (F, E) is soft open and $\overline{(F, E)}\tilde{\cap}\overline{(F, E)'} = \tilde{\Phi}$. Let x be a soft limit point of (F, E). Then $x \in \overline{(F, E)}$ but $x \notin \overline{(F, E)'}$. In particular, $x \notin (F, E)'$ and so $x \in (F, E)$. Thus (F, E) is soft closed and soft open. By Theorem 3.18, $(X, \mathcal{T}, E)$ is soft disconnected. This contradiction proves that (F, E) has a nonempty soft boundary.

Conversely, suppose that X is soft disconnected. Then by Theorem 3.18, $(X, \mathcal{T}, E)$ has a proper soft subset (F, E) which is both soft closed and soft open. Then $\overline{(F, A)} = (F, A)$, $\overline{(F, A)'} = (F, A)'$ and $\overline{(F, A)}\tilde{\cap}\overline{(F, A)'} = \tilde{\Phi}$. So (F, E) has empty soft boundary, a contradiction. Hence $(X, \mathcal{T}, E)$ is soft connected. This completes the proof. □

Theorem 3.21 *Let the pair (F, E) and (G, E) of soft sets be a soft disconnection in soft topological space $(X, \mathcal{T}, E)$ and (H, E) be a soft connected subspace of $(X, \mathcal{T}, E)$. Then (H, E) is contained in (F, E) or (G, E).*

Proof Contrarily suppose that (H, E) is neither contained in (F, E) nor in (G, E). Then $(H, E)\tilde{\cap}(F, E)$, $(H, E)\tilde{\cap}(G, E)$ are both nonempty soft open subsets of (H, E) such that $((H, E)\tilde{\cap}(F, E))\tilde{\cap}((H, E)\tilde{\cap}(G, E)) = \tilde{\Phi}$ and $((H, E)\tilde{\cap}(F, E))$ $\tilde{\cup}((H, E)\tilde{\cap}(G, E)) = (H, E)$. This gives that pair of $((H, E)\tilde{\cap}(F, E))$ and $((H, E)\tilde{\cap}(G, E))$ is a soft disconnection of (H, E). This contradiction proves the theorem. □

Theorem 3.22 *Let (G, E) be a soft connected subset of a soft topological space $(X, \mathcal{T}, E)$ and (F, E) be soft subset of X such that $(G, E)\tilde{\subseteq}(F, E)\tilde{\subseteq}\overline{(G, E)}$. Then (F, E) is soft connected.*

Proof It is sufficient to show that $\overline{(G, E)}$ is soft connected. On the contrary, suppose that $\overline{(G, E)}$ is soft disconnected. Then there exists a soft disconnection $((H, E), (K, E))$ of $\overline{(G, E)}$. That is, there are $((H, E)\tilde{\cap}(G, E))$, $((K, E)\tilde{\cap}(G, E))$ soft open sets in (G, E) such that $((H, E)\tilde{\cap}(G, E))\tilde{\cap}((K, E)\tilde{\cap}(G, E)) = ((H, E)\tilde{\cap}(K, E))\tilde{\cap}(G, E) = \tilde{\Phi}$, and $((H, E)\tilde{\cap}(G, E))\tilde{\cup}((K, E)\ \tilde{\cap}(G, E)) = ((H, E)\tilde{\cup}(K, E))\tilde{\cap}(G, E) = (G, E)$. This gives that pair $((H, E)\tilde{\cap}(G, E))$, $((K, E)\tilde{\cap}(G, E))$ is a soft disconnection of (G, E), a contradiction. This proves that $\overline{(G, E)}$ is soft connected. Hence the proof. □

Corollary 3.2 *If (F, E) is a soft connected soft subspace of a soft topological space $(X, \mathcal{T}, E)$, then $\overline{(F, E)}$ is soft connected.*

Theorem 3.23 *Let $(X, \mathcal{T}, E)$ be a soft regular space and $(Y, \mathcal{T}_Y, E)$ is a soft subspace of $(X, \mathcal{T}, E)$ such that $\mathcal{T}_Y = \{(Y_F, E)\,|(F, E) \in \mathcal{T}\}$ is soft relative topology on Y. Then $(Y, \mathcal{T}_Y, E)$ is soft regular space.*

Proof Let $(Y, \mathcal{T}_Y, E)$ be a soft subspace of soft regular space $(X, \mathcal{T}, E)$. Let $y \in Y$ and (G, E) be soft closed set in Y such that $y \notin (G, E)$. Now $\overline{(G, E)} \widetilde{\cap} Y = (G, E)$. Clearly, $y \notin \overline{(G, E)}$. Thus $\overline{(G, E)}$ is soft closed in X such that $y \notin \overline{(G, E)}$. Since X is soft regular, there exist soft open sets (F_1, E) and (F_2, E) such that $y \in (F_1, E)$, $\overline{(G, E)} \tilde{\subseteq} (F_2, E)$ and $(F_1, E) \cap (F_2, E) = \widetilde{\Phi}$. Then $Y \widetilde{\cap} (F_1, E)$, $Y \widetilde{\cap} (F_2, E)$ are soft disjoint soft open sets in Y such that $y \in Y \tilde{\cap} (F_1, E)$ and $(G, E) \tilde{\subseteq} Y \widetilde{\cap} (F_2, E)$. This completes the proof. □

3.5 Another Approach to Soft Topology

As mentioned in the beginning of the chapter, there are two approaches to soft topology in literature. Namely, one due to Shabir and Naz [125] and the other due to Cagman et al. [28]. All discussions in Sects. 3.1–3.4 was in the context of Shabir and Naz [125]. Now we will turn our attention to the second approach put forward by Cagman et al. [28]. To distinguish both, we adopt entirely different notations as introduced by Cagman et al. [28]. A brief account of some fundamental ideas in this line is provided in this section. Results provided are taken from Cagman et al. [28] and Babitha and Sunil [19]. A new notation is introduced for soft sets for the purpose of defining topology on soft sets by Cagman. This notation will be followed for the topological study of soft sets.

Definition 3.34 A soft set F_A on the universe U and parameter set E with $A \subseteq E$ is defined by the set of ordered pairs $F_A = \{(x, f_A(x)) : x \in E, f_A(x) \in P(U)\}$, where $f_A : E \to P(U)$ such that $f_A(x) = \emptyset$ if $x \notin A$. Here, f_A is called an approximate function of the soft set f_A. The value of $f_A(x)$ may be arbitrary. Some of them may be empty, some may have nonempty intersection.

Note that the set of all soft sets over U will be denoted by $S(U)_E$ or $S(U)$ if parameter set E is understood.

Definition 3.35 Let $F_A \in S(U)$. If $f_A(x) = \emptyset$ for all $x \in E$, then F_A is called an empty soft set, denoted by $F_\emptyset$.

Definition 3.36 Let $F_A, F_B \in S(U)$. Then, F_A is a soft subset of F_B denoted by $F_A \subseteq F_B$ if $f_A(x) \subseteq f_B(x)$ for all $x \in E$.

Definition 3.37 Let $F_A, F_B \in S(U)$. Then F_A and F_B are soft equal denoted by $F_A = F_B$ if and only if $f_A(x) = f_B(x)$ for all $x \in E$.

Definition 3.38 Let $F_A, F_B \in S(U)$. Then the soft union $F_A \cup F_B$, the soft intersection $F_A \cap F_B$ and the soft difference $F_A - F_B$ of F_A and F_B are defined by the approximate functions $f_{A \cup B}(x) = f_A(x) \cup f_B(x)$, $f_{A \cap B}(x) = f_A(x) \cap f_B(x)$, $f_{A-B}(x) = f_A(x) - f_B(x)$ respectively, and the soft complement F_A^c of F_A is defined by the approximate function $F_A^c(x) = f_A^c(x)$ where $f_A^C(x)$ is the complement of the set $f_A(x)$. i.e., $f_A^c(x) = U - f_A(x)$ for all $x \in E$.

Definition 3.39 Let $F_A \in S(U)$. A soft topology on F_A denoted by τ, is a collection of soft subsets of F_A having the following properties:

(i) $F_\emptyset, F_A \in \tau$.
(ii) $\{F_{A_i} \subseteq F_A : i \in I \subseteq \mathbb{N}\} \subseteq \tau \Rightarrow \cup_i F_{A_i} \in \tau$.
(iii) $\{F_{A_i} \subseteq F_A : 1 \leq i \leq n, n \in \mathbb{N}\} \subseteq \tau \Rightarrow \bigcap_{i=1}^n F_{A_i} \in \tau$. The pair (F_A, τ) is called a soft topological space.

Definition 3.40 Let (F_A, τ) be a soft topological space. Then, every element of τ is called a soft open set. Clearly, $F_\emptyset$ and F_A are soft open sets.

Definition 3.41 Let (F_A, τ) be a soft topological space and $\mathcal{B} \subseteq \tau$. If every element of τ can be written as the union of elements of $\mathcal{B}$, then $\mathcal{B}$ is called a soft basis for the soft topology τ. Each element of $\mathcal{B}$ is called a soft basis element.

Theorem 3.24 *Let (F_A, τ) be a soft topological space and $\mathcal{B}$ be a soft basis for τ. Then τ equals the collection of all soft unions of elements of $\mathcal{B}$.*

Definition 3.42 Let (F_A, τ) be soft topological space and $\mathcal{S}$ be a collection of soft subsets of F_A. Then $\mathcal{S}$ is called a soft sub base for a topology on F_A iff the soft union of finite soft intersection elements of $\mathcal{S}$ is a soft base for the topology. The soft topology generated by the soft sub basis $\mathcal{B}$ is defined to be the collection τ of soft set union of all finite soft set intersections of elements of $\mathcal{B}$.

Definition 3.43 Let (F_A, τ) be a soft topological space and $F_B \subseteq F_A$. Then, the collection

$$\tau_{F_B} = \{F_{A_i} \cap F_B : F_{A_i} \in \tau, i \in I \subseteq \mathbb{N}\}$$

is called a soft subspace topology on F_B

Definition 3.44 Let (F_A, τ) be a soft topological space and $F_B \subseteq F_A$. Then, F_B is said to be soft closed if the soft set F_B^C is soft open.

Theorem 3.25 *Let (F_A, τ) be a soft topological space. Then following conditions hold.*

(i) $F_\emptyset, F_A$ are soft closed sets.
(ii) Arbitrary soft intersections of the soft closed sets are soft closed.
(iii) Finite soft unions of the soft closed sets are soft closed.

Definition 3.45 Let (F_A, τ) be a soft topological space and $F_B \subseteq F_A$. Then, the soft closure of F_B, denoted by $\bar{F}_B$ is defined as the soft intersection of all soft closed supersets of F_B.

Theorem 3.26 *Let (F_A, τ) be a soft topological space and $F_B, F_C \subseteq F_A$. Then the following hold:*

(i) $\alpha \in \bar{F}_B$. If and only if every soft open set F_P containing α soft intersects F_B.

(ii) Supposing the soft topology of F_A is given by a soft basis, then $\alpha \in \bar{F}_B$ if and only if every soft basis element F_D containing α soft intersects F_B.

Theorem 3.27 Let (F_A, τ) *be a soft topological space and* $F_B \subseteq F_A$. *Then,* F_B *is a closed soft set if and only if* $F_B = \bar{F}_B$.

Definition 3.46 Let F_A be a soft set. Let τ be collection of all soft subsets of F_A. Then τ is called soft discrete topology on F_A. The space (F_A, τ) is called soft discrete topological space.

Definition 3.47 Let F_A be a soft set. Let τ be collection of F_A and F_ϕ. Then τ is called soft indiscrete topology on F_A. Then space (F_A, τ) is called soft indiscrete topological space.

Proposition 3.17 *Let* F_A *be a soft set over* U. *Then,*

(i) $F_A \cup F_A = F_A, F_A \cap F_A = F_A$
(ii) $F_A \cup F_U = F_A, F_A \cap F_\emptyset = F_\emptyset$
(iii) $F_A \cup F_E = F_E, F_A \cap F_E = F_A$
(iv) $F_A \cup F_{A^c} = F_E, F_A \cap F_{A^c} = F_\emptyset$.

Proposition 3.18 *Let* F_A, F_B, F_C *be soft sets over* U *Then,*

(i) $F_A \cup F_B = F_B \cup F_A, F_A \cap F_B = F_B \cap F_A$.
(ii) $(F_A \cup F_B)^c = F_A^c \cap F_B^c, (F_A \cap F_B)^c = F_A^c \cup F_B^c$.
(iii) $(F_A \cup F_B) \cup F_C = F_A \cup (F_B \cup F_C), (F_A \cap F_B) \cap F_C = F_A \cap (F_B \cap F_C)$.
(iv) $(F_A \cup F_B) \cap F_C = (F_A \cap F_C) \cup (F_B \cap F_C)$ *and* $(F_A \cap F_B) \cup F_C = (F_A \cup F_C) \cap (F_B \cup F_C)$.

Definition 3.48 Let (F_A, τ) be a soft topological space and $\alpha \in F_A$. If there exists a soft open set F_B such that $\alpha \in F_B$, then F_B is called a soft open neighborhood (or soft neighborhood) of α. The set of all soft neighborhoods of α, denoted $\tilde{v}(\alpha)$ is called the family of soft neighborhoods of α; that is,

$$\tilde{v}(\alpha) = \{F_B : F_B \in \tau, \alpha \in F_B\}.$$

3.5.1 Continuous Soft Set Functions

The notion of soft set functions considered in this subsection is in the sense of Babitha and Sunil [15] which is mentioned in Definition 1.33.

Theorem 3.28 Let (F_A, τ) *be a soft topological space and* $\mathcal{B}$ *be a collection of open sets of* F_A. *Then if for every* $\alpha \in F_A$ *and soft open set* F_{A_1} *containing* α *there exists* F_B *in* $\mathcal{B}$ *such that* $\alpha \in F_B$ *and* $F_B \subset F_{A_1}$ *then* B *is a soft basis for* τ.

Proof Suppose that given condition holds. Let $F_D \in \tau$. For each $\alpha \in F_D$ there exists F_{B_α} in $\mathcal{B}$ such that $\alpha \in F_{B_\alpha}$ and $F_{B_\alpha} \subseteq F_D$. Clearly $F_D = \cup_{\alpha \in F_D} F_{B_\alpha}$. Thus every member of τ can be expressed as union of some members of $\mathcal{B}$. So $\mathcal{B}$ is a soft basis for τ. □

Theorem 3.29 *Let (F_A, τ) be a soft topological space. Let $\mathcal{B}$ be a collection of open sets of* $\mathrm{F_A}$. *Then $\mathcal{B}$ is a soft basis for the soft-topology if and only if*
(i) F_A is a soft set union of members of $\mathcal{B}$ and
(ii) If F_B, F_C in $\mathcal{B}$ and $\alpha \in F_B \cap F_C$ then there exists $F_D \in \mathcal{B}$ such that $\alpha \in F_D$ and $F_D \subseteq F_B \cap F_C$.

Proof Assume that collection $\mathcal{B}$ satisfies (i) and (ii) and τ be the collection of unions of members of $\mathcal{B}$. Then $F_A \in \tau$ follows from (i). The empty union of members of $\mathcal{B}$ is F_ϕ and hence $F_\emptyset \in \tau$. Let $F_B, F_C \in \tau$. If $F_B \cap F_C = F_\emptyset$ then $F_B \cap F_C \in \tau$.

Otherwise, for each α in $F_B \cap F_C$, there exists $F_{D_\alpha} \in \mathcal{B}$ such that $\alpha \in F_{D_\alpha}$ and $F_{D_\alpha} \subseteq F_B \cap F_C$, from (ii), thus $F_B \cap F_C = \bigcup_{\alpha \in F_B \cap F_C} F_{D_\alpha}$ and each F_{D_α} is in $\mathcal{B}$. It follows that $F_B \cap F_C \in \tau$ by definition of τ. Let F_D be an arbitrary union of members of τ. Since each of the members of τ is union of members of $\mathcal{B}$, by definition of τ, $F_D \in \tau$. Then τ is a soft-topology on F_A and hence the collection $\mathcal{B}$ is a soft basis for a soft topology τ on $\boldsymbol{F_A}$.

For converse part suppose that $\mathcal{B}$ is a soft-basis for a soft-topology τ on F_A, then since $F_A \in \tau$, F_A is a union of members of $\mathcal{B}$ so that (i) hold. Let $F_B, F_C \in \mathcal{B}$ and $\alpha \in F_B \cap F_C$. Then $F_B \cap F_C \in \tau$ and take $F_D = F_B \cap F_C$ and (ii) holds. □

Definition 3.49 Let (F_A, τ) and (F_B, τ') be two soft topological spaces. A soft set function $g : F_A \to F_B$ is said to be continuous soft function if for each soft open subset F_{B_1} of F_B, the inverse image $g^{-1}(F_{B_1})$ is soft open subset of F_A.

Theorem 3.30 *Every soft set function from discrete soft topological space into any soft topological space is continuous soft set function.*

Proof Let (F_A, τ) and (F_B, τ') be two soft topological spaces. Suppose τ is discrete topology on F_A. Let $g : F_A \to F_B$ be a soft set function. Then for every soft open set $F_{B_1} \in \tau'$, the inverse image $g^{-1}(F_{B_1})$ is soft open with respect to discrete topology τ on F_A. Thus g is a continuous soft set function. □

Theorem 3.31 Let (F_A, τ) *and* (F_B, τ') *be two soft topological spaces, then the soft set function $g : F_A \to F_B$ is continuous soft function if the inverse image of every basis element is open.*

Proof Suppose that the inverse image of every basis element is open. Let $F_{B_1} \in \tau'$ be an arbitrary soft open set. By definition, F_{B_1} can be written as union of members of the basis $\mathcal{B}'$ of τ'. Hence $F_{B_1} = \cup_{j \in J} F_{B_j}$ Then

$$g^{-1}(F_{B_1}) = \cup_{j \in J} \; g^{-1}(F_{B_j})$$

since $g^{-1}(F_{B_j})$ is open and union of open sets are open, $g^{-1}(F_{B_1})$ is open. Hence g is continuous. □

Lemma 3.1 *Let F_A, F_B be two soft sets over U and $g : F_A \rightarrow F_B$ be a soft set function. Then*

(i) $gg^{-1}(F_{B_1}) \subseteq F_{B_1}$ *for every soft subset* F_{B_1} *of* F_B.
(ii) $g^{-1}(F_{B_1} - F_{B_2}) = g^{-1}(F_{B_1}) - g^{-1}(F_{B_2})$.

Proof (i) Let $\beta \in gg^{-1}(F(B_1))$. Then $g(\alpha) = \beta$ for $\alpha \in g^{-1}(F_{B_1})$ Now, $\alpha \in g^{-1}(F_{B_1}) \Rightarrow g(\alpha) \in F_{B_1} \Rightarrow \beta \in F_{B_1}$. Hence $gg^{-1}(F_{B_1}) \subseteq F_{B_1}$.

(ii) Let $\alpha \in g^{-1}(F_{B_1} - F_{B_2})$. Then $g(\alpha) \in F_{B_1} - F_{B_2} \Rightarrow g(\alpha) \in F_{B_1}$ and $g(\alpha) \notin F_{B_2}$. i.e., $\alpha \in g^{-1}(F_{B_1})$ and $\alpha \notin g^{-1}(F_{B_2})$ i.e., $\alpha \in g^{-1}(F_{(B_1)}) - g^{-1}(F_{B_2})$ So $g^{-1}(F_{B_1} - F_{B_2}) \subseteq g^{-1}(F_{B_1}) - g^{-1}(F_{B_2})$. Similarly we can prove the reverse implication $g^{-1}(F_{B_1}) - g^{-1}(F_{B_2}) \subseteq g^{-1}(F_{B_1} - F_{B_2})$ and hence the result. □

Theorem 3.32 (F_A, τ) *and* (F_B, τ') *be two soft topological spaces and* $g : F_A \rightarrow F_B$ *be a soft set function. Then the following are equivalent.*

(i) g *is continuous.*
(ii) For every soft subset F_{A_1} *of* F_A, $g(\overline{F_{A_1}}) \subseteq \overline{g(F_{A_1})}$.
(iii) For every soft closed set F_{B_1} *in* F_B *the set* $g^{-1}(F_{B_1})$ *is soft closed in* F_A.

Proof (i) $\Rightarrow$ (iii) Assume that g is continuous. Let F_{A_1} be a soft subset of F_A. It is enough to show that if $\alpha \in (\overline{F_{A_1}})$ then $g(\alpha) \in \overline{g(F_{A_1})}$. Let F_{B_1} be soft open set containing $g(\alpha)$. Then $g^{-1}(F_{B_1})$ is soft open set of F_A containing α and it must soft intersects F_{A_1} in some other element β. Then F_{B_1} soft intersects $g(F_{A_1})$ in some $g(\beta)$. Thus F_{B_1} being an arbitrary soft open set of $g(\alpha)$ intersects with $g(F_{A_1})$. Hence $g(\alpha) \in \overline{g(F_{A_1})}$. Hence $g(\overline{F_{A_1}}) \subseteq \overline{g(F_{A_1})}$.
(ii) $\Rightarrow$ (iii) Let F_{B_1} be a soft closed set of F_B. To show that $g^{-1}(F_{B_1})$ is soft closed in F_A, it suffices to show that the closure of $g^{-1}(F_{B_1})$ is contained in $g^{-1}(F_{B_1})$. Now $gg^{-1}(F_{B_1}) \subseteq F_{B_1}$ and $\alpha \in \overline{g^{-1}(F_{B_1})} \Rightarrow g(\alpha) \in \overline{g^{-1}(F_{B_1})}$. But

$$g(\overline{g^{-1}(F_{B_1})}) \subseteq \overline{gg^{-1}(F_{B_1})} \subseteq \overline{F_{B_1}} = F_{B_1}$$

as F_{B_1} is closed. Hence $\alpha \in g^{-1}(\bar{F}_{B_1})$ Thus $\overline{g^{-1}(F_{B_1})} \subseteq g^{-1}(F_{B_1})$. Hence $g^{-1}(F_{B_1})$ is soft closed in F_A.
(iii) $\Rightarrow$ (i) A soft set is closed if and only if its soft complement is open. Also we have $g^{-1}(F_A - F_B) = g^{-1}(F_A) - g^{-1}(F_B)$. Using these the implication follows clearly. □

Theorem 3.33 *Let* $(F_A, \tau), (F_B, \tau'), (F_C, \tau'')$ *be soft topological spaces. Then the following hold:*

(i) Any constant soft set function $g : F_A \rightarrow F_B$ *is continuous.*
(ii) If F_{A_1} *is soft subspace of* F_A, *then soft set function* $g : F_{A_1} \rightarrow F_B$ *defined by* $g(\alpha) = \alpha$ *is a continuous soft set function.*
(iii) If $g : F_A \rightarrow F_B, h : F_B \rightarrow F_C$ *are continuous soft set functions, then the composite function* $h \circ g$ *is also a continuous soft set function.*

Proof (i) For $\beta_0 \in F_B$, let g be the constant function defined by $g(\alpha) = \beta_0$ for every $\alpha \in F_A$. Then for every soft open set F_{B_1} in F_B, $g^{-1}(F_{B_1})$ is equal to either F_A or null soft set depending upon whether F_{B_1} contains β_0 or not. In either case $g^{-1}(F_{B_1})$ is soft open. Hence g is a continuous soft set function.

(ii) If F_{A_2} is soft open in F_B then $g^{-1}(F_{A_2}) = F_{A_2} \cap F_{A_1}$ which is soft open in F_{A_1} by definition of subspace topology. Hence g is a continuous soft set function.

(iii) If F_{C_1} is soft open in F_C, then since h is continuous soft function, $h^{-1}(F_{C_1})$ is soft open in F_B. Again as g is soft open, $g^{-1}(h^{-1}(F_{C_1}))$ is open in F_A. But we have $(h \circ g)^{-1} = g^{-1} \circ h^{-1}$. So $(h \circ g)^{-1}(F_{C_1})$ is soft open in F_A whenever F_{C_1} is soft open in F_C. Hence function $h \circ g$ is also a continuous soft set function. □

3.5.2 Soft Compactness

In this subsection, an attempt is made to extend the concept of compactness in general topology to the framework of soft topology.

Definition 3.50 A collection $\mathcal{C}$ of soft subsets of F_A is said to be a soft covering of F_A if the union of the elements of $\mathcal{C}$ is equal to F_A. It is a soft open covering if its elements are soft open subsets of soft topological space F_A.

Definition 3.51 A soft topological space is a soft compact space if every soft open covering $\mathcal{C}$ of F_A contains a finite subcollection that also covers F_A.

Theorem 3.34 *Let (F_A, τ) be a soft topological space and $F_B \subset F_A$ Then F_B is a soft compact subset of F_A iff the soft subspace (F_B, τ_{F_B}) is soft compact.*

Proof Suppose that F_B is a soft compact subset of F_A. Let $\mathcal{C}$ be a soft open cover of the space (F_B, τ_{F_B}). By definition each member F_C of $\mathcal{C}$ is of the form $F_B \cap F_{C'}$ for some $F_{C'} \in \tau$. Now fix for each F_C of $\mathcal{C}$ an element $F_{C'}$ in τ such that $F_C = F_B \cap F_{C'}$. Then the family $\{F_{C'} : F_C = F_B \cap F_C \text{ and } F_C \in \mathcal{C}\}$ is a soft open cover of F_B using soft open subsets of F_A. since F_B is a compact subset of F_A, This cover has a finite subcover say $\{F_{C'_1}, F_{C'_2}, \ldots, F_{C'_n}\}$ where $F_{C_1}, F_{C_2}, \ldots, F_{C_n} \in$ C. Clearly then $F_{C_1}, F_{C_2}, \ldots, F_{C_n}$ is a finite sub cover of $\mathcal{C}$. This shows that the subspace (F_B, τ_{F_B}) is compact.

Conversely suppose that the soft subspace (F_B, τ_{F_B}) is soft compact. Let $\mathcal{C}$ be a soft cover of F_B using soft open subsets of F_A. Then $\{F_B \cap F_C : F_C \in \mathcal{C}\}$ is an open cover of the space (F_B, τ_{F_B}). By soft compactness of the space (F_B, τ_{F_B}), this cover has a finite sub cover say $\{F_B \cap F_{C_i} : i = 1, 2, \ldots, n\}$ where $F_{C_i} \in \mathcal{C}$. Clearly this $\{Fc_i : i = 1, 2, \ldots, n\}$ is a finite subfamily of $\mathcal{C}$ covering the soft set F_B. Thus F_B is a soft compact subset of F_A. □

Theorem 3.35 *Let F_A be a compact space and suppose $g : F_A \to F_B$ is a soft continuous surjective function. Then F_B is compact. In other words every soft continuous image of a soft compact space is soft compact.*

Proof Let $\mathcal{C}$ be any soft open cover of F_B. Let $\mathcal{C}^{-1} = \{g^{-1}(C) : C \in \mathcal{C}\}$. Now since $\mathcal{C}^{-1}$ is a soft cover of F_A and g is soft continuous, it is a soft open cover of F_A. Now F_A is soft compact and hence some finitely many members of $\mathcal{C}^{-1}$ say $g^{-1}(C_1), g^{-1}(C_2), \ldots, g^{-1}(C_n)$ where $C_1, C_2, \ldots C_n \in \mathcal{C}$ cover F_A. But then $C_1, C_2, \ldots C_n$ cover F_B and g is onto. So F_B is soft compact. □

Theorem 3.36 *If F_A is a compact space and $F_B \subset F_A$ is closed in F_A, then F_B in its subspace topology is also soft compact.*

Proof Given F_A is a compact space and $F_B \subset F_A$ is closed. Let $\mathcal{C}$ be soft open cover of F_B in subspace topology on F_B. For each $F_C \in \mathcal{C}$ fix $F_{C'}$ in F_A such that $F_C = F_B \cap F_{C'}$. Then the family $\mathcal{C}' = \{F_{C'} : F_C = F_B \cap F_{C'}$ and $F_C \in \mathcal{C}\} \cup F_A - F_B$ is a soft open cover of F_A and hence admits a finite sub cover consists of say $F_{C'_1}, F_{C'_2}, \ldots, F_{C'_n}$, and possibly $F_A - F_B$. Then a finite sub cover of $\mathcal{C}$ is there that covers F_B. This shows that F_B in its subspace topology is compact. □

3.6 Soft Topologies Generated by Soft Set Relations

Here we observe the fact that soft set relations defined in the sense of Babitha and Sunil [15] give rise to soft topologies in the sense of Cagman et al. [28]. Discussions and examples provided are taken from Babitha and Sunil [20].

Definition 3.52 Let R be a soft set relation on F_A. The post soft set of $\alpha \in F_A$ in R is defined as $\alpha R = \{\beta \in F_A : \alpha R \beta\}$ and the pre soft set of $\alpha \in F_A$ is defined as $R\alpha = \{\beta \in F_A : \beta R \alpha\}$.

Definition 3.53 If R is a soft set relation defined on F_A, then the post class and pre class are defined respectively as $P_+ = \{\alpha R : \alpha \in F_A\}$ and $P_- = \{R\alpha : \alpha \in F_A\}$.

Theorem 3.37 *If R is a soft set relation on F_A, then the post class P_+ and pre class P_- form a sub basis.*

Proof Let β_+ be the collection of all finite intersection of members of P_+. Let $\mathcal{T}_1$ be the collection consists of subsets of F_A consisting of F_ϕ, F_A and all union of finite intersection of members of β_+. It is enough to show that $\mathcal{T}_1$ is a topology on F_A. For if $\mathcal{T}_1$ is a topology then β_+ is a soft basis and thus by definition, P_+ is a soft sub basis for F_A. Now β_+ is closed under finite intersection and $\mathcal{T}_1$ is also closed under finite intersection. Let $\{F_{A_i} : i \in I\}$ be a collection in $\mathcal{T}_1$. Then each F_{A_i} is a union of members of β_+. Thus $\bigcup_{i \in I} F_{A_i}$ is also a union of members of β_+. Then $\bigcup_{i \in I} F_{A_i}$ is in $\mathcal{T}_1$. By definition, F_A and F_ϕ are in $\mathcal{T}_1$. Thus $\mathcal{T}_1$ is a topology on F_A. Similarly we can prove for P_- by defining β_- and $\mathcal{T}_2$ corresponds to P_-. □

Note: The topologies $\mathcal{T}_1$ and $\mathcal{T}_2$ defined in Theorem 3.37 are called topologies generated by post class P_+ and pre class P_- respectively and β_+ and β_- are the soft basis generated by P_+ and P_- respectively.

Example 47 Let $F_A = \{a_1 = \{u_1, u_2\}, a_2 = \{u_1, u_4, u_7\}, a_3 = \{u_2, u_6\}, a_4 = \{u_3, u_7\}\}$ and R be soft set relation on F_A defined by
$R = \{F_A(a_1)RF_A(a_2), F_A(a_1)RF_A(a_4), F_A(a_2)RF_A(a_2)\}, \{F_A(a_2)RF_A(a_4), F_A(a_2)RF_A(a_3), F_A(a_3)RF_A(a_3), F_A(a_3)RF_A(a_4)\}$

Then the post soft set and pre soft sets are given by
$F_A(a_1)R = \{F_A(a_2), F_A(a_4)\}$
$F_A(a_2)R = \{F_A(a_2), F_A(a_3), F_A(a_4)\}$
$F_A(a_3)R = \{F_A(a_4), F_A(a_3)\}$
and
$RF_A(a_2) = \{F_A(a_2), F_A(a_1)\}$
$RF_A(a_3) = \{F_A(a_2), F_A(a_3)\}$
$RF_A(a_A) = \{F_A(a_3), F_A(a_2), F_A(a_1)\}$
The post class and pre class obtained from the the post soft set and pre soft sets is
$P_+ = \{\{F_A(a_2), F_A(a_4)\}, \{F_A(a_2), F_A(a_3), F_A(a_4)\}, \{F_A(a_4), F_A(a_3)\}\}$
and
$P_- = \{\{F_A(a_2), F_A(a_1)\}, \{F_A(a_2), F_A(a_3)\}, \{F_A(a_3), F_A(a_2), F_A(a_1)\}\}$
The soft bases $\mathcal{B}_+$ and $\mathcal{B}_-$ generated by P_+ and P_- respectively are given by
$\mathcal{B}_+ = \{\{F_A(a_4), F_A(a_3)\}, \{F_A(a_4)\}, \{F_A(a_2), F_A(a_4)\}, \{F_A(a_2), F_A(a_3), F_A(a_4)\}\}$
$\mathcal{B}_- = \{\{F_A(a_2), F_A(a_1)\}, \{F_A(a_2), F_A(a_3)\}, \{F_A(a_3), F_A(a_2), F_A(a_1)\}, \{F_A(a_2)\}\}$.
Then the topologies $\mathcal{T}_1$ and $\mathcal{T}_2$ generated by P_+ and P_- respectively are given by
$\mathcal{T}_1 = \{\{F_A(a_4), F_A(a_3)\}, \{F_A(a_4)\}, \{F_A(a_2), F_A(a_4)\}, \{F_A(a_2), F_A(a_3), F_A(a_4)\}, F_A, F_\phi\}$
$\mathcal{T}_2 = \{\{F_A(a_2), F_A(a_1)\}, \{F_A(a_2), F_A(a_3)\}, \{F_A(a_3), F_A(a_2), F_A(a_1)\}, \{F_A(a_2\}, F_A, F_\phi\}$

Definition 3.54 If $\mathcal{T}$ is a soft topology on F_A and the collection $\mathcal{T}^c = \{F_B^c : F_B \in \mathcal{T}\}$ is also a soft topology on F_A and $\mathcal{T}^c$ is called dual of the soft topology $\mathcal{T}$ on F_A.

Remark 3.10 For any arbitrary soft set relation, $\mathcal{T}_1$ is not the dual of $\mathcal{T}_2$.

Example 48 Let F_A be a soft set given by $F_A = \{F_A(a_1), F_A(a_2), F_A(a_3)\}$. And R be the reflexive relation defined on F_A given by

$R = \{F_A(a_1) \times F_A(a_1), F_A(a_2) \times F_A(a_2), F_A(a_3) \times F_A(a_3), F_A(a_1) \times F_A(a_3), F_A(a_3) \times F_A(a_2)\}$.
Then
$P_+ = \{\{F_A(a_1), F_A(a_3)\}, \{F_A(a_2)\}, \{F_A(a_3), F_A(a_2)\}\}$
$P_- = \{\{F_A(a_1)\}, \{F_A(a_2), F_A(a_3)\}, \{F_A(a_3), F_A(a_1)\}\}$
Then
$\mathcal{T}_1 = \{\{F_A(a_1), F_A(a_3)\}, \{F_A(a_2)\}, \{F_A(a_3), F_A(a_2)\}, \{F_A(a_3)\}, F_A, F_\varnothing\}$
$\mathcal{T}_2 = \{\{F_A(a_1)\}, \{F_A(a_2), F_A(a_3)\}, \{F_A(a_3), F_A(a_1)\}, \{F_A(a_3\}, F_A, F_\varnothing\}$
Clearly $\mathcal{T}_1$ is not the dual of $\mathcal{T}_2$. Also note that $\mathcal{T}_1 \neq \mathcal{T}_2$, Now consider a transitive relation S on F_A given by
$S = \{F_A(a_1) \times F_A(a_2), F_A(a_2) \times F_A(a_3), F_A(a_1) \times F_A(a_3)\}$
Then
$P_+ = \{\{F_A(a_2), F_A(a_3)\}, \{F_A(a_3)\}\}$
$P_- = \{\{F_A(a_1)\}, \{F_A(a_2), F_A(a_1)\}\}$.

Then
$\mathcal{T}_1 = \{\{F_A(a_2), F_A(a_3)\}, \{F_A(a_3)\}, F_A, F_\varnothing\}$
$\mathcal{T}_2 = \{\{F_A(a_1)\}, \{F_A(a_2), F_A(a_1)\}, F_A, F_\varnothing\}$
Clearly $\mathcal{T}_1$ is not the dual of $\mathcal{T}_2$. Also note that $\mathcal{T}_1 \neq \mathcal{T}_2$.

Lemma 3.2 *If R is a symmetric soft set relation on F_A, then $\mathcal{T}_1 = \mathcal{T}_2$.*

Proof Let R be a symmetric soft set relation on F_A. Then $\beta R\gamma$ iff $\gamma R\beta$. i.e., $\beta \in R\gamma$ iff $\beta \in \gamma R$. Thus $\gamma R = R\gamma$ for every γ in F_A. Then $P_+ = P_-$ and thus $\mathcal{T}_1 = \mathcal{T}_2$. □

Definition 3.55 A soft topological space is called a quasi-discrete soft topological space if every soft open set is closed set and vice versa.

Every discrete soft topology is a quasi discrete soft topology.
Theorem 3.38 *If R is a soft set relation on F_A and $\mathcal{T}_1, \mathcal{T}_2$ are soft topologies generated by R which are quasi discrete. Then $\mathcal{T}_1$ is the dual of $\mathcal{T}_2$ if $\mathcal{T}_1 = \mathcal{T}_2$.*

Proof Proof follows trivially. □

Definition 3.56 A soft set relation R is a pre order iff R is a reflexive and transitive soft set relation.

Lemma 3.3 *If R is a pre order soft set relation on F_A. Then*
(i) $F_B \in \mathcal{T}_1$ if and only if $F_B = \bigcup_{\alpha \in F_B} \alpha R$
(ii) $F_B \in \mathcal{T}_2$ if and only if $F_B = \bigcup_{\alpha \in F_B} R\alpha$

Proof (i) Assume that $F_B \in \mathcal{T}_1$. If $\beta \in F_B$, then by reflexivity we have $\beta \in \beta R$ and so $\beta \in \bigcup_{\alpha \in F_B} \alpha R$.
Now if $\beta \in \bigcup_{\alpha \in F_B} \alpha R$, there exists $\alpha \in F_B$ such that $\beta \in \alpha R$. Moreover if $\gamma \in \beta R$ then by transitivity, $\gamma \in \alpha R$. Hence $\beta R \subset \alpha R \subset \bigcup_{\alpha \in F_B} \alpha R$. Then $\beta \in F_B$. For, if $\beta \notin F_B$, then $\beta R \not\subset \bigcup_{\alpha \in F_B} \alpha R$. But this a contradiction. Hence $\beta \in F_B$. So $\bigcup_{\alpha \in F_B} \alpha R \subset F_B$.
Conversely suppose that R is a pre order soft set relation and $F_B = \bigcup_{\alpha \in F_B} \alpha R$. Then for every $\beta \in F_B$, there is $\beta R \in \mathcal{T}_1$ such that $\beta \in \beta R \subset \bigcup_{\alpha \in F_B} \alpha R = F_B$. Thus β is an interior point of F_B. Hence $F_B \in \mathcal{T}_1$.
(ii) can be proved in a similar way. □

Theorem 3.39 *If R is a pre order soft set relation on F_A, then the soft topology $\mathcal{T}_1$ is the dual of $\mathcal{T}_2$.*

Proof Assume that R is a pre order soft set relation on F_A and $F_B \in \mathcal{T}_1$. We will show that $F_B^c \in \mathcal{T}_1$. i.e., $F_B^c = \bigcup_{\alpha \in F_B^c} R\alpha$. Suppose that $F_B^c \neq \bigcup_{\alpha \in F_B^c} \alpha R$. Then there exists $\beta \in \bigcup_{\alpha \in F_B^c} \alpha R$ such that $\beta \notin F_B^c$. Hence there exists $\gamma \in F_B^c$ such that $\beta R\gamma$ and $\beta \in F_B$. Thus $\gamma \in \beta R \subset \bigcup_{\alpha \in F_B} \alpha R = F_B$. But this is a contradiction. Therefore $F_B^c = \bigcup_{\alpha \in F_B^c} R\alpha$ and $F_B^c \in \mathcal{T}_1$. Thus $\mathcal{T}_1$ is the dual of $\mathcal{T}_2$. □

Chapter 4
Soft Graphs, Soft Categories and Information Systems

Category theory paves the way to describe and compare objects with similar and different properties. It brings together the various branches of mathematics into a united whole. This chapter is an attempt to accommodate categorical concepts in the context of soft sets and soft graphs. Further, relationship between soft sets and classical information systems is also explored. Results, examples and discussions provided are taken from Ratheesh K. P [111] and D. Pei and D. Miao [102]. For Graph theory related terminologies see Harary [48]. Throughout this chapter, we alternatively use $\mathcal{F}_A(U)$ to represent soft set (F, A) over U.

4.1 Soft Relations and Equivalence Relations

Zhang and Yuan [164] defined soft relation as a parameterized family of crisp relations. Babitha and Sunil [15] explained soft set relation as a soft subset of cartesian product of two soft sets on the same universe. Qin et al. [109] redefined the concept of soft relation by using two different universes in the aforementioned soft set relation. Here, the concept of soft relation is discussed in a different way and soft equivalence relation and its characteristics are explored. Further soft partitions are introduced and establishes a one-to-one correspondence between soft equivalence relations and soft partitions.

4.1.1 Soft Relations

Definition 4.1 Let U be an initial universe set and let A and B be two sets of parameters. A soft relation from A to B is a soft set $\mathcal{G}_{A\times B}(U)$, which means $\mathcal{G}$:

S. J. John, *Soft Sets*, Studies in Fuzziness and Soft Computing 400,
https://doi.org/10.1007/978-3-030-57654-7_4

$A \times B \to \mathcal{P}(U)$ is a function. The domain of a soft relation $\mathcal{G}_{A\times B}(U)$ is the soft set $\mathcal{G}_D : A \to \mathcal{P}(U)$ defined by $\mathcal{G}_D(a) = \bigcup_{b\in B} \mathcal{G}(a, b)$ and the range is the soft set $\mathcal{G}_R : B \to \mathcal{P}(U)$ defined by $\mathcal{G}_R(b) = \bigcup_{a\in A} \mathcal{G}(a, b)$.

Example 49 Let $U = \{1, 2, \ldots, 10\}$ be a universe set and let $A = \{a_1, a_2, a_3\}$ and $B = \{b_1, b_2\}$ be the parameter sets. A soft relation $\mathcal{G}_{A\times B}(U)$ can be defined by

$$\begin{aligned} &\mathcal{G}(a_1, b_1) = \{1, 4, 5\} \quad \mathcal{G}(a_1, b_2) = \{1, 2, 9\} \\ &\mathcal{G}(a_2, b_1) = \{2, 6, 7\} \quad \mathcal{G}(a_2, b_2) = \{2, 4, 6, 8\} \\ &\mathcal{G}(a_3, b_1) = \{1, 3, 6\} \quad \mathcal{G}(a_3, b_2) = \{2, 9\}. \end{aligned}$$

The domain of $\mathcal{G}$ is the soft set $\mathcal{G}_D$ given by $\mathcal{G}_D(a_1) = \{1, 2, 4, 5, 9\}$, $\mathcal{G}_D(a_2) = \{2, 4, 6, 7, 8\}$, $\mathcal{G}_D(a_3) = \{1, 2, 3, 6, 9\}$.

The range of $\mathcal{G}$ is the soft set $\mathcal{G}_R$ given by $\mathcal{G}_R(b_1) = \{1, 2, 3, 4, 5, 6, 7\}$, $\mathcal{G}_R(b_2) = \{1, 2, 4, 6, 8, 9\}$.

Definition 4.2 (Operations on soft relations) Let $\mathcal{G}_{A\times B}(U)$ and $\mathcal{H}_{A\times B}(U)$ be two soft relations and let $(a, b) \in A \times B$, then

1. Union : the soft set $(\mathcal{G} \cup \mathcal{H})_{A\times B}$ defined by $(\mathcal{G} \cup \mathcal{H})(a, b) = \mathcal{G}(a, b) \cup \mathcal{H}(a, b)$
2. Intersection : the soft set $(\mathcal{G} \cap \mathcal{H})_{A\times B}$ defined by $(\mathcal{G} \cap \mathcal{H})(a, b) = \mathcal{G}(a, b) \cap \mathcal{H}(a, b)$
3. Transpose : the soft set $\mathcal{G}^t_{B\times A}$ is the transpose of $\mathcal{G}$ defined as $\mathcal{G}^t(b, a) = \mathcal{G}(a, b)$
4. Composition of soft relations: Let $\mathcal{G}_{A\times B}(U)$ and $\mathcal{H}_{B\times C}(U)$ be two soft relations, the composite of $\mathcal{G}$ and $\mathcal{H}$ is a soft set $(\mathcal{G} \circ \mathcal{H})_{A\times C}$ defined by

$$(\mathcal{G} \circ \mathcal{H})(a, c) = \bigcup_{b\in B} [\mathcal{G}(a, b) \cap \mathcal{H}(b, c)]$$

for all $(a, c) \in A \times C$.

Proposition 4.1 *Let $\mathcal{G}_{E\times E}(U)$, $\mathcal{H}_{E\times E}(U)$ and $\mathcal{I}_{E\times E}(U)$ be soft relations on a set E, then*

(a) if $\mathcal{G} \subseteq \mathcal{H}$, then $\mathcal{G} \circ \mathcal{I} \subseteq \mathcal{H} \circ \mathcal{I}$ and $\mathcal{G}^t \subseteq \mathcal{H}^t$
(b) $\mathcal{G} \circ (\mathcal{H} \cup \mathcal{I}) = (\mathcal{G} \circ \mathcal{H}) \cup (\mathcal{G} \circ \mathcal{I})$
(c) $(\mathcal{G} \cup \mathcal{H})^t = \mathcal{G}^t \cup \mathcal{H}^t$
(d) $(\mathcal{G} \cap \mathcal{H})^t = \mathcal{G}^t \cap \mathcal{H}^t$

Proof For any $e_1, e_2, e_3 \in E$,

(a)

$$\begin{aligned}(\mathcal{G} \circ \mathcal{I})(e_1, e_3) &= \bigcup_{e_2 \in E} [\mathcal{G}(e_1, e_2) \cap \mathcal{I}(e_2, e_3)] \\ &\subseteq \bigcup_{e_2 \in E} [\mathcal{H}(e_1, e_2) \cap \mathcal{I}(e_2, e_3)] \\ &= (\mathcal{H} \circ \mathcal{I})(e_1, e_3)\end{aligned}$$

and

$$\begin{aligned}\mathcal{G}^t(e_2, e_1) &= \mathcal{G}(e_1, e_2) \\ &\subseteq \mathcal{H}(e_1, e_2) \\ &= \mathcal{H}^t(e_2, e_1)\end{aligned}$$

(b)

$$\begin{aligned}[\mathcal{G} \circ (\mathcal{H} \cup \mathcal{I})](e_1, e_3) &= \bigcup_{e_2 \in E} [\mathcal{G}(e_1, e_2) \cap (\mathcal{H} \cup \mathcal{I})(e_2, e_3)] \\ &= \bigcup_{e_2 \in E} [\mathcal{G}(e_1, e_2) \cap (\mathcal{H}(e_2, e_3) \cup \mathcal{I}(e_2, e_3))] \\ &= \bigcup_{e_2 \in E} [(\mathcal{G}(e_1, e_2) \cap \mathcal{H}(e_2, e_3)) \cup (\mathcal{G}(e_1, e_2) \cap \mathcal{I}(e_2, e_3))] \\ &= [(\mathcal{G} \circ \mathcal{H}) \cup (\mathcal{G} \circ \mathcal{I})](e_1, e_3)\end{aligned}$$

(c)

$$\begin{aligned}(\mathcal{G} \cup \mathcal{H})^t(e_1, e_2) &= (\mathcal{G} \cup \mathcal{H})(e_2, e_1) \\ &= \mathcal{G}(e_2, e_1) \cup \mathcal{H}(e_2, e_1) \\ &= \mathcal{G}^t(e_1, e_2) \cup \mathcal{H}^t(e_1, e_2) \\ &= (\mathcal{G}^t \cup \mathcal{H}^t)(e_1, e_2)\end{aligned}$$

(d)

$$\begin{aligned}(\mathcal{G} \cap \mathcal{H})^t(e_1, e_2) &= (\mathcal{G} \cap \mathcal{H})(e_2, e_1) \\ &= \mathcal{G}(e_2, e_1) \cap \mathcal{H}(e_2, e_1) \\ &= \mathcal{G}^t(e_1, e_2) \cap \mathcal{H}^t(e_1, e_2) \\ &= (\mathcal{G}^t \cap \mathcal{H}^t)(e_1, e_2)\end{aligned}$$

Hence the proof. □

Definition 4.3 A soft relation $\mathcal{G}_{E\times E}(U)$ on a set E is said to be

$$\begin{aligned} \text{reflexive if } \mathcal{G}(a,a) &= U, \ \ for\ all\ a \in E \\ \text{symmetric if } \mathcal{G}(a,b) &= \mathcal{G}(b,a), \ \ for\ all\ a,b \in E \\ \text{transitive if } (\mathcal{G}\circ\mathcal{G})(a,b) &\subseteq \mathcal{G}(a,b), \ \ for\ all\ a,b \in E \end{aligned}$$

A soft relation is said to be a soft equivalence relation if it is reflexive, symmetric and transitive.

Example 50 Let $U = \{1, 2, 3, 4, 5\}$ and $E = \{a, b, c\}$ be an universe set and a parameter set respectively. Consider a soft relation $\mathcal{G}_{E\times E}(U)$ defined by

$$\begin{array}{lll} \mathcal{G}(a,a) = U & \mathcal{G}(a,b) = \{1,3,4\} & \mathcal{G}(a,c) = \{1,3,4,5\} \\ \mathcal{G}(b,a) = \{1,3,4\} & \mathcal{G}(b,b) = U & \mathcal{G}(b,c) = \{2,3,4\} \\ \mathcal{G}(c,a) = \{1,3,4,5\} & \mathcal{G}(c,b) = \{2,3,4\} & \mathcal{G}(c,c) = U \end{array}$$

The reflexivity and symmetry of $\mathcal{G}$ follows from the definition. Moreover, $\mathcal{G}(a,b)\cap\mathcal{G}(b,c)\subseteq\mathcal{G}(a,c)$, for all $a,b,c \in E$. Thus $\mathcal{G}$ is transitive and hence a soft equivalence relation.

Proposition 4.2 *Let $\mathcal{F}_E(U)$ be a soft set, then the soft relation $\mathcal{G}_{E\times E}(U)$ defined by*

$$\mathcal{G}(a,b) = \begin{cases} U & \textit{if } a = b; \\ \mathcal{F}(a)\cap\mathcal{F}(b) & \textit{otherwise} \end{cases}$$

is a soft equivalence relation

Proof From the definition of $\mathcal{G}$, it is noted that $\mathcal{G}(a,a) = U$ and $\mathcal{G}(a,b) = \mathcal{G}(b,a)$, for all $a,b \in E$. Thus $\mathcal{G}$ is reflexive and symmetric.
For transitivity, we have to show that $\mathcal{G}(a,c)\cap\mathcal{G}(c,b)\subseteq\mathcal{G}(a,b)$, for all $a,b,c \in E$.

Case (I): If $a = c, a = b$ or $c = b$, then $\mathcal{G}(a,c) = U, \mathcal{G}(a,b) = U$ or $\mathcal{G}(c,b) = U$ respectively. It implies that $\mathcal{G}(a,c)\cap\mathcal{G}(c,b)\subseteq\mathcal{G}(a,b)$.

Case (II): If a, b and c are distinct, then

$$\begin{aligned} \mathcal{G}(a,c)\cap\mathcal{G}(c,b) &= [\mathcal{F}(a)\cap\mathcal{F}(c)]\cap[\mathcal{F}(c)\cap\mathcal{F}(b)] \\ &= \mathcal{F}(a)\cap\mathcal{F}(c)\cap\mathcal{F}(b) \\ &\subseteq \mathcal{F}(a)\cap\mathcal{F}(b) = \mathcal{G}(a,b) \end{aligned}$$

So $(\mathcal{G}\circ\mathcal{G})(a,b) = \bigcup_{c\in B}[\mathcal{G}(a,c)\cap\mathcal{G}(c,b)] \subseteq \bigcup_{c\in B}\mathcal{G}(a,b) = \mathcal{G}(a,b), \ \forall a,b \in E.$
It follows that $\mathcal{G}$ is transitive and hence $\mathcal{G}$ is a soft equivalence relation. □

Remark 4.1 Let $\mathcal{F}_E(U)$ be a soft set, then the soft relation $\mathcal{H}_{E\times E}(U)$ defined by

$$\mathcal{H}(a,b) = \begin{cases} U & \text{if } a = b; \\ \mathcal{F}(a)\cup\mathcal{F}(b) & \text{if otherwise} \end{cases}$$

is not a soft equivalence relation on E.

Example 51 Let $U = \{1, 2, 3, 4\}$ and $E = \{a, b, c\}$ be universe and parameter set respectively. A soft set $\mathcal{F}_E(U)$ is defined by

$$\mathcal{F}(a) = \{1, 2\}, \ \mathcal{F}(b) = \{2, 3\} \ \text{and} \ \mathcal{F}(c) = \{1, 2, 4\}.$$

By definition, $\mathcal{H}(a, b) = \mathcal{F}(a) \cup \mathcal{F}(b) = \{1, 2, 3\}$ and $(\mathcal{H} \circ \mathcal{H})(a, b) = \{1, 2, 3, 4\}$. So $(\mathcal{H} \circ \mathcal{H})(a, b) \nsubseteq \mathcal{H}(a, b)$. Hence it is not transitive.

Proposition 4.3 *If $\mathcal{G}_{E\times E}(U)$ is a soft equivalence relation, then so is $\mathcal{G}^t$.*

Proof Since $\mathcal{G}$ reflexive, $\mathcal{G}^t(a, a) = \mathcal{G}(a, a) = U$ for all $a \in E$. So $\mathcal{G}^t$ is reflexive.

Also since $\mathcal{G}$ symmetric, $\mathcal{G}^t(a, b) = \mathcal{G}(b, a) = \mathcal{G}(a, b) = \mathcal{G}^t(b, a)$. Hence $\mathcal{G}^t$ is symmetric.

To prove transitivity, let $a, b, c \in E$

$$\begin{aligned}\mathcal{G}^t(a, b) \cap \mathcal{G}^t(b, c) &= \mathcal{G}(b, a) \cap \mathcal{G}(c, b) \\ &= \mathcal{G}(c, b) \cap \mathcal{G}(b, a) \\ &\subseteq \mathcal{G}(c, a) = \mathcal{G}^t(a, c), \quad \text{since } \mathcal{G} \text{ is transitive}\end{aligned}$$

Therefore $\mathcal{G}^t$ is transitive and it follows that $\mathcal{G}^t$ is a soft equivalence relation. □

Remark 4.2 If $\{\mathcal{G}_j\}_{j\in J}$ is a family of soft equivalence relations on a set E, then $\mathcal{G} = \bigcap\limits_{j\in J} \mathcal{G}_j$ is a soft equivalence relation on E.

Remark 4.3 Let $\mathcal{G}_{E\times E}(U)$ be a soft equivalence relation on a set E, then any $a \in E$, $\mathcal{G}_a : E \to \mathcal{P}(U)$ defined by $\mathcal{G}_a(e) = \mathcal{G}(a, e)$ is a soft set on E.

The soft set $\mathcal{G}_a$ is called soft equivalence class of $\mathcal{G}$ on E and is denoted by $[\mathcal{G}_a]$. The soft equivalence classes of a soft equivalence relation need not be disjoint.

Example 52 The soft equivalence classes of Example 50 are

$[\mathcal{G}_a] = \{(a, U), (b, \{1, 3, 4\}), (c, \{1, 3, 4, 5\})\}$
$[\mathcal{G}_b] = \{(a, \{1, 3, 4\}), (b, U), (c, \{2, 3, 4\})\}$
$[\mathcal{G}_c] = \{(a, \{1, 3, 4, 5\}), (b, \{2, 3, 4\}), (c, U)\}$

Now $[\mathcal{G}_a](b) \cap [\mathcal{G}_c](b) = \{1, 3, 4\} \cap \{2, 3, 4\} = \{3, 4\} \neq \emptyset$. Therefore the soft equivalence classes need not be disjoint.

Proposition 4.4 *Let $[\mathcal{G}_a]$ and $[\mathcal{G}_b]$ be two soft equivalence classes of a soft equivalence relation $\mathcal{G}_{E\times E}(U)$, then*

(a) $[\mathcal{G}_a](e) \cap [\mathcal{G}_b](e) \subseteq \mathcal{G}(a, b)$, *for all* $e \in E$
(b) $[\mathcal{G}_a] = [\mathcal{G}_b]$ *if and only if* $\mathcal{G}(a, b) = U$
(c) *if* $[\mathcal{G}_a] \neq [\mathcal{G}_b]$, *then* $[\mathcal{G}_a](e) \cap [\mathcal{G}_b](e) \subsetneq U$, *for all* $e \in E$

Proof Let $[\mathcal{G}_a]$ and $[\mathcal{G}_b]$ be equivalence classes of $\mathcal{G}$.

(a) For $e \in E$, we have

$$\begin{aligned}[\mathcal{G}_a](e) \cap [\mathcal{G}_b](e) &= \mathcal{G}(a,e) \cap \mathcal{G}(b,e)\\ &= \mathcal{G}(a,e) \cap \mathcal{G}(e,b), \quad \text{since } \mathcal{G} \text{ is symmetric}\\ &\subseteq \mathcal{G}(a,b), \quad \text{since } \mathcal{G} \text{ is transitive.}\end{aligned}$$

(b) Suppose $[\mathcal{G}_a] = [\mathcal{G}_b]$, then $\mathcal{G}(a,b) = [\mathcal{G}_a](b) = [\mathcal{G}_b](b) = U$
Conversely, suppose that $\mathcal{G}(a,b) = U$, then

$$\begin{aligned}[\mathcal{G}_a](e) = \mathcal{G}(a,e) &\supseteq \mathcal{G}(a,b) \cap \mathcal{G}(b,e), \quad \text{since } \mathcal{G} \text{ is symmetric}\\ &= U \cap \mathcal{G}(b,e)\\ &= \mathcal{G}(b,e) = [\mathcal{G}_b](e)\end{aligned}$$

Thus $[\mathcal{G}_a](e) \supseteq [\mathcal{G}_b](e)$, for all $e \in E$. Similarly, for any $e \in E$, we can prove that $[\mathcal{G}_b](e) \supseteq [\mathcal{G}_a](e)$. Hence $[\mathcal{G}_a](e) = [\mathcal{G}_b](e)$.

(c) Suppose $[\mathcal{G}_a](e) \cap [\mathcal{G}_b](e) = U$, then

$$\begin{aligned}\mathcal{G}(a,b) &\supseteq \mathcal{G}(a,e) \cap \mathcal{G}(e,b), \quad \text{since } \mathcal{G} \text{ is transitive}\\ &= \mathcal{G}(a,e) \cap \mathcal{G}(b,e), \quad \text{since } \mathcal{G} \text{ is symmetric}\\ &= [\mathcal{G}_a](e) \cap [\mathcal{G}_b](e) = U\end{aligned}$$

Therefore $\mathcal{G}(a,b) = U$. Also, from (b), we get $[\mathcal{G}_a](e) = [\mathcal{G}_b](e)$.
Hence $[\mathcal{G}_a](e) \cap [\mathcal{G}_b](e) \subsetneq U$. □

Proposition 4.5 *Let $\mathcal{G}_{E\times E}(U)$ be a soft equivalence relation on E and let $S \in \mathcal{P}(U)$. Then the S-cut of $\mathcal{G}$, $\mathcal{G}^S = \{(a,b) \in E \times E : S \subseteq \mathcal{G}(a,b)\}$ is a crisp equivalence relation on E.*

Proof Let $S \in \mathcal{P}(U)$. For each $a \in E$, $\mathcal{G}(a,a) = U \supseteq S$. Hence $(a,a) \in \mathcal{G}^S$, it follows that $\mathcal{G}^S$ is reflexive for all $a \in E$. Suppose $(a,b) \in \mathcal{G}^S$, then $S \subseteq \mathcal{G}(a,b)$ implies that $S \subseteq \mathcal{G}(b,a)$ for all $a, b \in E$, since $\mathcal{G}$ is symmetric. Finally, assume that $(a,b), (b,c) \in \mathcal{G}^S$, then

$$\begin{aligned}\mathcal{G}(a,c) &\supseteq (\mathcal{G} \circ \mathcal{G})(a,c), \quad \text{since } \mathcal{G} \text{ is transitive}\\ &= \bigcup_{e\in E} [\mathcal{G}(a,e) \cap \mathcal{G}(e,c)]\\ &\supseteq \mathcal{G}(a,b) \cap \mathcal{G}(b,a)\\ &\supseteq S \cap S = S\end{aligned}$$

Therefore $(a,c) \in \mathcal{G}^S$. This completes the proof. □

4.1.2 Soft Partitions

This subsection deals with soft partitions and its connection with soft equivalence relation. In [15], Babitha and Sunil defines soft set partition as a disjoint collection of nonempty soft sets having some properties, which is entirely different from this.

Definition 4.4 A collection P of soft sets over U with parameter set E is said to be a soft partition on E, if

1. for all $\mathcal{F} \in P$, there is some $a \in E$ such that $\mathcal{F}(a) = U$
2. for all $a \in E$, there is exactly one $\mathcal{F} \in P$ such that $\mathcal{F}(a) = U$
3. $\mathcal{F}, \mathcal{G} \in P$ such that $\mathcal{F}(a) = \mathcal{G}(b) = U$ for some $a, b \in E$, then $\mathcal{F}(b) \cap \mathcal{F}(e) \subseteq \mathcal{G}(e)$, for all $e \in E$.

Example 53 Let $U = \{1, 2, 3, 4, 5, 6\}$ and $E = \{a, b, c\}$ be universal and parameter set respectively. The soft sets on E defined as

$$\begin{aligned}
\mathcal{F} &= \{(a, U), (b, \{2, 4, 5, 6\}), (c, \{1, 2, 4, 5, 6\})\} \\
\mathcal{G} &= \{(a, \{2, 4, 5, 6\}), (b, U), (c, \{2, 3, 4, 5, 6\})\} \\
\mathcal{H} &= \{(a, \{1, 2, 4, 5, 6\}), (b, \{2, 3, 4, 5, 6\}), (c, U)\}.
\end{aligned}$$

Then $\{\mathcal{F}, \mathcal{G}, \mathcal{H}\}$ form a soft partition on E.

Proposition 4.6 *Let P be a soft partition on E. If $\mathcal{F}, \mathcal{G} \in P$ such that $\mathcal{F}(a) = \mathcal{G}(b) = U$ for some $a, b \in E$, then $\mathcal{F}(b) = \mathcal{G}(a)$.*

Proof Given $\mathcal{F}(a) = \mathcal{G}(b) = U$ and since P is a partition, we have

$$\begin{aligned}
\mathcal{F}(b) \cap \mathcal{F}(a) &\subseteq \mathcal{G}(a) \\
\mathcal{F}(b) \cap U &\subseteq \mathcal{G}(a) \\
\text{which implies} \qquad \mathcal{F}(b) &\subseteq \mathcal{G}(a)
\end{aligned}$$

Also, from $\mathcal{G}(a) \cap \mathcal{G}(b) \subseteq \mathcal{F}(b)$, we get $\mathcal{G}(a) \subseteq \mathcal{F}(b)$ and the proof is complete. □

Theorem 4.1 *Let P be a soft partition on E. Then the soft relation $\mathcal{H}_{E \times E}(U)$ defined by $\mathcal{H}(a, b) = F(b)$, where $\mathcal{F} \in P$ with $\mathcal{F}(a) = U$ is a soft equivalence relation.*

Proof By definition of soft partition, $\mathcal{H}$ is well defined. From the definition of $\mathcal{H}$, $\mathcal{H}(a, a) = \mathcal{F}(a) = U$, for all $a \in E$. That is $\mathcal{H}$ is reflexive.

For symmetry, if $\mathcal{H}(a, b) = \mathcal{F}(b)$ and $\mathcal{H}(b, a) = \mathcal{G}(a)$, where $\mathcal{F}, \mathcal{G} \in P$ with $\mathcal{F}(a) = \mathcal{G}(b) = U$. By proposition 4.6, $\mathcal{F}(b) = \mathcal{G}(a)$ which implies
$\mathcal{H}(a, b) = \mathcal{H}(b, a)$, for all $a, b \in E$

For transitivity, consider $a, b, c \in E$ and

$$\begin{aligned}
&\mathcal{H}(b, a) = \mathcal{F}(a) \text{ with } \mathcal{F}(b) = U \\
&\mathcal{H}(a, c) = \mathcal{G}(c) \text{ with } \mathcal{G}(a) = U
\end{aligned}$$

$\mathcal{H}(b, c) = \mathcal{F}(c)$ with $\mathcal{F}(b) = U$

Since $\mathcal{F}(b) = \mathcal{G}(a) = U$. By the Definition 4.4 of soft partition, we have

$\mathcal{F}(a) \cap \mathcal{F}(c) \subseteq \mathcal{G}(c)$
i.e., $\mathcal{H}(b, a) \cap \mathcal{H}(b.c) \subseteq \mathcal{H}(a, c)$
i.e., $\mathcal{H}(a, b) \cap \mathcal{H}(b.c) \subseteq \mathcal{H}(a, c)$, since $\mathcal{H}$ is symmetric.

It follows that $\mathcal{H}$ is transitive. Hence the theorem. □

Remark 4.4 The collection of soft equivalence classes of the soft equivalence relation $\mathcal{H}$ defined in Theorem 4.1 is precisely the soft partition P.

Theorem 4.2 *If $\mathcal{G}$ is a soft equivalence relation on E, then the collection $\mathcal{C}$ of soft equivalence classes form a soft partition on E.*

Proof Suppose that $\mathcal{G}$ is a soft equivalence relation on E. Then the soft equivalence classes of $\mathcal{G}$ is the collection $\mathcal{C} = \{[\mathcal{G}_a] : a \in E\}$, where $\mathcal{G}_a$ is a soft set and is defined by $\mathcal{G}_a(e) = \mathcal{G}(a, e)$, for all $e \in E$.

For any $\mathcal{G}_a \in \mathcal{C}$, $\mathcal{G}_a(a) = \mathcal{G}(a, a) = U$. Hence condition (1) of soft partition is satisfied.

Now any $a \in E, \mathcal{G}_a(a) = U$. Suppose there exist an $x \in E$ such that $\mathcal{G}_x(a) = U$, then by Proposition 4.4(b) $[\mathcal{G}_x] = [\mathcal{G}_a]$. Which establish the condition (2) of soft partition.

Finally, if $[\mathcal{G}_a], [\mathcal{G}_b] \in \mathcal{C}$ such that $\mathcal{G}_a(x) = \mathcal{G}_b(y) = U$. For any $e \in E$,

$$\begin{aligned}
\mathcal{G}_a(y) \cap \mathcal{G}_a(e) &= \mathcal{G}(a, y) \cap \mathcal{G}(a, e) \\
&= \mathcal{G}(y, a) \cap \mathcal{G}(a, e) \\
&\subseteq \mathcal{G}(y, e) = \mathcal{G}_y(e) \\
&= \mathcal{G}_b(e), \qquad \text{since } \mathcal{G}_b(y) = U
\end{aligned}$$

Which completes the condition (3) and hence $\mathcal{C}$ form a soft partition on E. □

Remark 4.5 The soft equivalence relation defined by using the soft partition $\mathcal{C}$ in Theorem 4.2 is precisely the $\mathcal{G}$. This establishes a one to one correspondence between soft equivalence relations and soft partitions.

Proposition 4.7 *Let P be a soft partition on a set E, then the soft relation $\mathcal{G}$ on E defined by $\mathcal{G}(a, b) = \bigcup_{\mathcal{F} \in P} (\mathcal{F}(a) \cap \mathcal{F}(b))$ is a soft equivalence relation on E.*

Proof For reflexivity, any $a \in E$

$$\begin{aligned}
\mathcal{G}(a, a) &= \bigcup_{\mathcal{F} \in P} (\mathcal{F}(a) \cap \mathcal{F}(a)) \\
&= \bigcup_{\mathcal{F} \in P} \mathcal{F}(a) = U, \quad \text{by definition } 4.4(2)
\end{aligned}$$

For symmetry, any $a, b \in E$

$$\begin{aligned}\mathcal{G}(a,b) &= \bigcup_{\mathcal{F}\in P}(\mathcal{F}(a)\cap\mathcal{F}(b))\\ &= \bigcup_{\mathcal{F}\in P}(\mathcal{F}(b)\cap\mathcal{F}(a)) = \mathcal{G}(b,a)\end{aligned}$$

For transitivity, let $a, b, c \in E$

$$\begin{aligned}\mathcal{G}(a,b)\cap\mathcal{G}(b,c) &= [\bigcup_{\mathcal{F}\in P}(\mathcal{F}(a)\cap\mathcal{F}(b))]\cap[\bigcup_{\mathcal{F}\in P}(\mathcal{F}(b)\cap\mathcal{F}(c))]\\ &= \bigcup_{\mathcal{F}\in P}[\mathcal{F}(a)\cap\mathcal{F}(b)\cap\mathcal{F}(c)]\\ &\subseteq \bigcup_{\mathcal{F}\in P}[\mathcal{F}(a)\cap\mathcal{F}(c)] = \mathcal{G}(a,c)\end{aligned}$$

□

Proposition 4.8 *Let P be a soft partition on E. If $|E| = n$, then P contains at most n soft sets.*

Proof Suppose P is a soft partition and $|P| > n$, then there exists two soft sets $\mathcal{P}_1, \mathcal{P}_2 \in P$ and $a \in E$ such that $\mathcal{P}_1(a) = \mathcal{P}_2(a) = U$. It implies that P is not a soft partition. □

4.2 Soft Graphs and Chained Soft Graphs

Graph theoretical concepts are very prominent in decision-making problems. Rosenfeld [115] proposed the notion of fuzzy graphs using symmetric fuzzy relation on a fuzzy set. He presented fuzzy graph as a weighted graph having values from [0, 1], which made the growth of graph theory in a different level. Thumbakara et al. [142] introduced the concept of soft graphs as a parameterized family of connected subgraphs of a simple graph. Akram et al.[4] focus their discussion on vertex induced soft graph and edge induced soft graph. Further, some work related to rough graph can be seen in Mathew, B. et al. [84]. In this context, the aim of this section is to focus on the study of the behaviour and the structure of soft graphs, using soft relation on a soft set.

4.2.1 Soft Relation on a Soft Set

Definition 4.5 Let $\mathcal{F}_E(U)$ be a soft set over U and $\mathcal{G}_{E\times E}(U)$ is a soft relation on E. Then $\mathcal{G}_{E\times E}(U)$ is said to be a soft relation on a soft set $\mathcal{F}_E(U)$, if $\mathcal{G}(a,b) \subseteq \mathcal{F}(a) \cap \mathcal{F}(b)$, for all $a, b \in E$.

Example 54 Let $U = \{u_1, u_2, u_3, u_4\}$ and $E = \{a, b, c\}$ be universe and parameter set respectively. A soft set $\mathcal{F}_E$ is defined by $\mathcal{F} : E \to \mathcal{P}(U)$ as

$$\mathcal{F}(a) = \{u_1, u_3, u_4\},\ \mathcal{F}(b) = \{u_2, u_3\} \text{ and } \mathcal{F}(c) = \{u_1, u_2, u_3\}.$$

Define $\mathcal{G} : E \times E \to \mathcal{P}(U)$ as

$$\begin{array}{lll} \mathcal{G}(a,a) = \{u_1, u_4\} & \mathcal{G}(a,b) = \{u_3\} & \mathcal{G}(a,c) = \{u_1, u_3\} \\ \mathcal{G}(b,a) = \{u_3\} & \mathcal{G}(b,b) = \{u_2, u_3\} & \mathcal{G}(b,c) = \{u_2\} \\ \mathcal{G}(c,a) = \{u_1\} & \mathcal{G}(c,b) = \{u_3\} & \mathcal{G}(c,c) = \{u_2, u_3\} \end{array}$$

Then $\mathcal{G}(a,b) \subseteq \mathcal{F}(a) \cap \mathcal{F}(b)$ for all $a, b \in E$, so $\mathcal{G}_{E\times E}(U)$ is a soft relation on $\mathcal{F}_E(U)$.

Proposition 4.9 *Let $\mathcal{G}_{E\times E}(U)$ and $\mathcal{H}_{E\times E}(U)$ be two soft relations on a soft set $\mathcal{F}_E(U)$ and for $a, b, c \in E$, then*

(a) Union : the soft set $(\mathcal{G} \cup \mathcal{H})_{E\times E}$ defined by $(\mathcal{G} \cup \mathcal{H})(a,b) = \mathcal{G}(a,b) \cup \mathcal{H}(a,b)$
(b) Intersection : the soft set $(\mathcal{G} \cap \mathcal{H})_{E\times E}$ defined by $(\mathcal{G} \cap \mathcal{H})(a,b) = \mathcal{G}(a,b) \cap \mathcal{H}(a,b)$
(c) Transpose : the soft set $\mathcal{G}^t_{E\times E}$ is the transpose of $\mathcal{G}$ defined by $\mathcal{G}^t(b,a) = \mathcal{G}(a,b)$
(d) Composition of soft relations: the composite of $\mathcal{G}$ and $\mathcal{H}$ is a soft set $(\mathcal{G} \circ \mathcal{H})_{E\times E}$ defined by

$$(\mathcal{G} \circ \mathcal{H})(a,c) = \bigcup_{b\in B} [\mathcal{G}(a,b) \cap \mathcal{H}(b,c)]$$

are soft relations on the soft set $\mathcal{F}_E(U)$.

Proof For $a, b \in E$, $\mathcal{G}(a,b) \subseteq \mathcal{F}(a) \cap \mathcal{F}(b)$ and $\mathcal{H}(a,b) \subseteq \mathcal{F}(a) \cap \mathcal{F}(b)$

(a) $(\mathcal{G} \cup \mathcal{H})(a,b) = \mathcal{G}(a,b) \cup \mathcal{H}(a,b) \subseteq \mathcal{F}(a) \cap \mathcal{F}(b)$
(b) $(\mathcal{G} \cap \mathcal{H})(a,b) = \mathcal{G}(a,b) \cap \mathcal{H}(a,b) \subseteq \mathcal{F}(a) \cap \mathcal{F}(b)$
(c) $\mathcal{G}^t(b,a) = \mathcal{G}(a,b) \subseteq \mathcal{F}(a) \cap \mathcal{F}(b)$
(d)

$$\begin{aligned} (\mathcal{G} \circ \mathcal{H})(a,c) &= \bigcup_{b\in E} [\mathcal{G}(a,b) \cap \mathcal{H}(b,c)] \\ &\subseteq \bigcup_{b\in E} [(\mathcal{F}(a) \cap \mathcal{F}(b)) \cap (\mathcal{F}(b) \cap \mathcal{F}(c))] \end{aligned}$$

$$= \bigcup_{b \in E} [\mathcal{F}(a) \cap \mathcal{F}(b) \cap \mathcal{F}(c)]$$
$$\subseteq \bigcup_{b \in E} [\mathcal{F}(a) \cap \mathcal{F}(c)]$$
$$= \mathcal{F}(a) \cap \mathcal{F}(c).$$

Thus $\mathcal{G} \cup \mathcal{H},\ \mathcal{G} \cap \mathcal{H},\ \mathcal{G}^t$ and $\mathcal{G} \circ \mathcal{H}$ are soft relations on the soft set $\mathcal{F}_E(U)$.

□

Definition 4.6 A soft relation $\mathcal{G}_{E \times E}(U)$ on a soft set $\mathcal{F}_E(U)$ is said to be

reflexive	if $\mathcal{G}(a, a) = \mathcal{F}(a)$, for all $a \in E$
symmetric	if $\mathcal{G}(a, b) = \mathcal{G}(b, a)$, for all $a, b \in E$
transitive	if $(\mathcal{G} \circ \mathcal{G})(a, b) \subseteq \mathcal{G}(a, b)$, for all $a, b \in E$.

A soft relation on a soft set is said to be a soft equivalence relation on a soft set, if it is reflexive, symmetric and transitive.

Proposition 4.10 *If $\mathcal{G}_{E \times E}(U)$ and $\mathcal{H}_{E \times E}(U)$ are reflexive soft relations on a soft set $\mathcal{F}_E(U)$, then*

(a) $\mathcal{G}(a, b) \subseteq \mathcal{G}(a, a)$ and $\mathcal{G}(b, a) \subseteq \mathcal{G}(a, a)$ for all $a, b \in E$
(b) $\mathcal{G} \circ \mathcal{H}$ is reflexive
(c) $\mathcal{H} \subseteq \mathcal{G} \circ \mathcal{H}$ and $\mathcal{H} \subseteq \mathcal{H} \circ \mathcal{G}$, for any soft relation $\mathcal{H}$ on $\mathcal{F}$.

Proof Let $a, b, c \in E$, we have

(a) $\mathcal{G}(a, b) \subseteq \mathcal{F}(a) \cap \mathcal{F}(b) \subseteq \mathcal{F}(a) = \mathcal{G}(a, a)$ and $\mathcal{G}(b, a) \subseteq \mathcal{F}(b) \cap \mathcal{F}(a) \subseteq \mathcal{F}(a) = \mathcal{G}(a, a)$

(b)

$$(\mathcal{G} \circ \mathcal{H})(a, a) = \bigcup_{b \in E} [\mathcal{G}(a, b) \cap \mathcal{H}(b, a)]$$
$$= (\mathcal{G}(a, a) \cap \mathcal{H}(a, a)) \cup \Big(\bigcup_{\substack{b \in E \\ b \neq a}} [\mathcal{G}(a, b) \cap \mathcal{H}(b, a)] \Big)$$
$$= \mathcal{F}(a) \cup S = \mathcal{F}(a)$$

where $S = \bigcup_{\substack{b \in E \\ b \neq a}} [\mathcal{G}(a, b) \cap \mathcal{H}(b, a)] \subseteq \mathcal{F}(a) \cap \mathcal{F}(b) \subseteq \mathcal{F}(a)$.

(c)

$$\begin{aligned}(\mathcal{G} \circ \mathcal{H})(a, c) &= \bigcup_{b \in E}[\mathcal{G}(a, b) \cap \mathcal{H}(b, c)] \\ &\supseteq \mathcal{G}(a, a) \cap \mathcal{H}(a, c) \\ &= \mathcal{F}(a) \cap \mathcal{H}(a, c) \\ &= \mathcal{H}(a, c), \qquad \text{since } \mathcal{H}(a, c) \subseteq \mathcal{F}(a) \cap \mathcal{F}(c).\end{aligned}$$

Thus $\mathcal{H} \subseteq \mathcal{G} \circ \mathcal{H}$. Similarly $\mathcal{H} \subseteq \mathcal{H} \circ \mathcal{G}$ hold.

□

4.2.2 *Soft Graphs*

Definition 4.7 A soft graph G is a pair $(\mathcal{F}, \mathcal{G})$, where $\mathcal{F} = \mathcal{F}_E(U)$ is a soft set and $\mathcal{G} = \mathcal{G}_{E\times E}(U)$ is a symmetric soft relation on $\mathcal{F}_E(U)$ and is denoted by $G = (\mathcal{F}, \mathcal{G})_E$.

Remark 4.6 A soft graph $G = (\mathcal{F}, \mathcal{G})_E$ is a pair of soft sets satisfying $\mathcal{G}(a, b) \subseteq \mathcal{F}(a) \cap \mathcal{F}(a)$, for all $a, b \in E$. The underlying crisp graph of a soft graph $G = (\mathcal{F}, \mathcal{G})_E$ is denoted by $G^* = (\mathcal{F}^*, \mathcal{G}^*)$ where $\mathcal{F}^* = \{a \in E : \mathcal{F}(a) \neq \emptyset\}$ and $\mathcal{G}^* = \{(a, b) \in E \times E : \mathcal{G}(a, b) \neq \emptyset\}$.

We consider only loopless graphs unless otherwise specified.

Example 55 Let $U = \{1, 2, 3, 4, 5, 6\}$ and $E = \{a, b, c\}$ be universe and parameter set respectively. A soft set $\mathcal{F}_E(U)$ is defined by

$$\mathcal{F}(a) = \{1, 2, 3, 4, 6\} \; \mathcal{F}(b) = \{1, 2, 4, 5\} \; \mathcal{F}(c) = \{2, 4, 5, 6\}$$

and a symmetric soft relation defined by $\mathcal{G}_{E\times E}(U)$ as (Fig. 4.1)

$$\begin{aligned}&\mathcal{G}(a, b) = \mathcal{G}(b, a) = \{2\} \; \mathcal{G}(a, c) = \mathcal{G}(c, a) = \{4, 6\} \\ &\mathcal{G}(b, c) = \mathcal{G}(c, b) = \{4\}.\end{aligned}$$

Definition 4.8 A soft graph $G = (\mathcal{F}, \mathcal{G})_A$ is said to be an absolute soft graph, if $\mathcal{F}_A$ and $\mathcal{G}_{A\times A}$ are absolute soft sets.

Definition 4.9 For a soft graph $G = (\mathcal{F}, \mathcal{G})_E$, order of G is defined and denoted by the set $P = \bigcup_{a\in E} \mathcal{F}(a)$ and size of G is defined and denoted by the set $Q = \bigcup_{a,b\in E} \mathcal{G}(a, b)$.

Example 56 In the previous Example 55, the order and size of the soft graph $G = (\mathcal{F}, \mathcal{G})_E$ is $P = \{1, 2, 3, 4, 5, 6\}$ and $Q = \{2, 4, 6\}$ respectively.

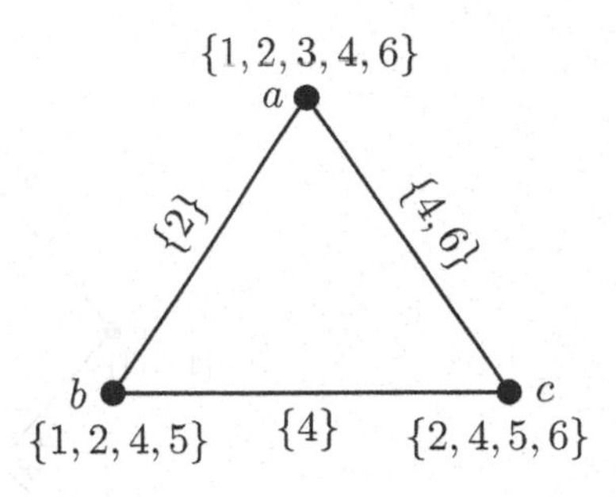

Fig. 4.1 Soft graph $G = (\mathcal{F}, \mathcal{G})_E$

Definition 4.10 Let $G = (\mathcal{F}, \mathcal{G})_E$ be a soft graph. The degree of a vertex $a \in E$ is $d_G(a) = \bigcup_{b \in E} \mathcal{G}(a, b)$.

Example 57 In Example 55, $d_G(a) = \{2, 4, 6\}$, $d_G(b) = \{2, 4\}$ and $d_G(c) = \{4, 6\}$.

A fuzzy graph is a pair $G = (\sigma, \mu)_A$, where $\sigma : A \to [0, 1]$ is a fuzzy set on A and $\mu : A \times A \to [0, 1]$ is a fuzzy relation on σ satisfy $\mu(a, b) \leq \sigma(a) \wedge \sigma(a)$.

Proposition 4.11 *Every fuzzy graph can be represented by a soft graph.*

Proof Let $(\sigma, \mu)_E$ be a fuzzy graph on E, where $\sigma : E \to [0, 1]$ is a fuzzy set and $\mu : E \times E \to [0, 1]$ is a symmetric fuzzy relation on σ. Define a soft set $\mathcal{F}$ on E (i.e., $\mathcal{F} : E \to \mathcal{P}([0, 1])$) by $\mathcal{F}(a) = [0, a_i]$, where $a_i = \sigma(a)$ and note that $\sigma_E(x) = \sup \mathcal{F}(x)$.

Define a soft relation $\mathcal{G} : E \times E \to \mathcal{P}([0, 1])$ by $\mathcal{G}(a, b) = [0, b_i]$, where $\mu(a, b) = b_i$ and note that $\mu_{E \times E}(x, y) = \sup \mathcal{G}(x, y)$. Since $\mu(a, b) \leq \sigma(a) \wedge \sigma(b)$, we get $\mathcal{G}(a, b) \subseteq \mathcal{F}(a) \cap \mathcal{F}(b)$. Thus the fuzzy graph $(\sigma, \mu)_E$ can be represented as the soft graph $(\mathcal{F}, \mathcal{G})_E$. □

Definition 4.11 Let $G = (\mathcal{F}, \mathcal{G})_E$ be a soft graph. Then G is said to be strong if $\mathcal{G}(a, b) = \mathcal{F}(a) \cap \mathcal{F}(b)$, for all $(a, b) \in \mathcal{G}^*$ and is a complete soft graph if $\mathcal{G}(a, b) = \mathcal{F}(a) \cap \mathcal{F}(b)$, for all $a, b \in \mathcal{F}^*$.

Proposition 4.12 *Every complete soft graph is a strong soft graph.*

Proof Let $G = (\mathcal{F}, \mathcal{G})_E$ be a complete soft graph, then $\mathcal{G}(a, b) = \mathcal{F}(a) \cap \mathcal{F}(b)$, for all $a, b \in \mathcal{F}^*$. For any $(x, y) \in \mathcal{G}^*$, $\mathcal{G}(x, y) \neq \emptyset$. It follows that $\mathcal{F}(x) \neq \emptyset$ and $\mathcal{F}(y) \neq \emptyset$, since G is a soft graph. Thus $x, y \in \mathcal{F}^*$. Then $\mathcal{G}(x, y) = \mathcal{F}(x) \cap \mathcal{F}(y)$, for all $(x, y) \in \mathcal{G}^*$. Hence G is a strong soft graph. □

Remark 4.7 The converse of above proposition is not true in general i.e., every strong soft graph need not be complete.

Example 58 Let $U = \{1, 2, 3, 4, 5\}$ and $E = \{a, b, c, d\}$ be universe and parameter sets respectively (Fig. 4.2).

Here $a, c \in \mathcal{F}^*$, but $\mathcal{G}(a, c) = \emptyset$. So it is not a complete soft graph.

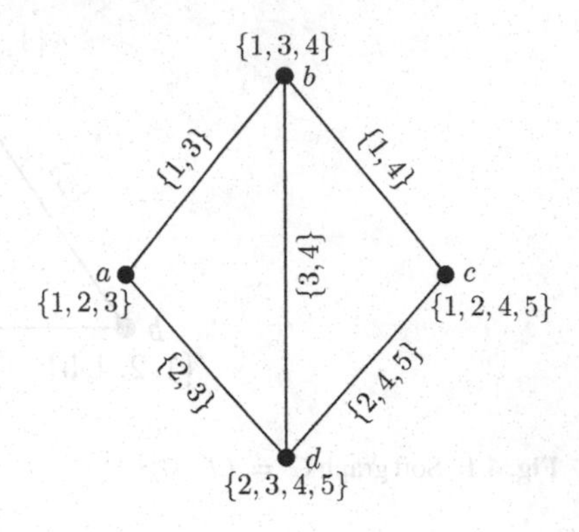

Fig. 4.2 Strong soft graph, but not complete

Definition 4.12 Let U be an universal set and let E and E' be parameter sets. An isomorphism between two soft graphs $(\mathcal{F}, \mathcal{G})_E$ and $(\mathcal{F}', \mathcal{G}')_{E'}$ is a map h, where $h : E \to E'$ is a bijection such that

1. $\mathcal{F}(a) = \mathcal{F}'(h(a))$, for all $a \in E$
2. $\mathcal{G}(a, b) = \mathcal{G}'(h(a), h(b))$, for all $a, b \in E$.

Proposition 4.13 *An isomorphism between two soft graphs preserves their order, size and degree of a vertex.*

Proof Let $G = (\mathcal{F}, \mathcal{G})_E$ and $G' = (\mathcal{F}', \mathcal{G}')_{E'}$ be two isomorphic soft graphs and let $h : G \to G'$ be an isomorphism between G and G'.

Let P and Q be the order and size of G and P' and Q' be the order and size of G' respectively. Then

$$P = \bigcup_{a\in E} \mathcal{F}(a) = \bigcup_{a\in E} \mathcal{F}'(h(a)) = P'$$

and

$$Q = \bigcup_{a\in E} \mathcal{G}(a, b) = \bigcup_{a\in E} \mathcal{G}'(h(a), h(b)) = Q'.$$

Let $h(a) = a'$ and $h(b) = b'$, then $\mathcal{G}(a, b) = \mathcal{G}'(h(a), h(b)) = \mathcal{G}'(a', b')$

$$d_G(a) = \bigcup_{b\in E} \mathcal{G}(a, b) = \bigcup_{b\in E} \mathcal{G}'(h(a), h(b)) = \bigcup_{b'\in E'} \mathcal{G}'(a', b') = d_{G'}(a')$$

Hence the result. □

4.2.3 Operations on Soft Graphs

Throughout this subsection an edge $e = (a, b)$ will be denoted as ab.

Definition 4.13 Let $G_1 = (\mathcal{F}_1, \mathcal{G}_1)$ and $G_2 = (\mathcal{F}_2, \mathcal{G}_2)$ be two soft graphs with $G_1^* = (V_1, E_1)$ and $G_2^* = (V_2, E_2)$ as underlying crisp graphs respectively. Define soft sets $\mathcal{F}_1 \cup \mathcal{F}_2$ on $V_1 \cup V_2$ and $\mathcal{G}_1 \cup \mathcal{G}_2$ on $E_1 \cup E_2$ as follows,

$$(\mathcal{F}_1 \cup \mathcal{F}_2)(a) = \begin{cases} \mathcal{F}_1(a) & \text{if } a \in V_1 \setminus V_2; \\ \mathcal{F}_2(a) & \text{if } a \in V_2 \setminus V_1; \\ \mathcal{F}_1(a) \cup \mathcal{F}_2(a) & \text{if } a \in V_1 \cap V_2 \end{cases}$$

$$(\mathcal{G}_1 \cup \mathcal{G}_2)(e) = \begin{cases} \mathcal{G}_1(e) & \text{if } e \in E_1 \setminus E_2; \\ \mathcal{G}_2(e) & \text{if } e \in E_2 \setminus E_1; \\ \mathcal{G}_1(e) \cup \mathcal{G}_2(e) & \text{if } e \in E_1 \cap E_2 \end{cases}$$

Proposition 4.14 *Let $(\mathcal{F}_1, \mathcal{G}_1)$ and $(\mathcal{F}_2, \mathcal{G}_2)$ be two soft graphs with underlying crisp graphs $G_1^* = (V_1, E_1)$ and $G_2^* = (V_2, E_2)$ respectively. Then $(\mathcal{F}_1 \cup \mathcal{F}_2, \mathcal{G}_1 \cup \mathcal{G}_2)$ is a soft graph.*

Proof Suppose $e = uv \in E_1 \setminus E_2$, then
Case I: $u, v \in V_1 \cap V_2$

$$\begin{aligned}(\mathcal{G}_1 \cup \mathcal{G}_2)(uv) &= \mathcal{G}_1(uv) \\ &\subseteq \mathcal{F}_1(u) \cap \mathcal{F}_1(v) \\ &\subseteq [\mathcal{F}_1(u) \cup \mathcal{F}_2(u)] \cap [\mathcal{F}_1(v) \cup \mathcal{F}_2(v)] \\ &= (\mathcal{F}_1 \cup \mathcal{F}_2)(u) \cap (\mathcal{F}_1 \cup \mathcal{F}_2)(v)\end{aligned}$$

Case II: $u \in V_1 \setminus V_2, v \in V_1 \cap V_2$

$$\begin{aligned}(\mathcal{G}_1 \cup \mathcal{G}_2)(uv) &= \mathcal{G}_1(uv) \\ &\subseteq \mathcal{F}_1(u) \cap \mathcal{F}_1(v) \\ &\subseteq (\mathcal{F}_1 \cup \mathcal{F}_2)(u) \cap [\mathcal{F}_1(v) \cup \mathcal{F}_2(v)] \\ &= (\mathcal{F}_1 \cup \mathcal{F}_2)(u) \cap (\mathcal{F}_1 \cup \mathcal{F}_2)(v)\end{aligned}$$

Case III: $u, v \in V_1 \setminus V_2$

$$\begin{aligned}(\mathcal{G}_1 \cup \mathcal{G}_2)(uv) &= \mathcal{G}_1(uv) \\ &\subseteq \mathcal{F}_1(u) \cap \mathcal{F}_1(v) \\ &= (\mathcal{F}_1 \cup \mathcal{F}_2)(u) \cap (\mathcal{F}_1 \cup \mathcal{F}_2)(v)\end{aligned}$$

Similarly, if $uv \in E_2 \setminus E_1$, then $(\mathcal{G}_1 \cup \mathcal{G}_2)(uv) \subseteq (\mathcal{F}_1 \cup \mathcal{F}_2)(u) \cap (\mathcal{G}_1 \cup \mathcal{G}_2)(v)$.
Suppose that $uv \in E_1 \cap E_2$, then

$$\begin{aligned}(\mathcal{G}_1 \cup \mathcal{G}_2)(uv) &= \mathcal{G}_1(uv) \cup \mathcal{G}_2(uv)\\ &\subseteq [\mathcal{F}_1(u) \cap \mathcal{F}_1(v)] \cup [\mathcal{F}_2(u) \cap \mathcal{F}_2(v)]\\ &\subseteq [\mathcal{F}_1(u) \cup \mathcal{F}_2(u)] \cap [\mathcal{F}_1(v) \cup \mathcal{F}_2(v)]\\ &= (\mathcal{F}_1 \cup \mathcal{F}_2)(u) \cap (\mathcal{F}_1 \cup \mathcal{F}_2)(v)\end{aligned}$$

Thus $(\mathcal{G}_1 \cup \mathcal{G}_2)(uv) \subseteq (\mathcal{F}_1 \cup \mathcal{F}_2)(u) \cap (\mathcal{F}_1 \cup \mathcal{F}_2)(v)$, for all $e = uv \in E_1 \cup E_2$. □

Remark 4.8 The soft graph $(\mathcal{F}_1 \cup \mathcal{F}_2, \mathcal{G}_1 \cup \mathcal{G}_2)$ in Proposition 4.14 is called the union of $(\mathcal{F}_1, \mathcal{G}_1)$ and $(\mathcal{F}_2, \mathcal{G}_2)$.

Definition 4.14 Let $G_1 = (\mathcal{F}_1, \mathcal{G}_1)$ and $G_2 = (\mathcal{F}_2, \mathcal{G}_2)$ be two soft graphs with underlying crisp graph $G_1^* = (V_1, E_1)$ and $G_2^* = (V_2, E_2)$ respectively. Assume $V_1 \cap V_2 = \emptyset$. Define soft sets $\mathcal{F}_1 + \mathcal{F}_2$ on $V_1 \cup V_2$ and $\mathcal{G}_1 + \mathcal{G}_2$ on $E_1 \cup E_2 \cup E'$, where E' is the set of all edges joining vertices of V_1 and V_2, as follows:

$$(\mathcal{F}_1 + \mathcal{F}_2)(a) = \begin{cases}\mathcal{F}_1(a) \text{ if } a \in V_1;\\ \mathcal{F}_2(a) \text{ if } a \in V_2\end{cases}$$

$$(\mathcal{G}_1 + \mathcal{G}_2)(ab) = \begin{cases}\mathcal{G}_1(ab) & \text{if } ab \in E_1;\\ \mathcal{G}_2(ab) & \text{if } ab \in E_2;\\ \mathcal{F}_1(a) \cap \mathcal{F}_2(b) & \text{if } ab \in E', a \in V_1, b \in V_2\end{cases}$$

Proposition 4.15 *Let $(\mathcal{F}_1, \mathcal{G}_1)$ and $(\mathcal{F}_2, \mathcal{G}_2)$ be two soft graphs with underlying crisp graphs $G_1^* = (V_1, E_1)$ and $G_2^* = (V_2, E_2)$ respectively. Then $(\mathcal{F}_1 + \mathcal{F}_2, \mathcal{G}_1 + \mathcal{G}_2)$ is a soft graph.*

Proof Case I: Assume that $ab \in E_1$. It follows that $a, b \in V_1$, then

$$\begin{aligned}(\mathcal{G}_1 + \mathcal{G}_2)(ab) &= \mathcal{G}_1(ab)\\ &\subseteq \mathcal{F}_1(a) \cap \mathcal{F}_1(b)\\ &= (\mathcal{F}_1 + \mathcal{F}_2)(a) \cap (\mathcal{F}_1 + \mathcal{F}_2)(b)\end{aligned}$$

Case II: Assume that $ab \in E_2$. It follows that $a, b \in V_2$, then

$$\begin{aligned}(\mathcal{G}_1 + \mathcal{G}_2)(ab) &= \mathcal{G}_2(ab)\\ &\subseteq \mathcal{F}_2(a) \cap \mathcal{F}_2(b)\\ &= (\mathcal{F}_1 + \mathcal{F}_2)(a) \cap (\mathcal{F}_1 + \mathcal{F}_2)(b)\end{aligned}$$

Case II: Assume that $ab \in E'$ and $a \in V_1, b \in V_2$, then

$$\begin{aligned}(\mathcal{G}_1 + \mathcal{G}_2)(ab) &= \mathcal{F}_1(a) \cap \mathcal{F}_2(b)\\ &= (\mathcal{F}_1 + \mathcal{F}_2)(a) \cap (\mathcal{F}_1 + \mathcal{F}_2)(b)\end{aligned}$$

In all cases, $(\mathcal{G}_1 + \mathcal{G}_2)(ab) \subseteq (\mathcal{F}_1 + \mathcal{F}_2)(a) \cap (\mathcal{F}_1 + \mathcal{F}_2)(b)$, for all $e = ab \in E_1 \cup E_2 \cup E'$. □

Remark 4.9 The soft graph $(\mathcal{F}_1 + \mathcal{F}_2, \mathcal{G}_1 + \mathcal{G}_2)$ in Proposition 4.15 is called the join of $(\mathcal{F}_1, \mathcal{G}_1)$ and $(\mathcal{F}_2, \mathcal{G}_2)$.

Definition 4.15 Let $G = (\mathcal{F}, \mathcal{G})_E$ be a soft graph. The complement of G is defined as $\overline{G} = (\overline{\mathcal{F}}, \overline{\mathcal{G}})$, where $\overline{\mathcal{F}}(a) = \mathcal{F}(a)$ and $\overline{\mathcal{G}}(ab) = (\mathcal{F}(a) \cap \mathcal{F}(b)) \setminus \mathcal{G}(ab)$, for all $a, b \in E$.

Proposition 4.16 *Let $G = (\mathcal{F}, \mathcal{G})_E$ be a soft graph, then $\overline{\overline{G}} = G$.*

Proof From the definition of complement, $\overline{\overline{\mathcal{F}}}(a) = \overline{\mathcal{F}}(a) = \mathcal{F}(a)$, for all $a \in E$.
Now, for all $a, b \in E$,

$$\begin{aligned}\overline{\overline{\mathcal{G}}}(ab) &= (\overline{\mathcal{F}}(a) \cap \overline{\mathcal{F}}(b)) \setminus \overline{\mathcal{G}}(ab)\\ &= (\mathcal{F}(a) \cap \mathcal{F}(b)) \setminus [(\mathcal{F}(a) \cap \mathcal{F}(b)) \setminus \mathcal{G}(ab)]\\ &= \mathcal{G}(ab) \quad \text{Since } \ \mathcal{G}(ab) \subseteq \mathcal{F}(a) \cap \mathcal{F}(b)\end{aligned}$$

Hence $\overline{\overline{G}} = G$. □

Proposition 4.17 *Let $G_1 = (\mathcal{F}_1, \mathcal{G}_1)$ and $G_2 = (\mathcal{F}_2, \mathcal{G}_2)$ be two soft graphs with underlying crisp graphs $G_1^* = (V_1, E_1)$ and $G_2^* = (V_2, E_2)$ respectively and $V_1 \cap V_2 = \emptyset$, then*

(a) $\overline{G_1 \cup G_2} = \overline{G_1} + \overline{G_2}$
(b) $\overline{G_1 + G_2} = \overline{G_1} \cup \overline{G_2}$

Proof (a) Now $\overline{G_1 \cup G_2} = (\overline{\mathcal{F}_1 \cup \mathcal{F}_2}, \overline{\mathcal{G}_1 \cup \mathcal{G}_2})$

$$\begin{aligned}(\overline{\mathcal{F}_1 \cup \mathcal{F}_2})(a) &= (\mathcal{F}_1 \cup \mathcal{F}_2)(a)\\ &= \begin{cases}\mathcal{F}_1(a) \text{ if } a \in V_1;\\ \mathcal{F}_2(a) \text{ if } a \in V_2\end{cases}\\ &= \begin{cases}\overline{\mathcal{F}_1}(a) \text{ if } a \in V_1;\\ \overline{\mathcal{F}_2}(a) \text{ if } a \in V_2\end{cases}\\ &= (\overline{\mathcal{F}_1} + \overline{\mathcal{F}_2})(a), \ \text{for all } a \in V_1 \cup V_2\end{aligned}$$

Now $(\overline{\mathcal{G}_1 \cup \mathcal{G}_2})(ab) = [(\mathcal{F}_1 \cup \mathcal{F}_2)(a) \cap (\mathcal{F}_1 \cup \mathcal{F}_2)(b)] \setminus (\mathcal{G}_1 \cup \mathcal{G}_2)(ab)$

Case I. If $a, b \in V_1$ i.e., $ab \in E_1$
$(\overline{\mathcal{G}_1 \cup \mathcal{G}_2})(ab) = [\mathcal{F}_1(a) \cap \mathcal{F}_1(b)] \setminus \mathcal{G}_1(ab) = \overline{\mathcal{G}_1}(ab)$
Case II. If $a, b \in V_2$ i.e., $ab \in E_2$
$(\overline{\mathcal{G}_1 \cup \mathcal{G}_2})(ab) = [\mathcal{F}_2(a) \cap \mathcal{F}_2(b)] \setminus \mathcal{G}_2(ab) = \overline{\mathcal{G}_2}(ab)$
Case III. If $a \in V_1$ and $b \in V_2$ i.e., $ab \in E'$
$(\overline{\mathcal{G}_1 \cup \mathcal{G}_2})(ab) = [\mathcal{F}_1(a) \cap \mathcal{F}_2(b)] = \overline{\mathcal{F}_1}(a) \cap \overline{\mathcal{F}_2}(b)$
Therefore $(\overline{\mathcal{G}_1 \cup \mathcal{G}_2})(e) = (\overline{\mathcal{G}_1} + \overline{\mathcal{G}_2})(e)$, for all $e = ab \in E_1 \cup E_2 \cup E'$. Hence $\overline{G_1 \cup G_2} = \overline{G_1} + \overline{G_2}$.

(b) Now $\overline{G_1 + G_2} = (\overline{\mathcal{F}_1 + \mathcal{F}_2}, \overline{\mathcal{G}_1 + \mathcal{G}_2})$

$$\begin{aligned}
(\overline{\mathcal{F}_1 + \mathcal{F}_2})(a) &= (\mathcal{F}_1 + \mathcal{F}_2)(a) \\
&= \begin{cases} \mathcal{F}_1(a) \text{ if } a \in V_1; \\ \mathcal{F}_2(a) \text{ if } a \in V_2 \end{cases} \\
&= \begin{cases} \overline{\mathcal{F}_1}(a) \text{ if } a \in V_1; \\ \overline{\mathcal{F}_2}(a) \text{ if } a \in V_2 \end{cases} \\
&= (\overline{\mathcal{F}_1} \cup \overline{\mathcal{F}_2})(a), \text{ for all } a \in V_1 \cup V_2
\end{aligned}$$

□

Now, $(\overline{\mathcal{G}_1 + \mathcal{G}_2})(ab) = [(\mathcal{F}_1 + \mathcal{F}_2)(a) \cap (\mathcal{F}_1 + \mathcal{F}_2)(b)] \setminus (\mathcal{G}_1 + \mathcal{G}_2)(ab)$.

Case I. If $a, b \in V_1$ i.e., $ab \in E_1$ $(\overline{\mathcal{G}_1 + \mathcal{G}_2})(ab) = [\mathcal{F}_1(a) \cap \mathcal{F}_1(b)] \setminus \mathcal{G}_1(ab) = \overline{\mathcal{G}_1}(ab)$

Case II. If $a, b \in V_2$ i.e. $ab \in E_2$ $(\overline{\mathcal{G}_1 + \mathcal{G}_2})(ab) = [\mathcal{F}_2(a) \cap \mathcal{F}_2(b)] \setminus \mathcal{G}_2(ab) = \overline{\mathcal{G}_2}(ab)$

Case III. If $a \in V_1$ and $b \in V_2$ i.e. $ab \in E'$ $(\overline{\mathcal{G}_1 + \mathcal{G}_2})(ab) = [\mathcal{F}_1(a) \cap \mathcal{F}_2(b)] \setminus [\mathcal{F}_1(a) \cap \mathcal{F}_2(b)] = \emptyset$

Therefore $(\overline{\mathcal{G}_1 + \mathcal{G}_2})(e) = (\overline{\mathcal{G}_1} \cup \overline{\mathcal{G}_2})(e)$, for all $e = ab \in E_1 \cup E_2 \cup E'$. Hence $\overline{G_1 + G_2} = \overline{G_1} \cup \overline{G_2}$.

4.3 Chained Soft Graphs

A path P of length n in a soft graph $G = (\mathcal{F}, \mathcal{G})_E$ is a sequence of distinct vertices $a_0, a_1, a_2, \ldots, a_n$ such that $\mathcal{G}(a_{i-1}, a_i) \neq \emptyset$ for $1 \leq i \leq n$. A closed path with $n \geq 3$ is called a cycle. The pairs (a_{i-1}, a_i) are called the arcs of the path P. The set $\bigcap_{i=1}^{n} \mathcal{G}(a_{i-1}, a_i)$ is called the strength of the path P and is denoted by $S(P)$. A path with empty strength is called a null path and otherwise is called a flow path. A soft

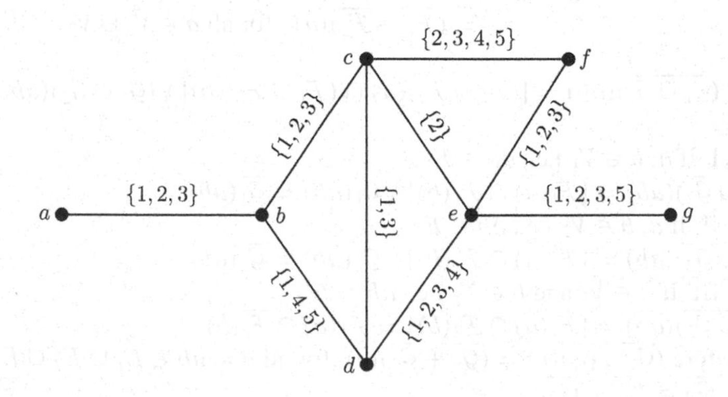

Fig. 4.3 Null path and flow path in a soft graph

graph $G = (\mathcal{F}, \mathcal{G})_A$ is connected if there is a path between any pair of vertices. A strongest path between two vertices a and b is a path P with strength $S(P)$ such that $S(P_i) \subseteq S(P)$, where P_i is a path between a and b for all i. Any pair of vertices in a soft graph connected by a flow path is called strongly connected soft graph.

Example 59 Let $U = \{1, 2, 3, 4, 5\}$ and $E = \{a, b, c, d, e, f, g\}$ be universe and parameter sets respectively. Consider the soft graph given in Fig. 4.3,

All $a - g$ paths in the soft graph are given by

	Path	Strength	Type
P_1	$a - b - d - e - g$	$\{1\}$	Flow path
P_2	$a - b - d - c - e - g$	$\emptyset$	Null path
P_3	$a - b - d - c - f - e - g$	$\emptyset$	Null path
P_4	$a - b - c - e - g$	$\{2\}$	Flow path
P_5	$a - b - c - f - e - g$	$\{2, 3\}$	Flow path
P_6	$a - b - c - d - e - g$	$\{1, 3\}$	Flow path

From the above table, paths P_1 and P_5 are two $a - g$ paths with different strength and there exist no strongest path between a and g. Thus, it is very challenging to find a strongest path between any two vertices in general. To resolve this difficulty, we introduce the chain concept in the context of soft graph.

Definition 4.16 A soft graph $G = (\mathcal{F}, \mathcal{G})_E$ is a vertex chained soft graph if $\mathcal{F}(E)$ is a chain under set inclusion and is called edge chained soft graph if $\mathcal{G}(E \times E)$ is a chain under set inclusion. If G is both vertex and edge chained soft graph, then G is called a chained soft graph.

Example 60 Let $U = \{1, 2, 3, 4\}$ and $E = \{a, b, c, d\}$ be universe set and parameter set respectively (Fig. 4.4 is a Chained soft graph).

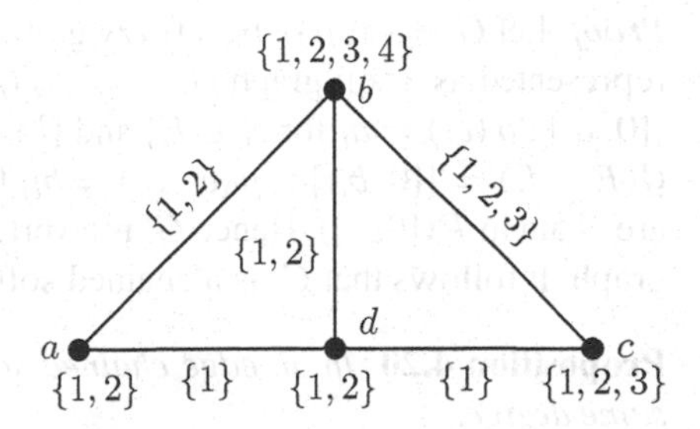

Fig. 4.4 Chained soft graph

Remark 4.10 From the Example 60, we can see that there are four $a-c$ paths as shown in the table.

	Path	Strength
P_1	$a-d-c$	$\{1\}$
P_2	$a-d-b-c$	$\{1\}$
P_3	$a-b-c$	$\{1,2\}$
P_4	$a-b-d-c$	$\{1\}$

Also, the strength of the path P_3 is $S(P_3)=\{1,2\}$ and $S(P_3)\supseteq S(P_i)$, for $i=1,2,4$. Thus P_3 is the strongest path between a and c.

Proposition 4.18 *Every complete vertex chained soft graph is an edge chained soft graph.*

Proof Let $G=(\mathcal{F},\mathcal{G})$ be a complete vertex chained soft graph. Let (a_1,b_1), (a_2,b_2) be any two edges in G,

$$\mathcal{G}(a_1,b_1)=\mathcal{F}(a_1)\cap\mathcal{F}(b_1)$$
$$\mathcal{G}(a_2,b_2)=\mathcal{F}(a_2)\cap\mathcal{F}(b_2)$$

Since $\mathcal{F}(A)$ is a chain, it follows that

$$\mathcal{G}(a_1,b_1)=\mathcal{F}(a_1)\text{ or }\mathcal{F}(b_1)$$
$$\mathcal{G}(a_2,b_2)=\mathcal{F}(a_2)\text{ or }\mathcal{F}(b_2)$$

Thus $\mathcal{G}(a_1,b_1)\subseteq\mathcal{G}(a_2,b_2)$ or $\mathcal{G}(a_2,b_2)\subseteq\mathcal{G}(a_1,b_1)$. Hence G is an edge chained soft graph. □

Proposition 4.19 *Every fuzzy graph can be represented by a chained soft graph.*

Proof Let $G=(\sigma,\mu)_E$ be a fuzzy graph on E. Then by Proposition 4.11, G can be represented as a soft graph $G'=(\mathcal{F},\mathcal{G})_E$, where $\mathcal{F}$ is a soft set defined by $\mathcal{F}(E)=\{[0,a_i]:\sigma(e_i)=a_i\text{ for }e_i\in E\}$ and $\mathcal{G}$ is a soft relation on the soft set $\mathcal{F}$ defined as $\mathcal{G}(E\times E)=\{[0,b_{ij}]:\ \mu(e_i,e_j)=b_{ij}\text{ for }e_i,e_j\in E\ \}$. Then $\mathcal{F}(E)$ and $\mathcal{G}(E\times E)$ are chain in $\mathcal{P}([0,1])$. Hence G' is a vertex chained soft graph and edge chained soft graph. It follows that G' is a chained soft graph. □

Proposition 4.20 *In an edge chained soft graph, there are at least two vertices of same degree.*

Proof Let $G=(\mathcal{F},\mathcal{G})_E$ be an edge chained soft graph, then $\mathcal{G}(E\times E)$ is a chain and it has a maximal element, say $\mathcal{G}(a,b)$. Hence $d(a)=d(b)=\mathcal{G}(a,b)$, which proves the result. □

Proposition 4.21 *Every connected edge chained soft graph is a strongly connected soft graph.*

Proof Let $G = (\mathcal{F}, \mathcal{G})$ be a connected edge chained soft graph and let a and b be any two vertices in G. Since G is connected soft graph, there exists a path P $(a = a_0, a_1, a_2, \ldots, a_n = b)$ with $\mathcal{G}(a_{i-1}, a_i) \neq \emptyset$, for all $i = 1, 2, \ldots, n$. Since $\mathcal{G}(E \times E)$ is a chain, $\bigcap_{i=1}^{n} \mathcal{G}(a_{i-1}, a_i) \neq \emptyset$. Thus P is a flow path and hence G is a strongly connected soft graph. □

4.4 Category of Soft Sets

Many researchers studied the category of soft sets using different definitions of morphisms. Borzooei et al.[24] explored the concept of the category of soft sets using a pair of functions as morphisms. Zhou et al. [167] studied category of soft sets with the help of a single function as morphism. In [121], Sardar and Gupta came forward with soft category using the parameterization technique.

4.4.1 Basic Notions of Category Theory

The fundamentals of the category theory can be found in [2, 59, 77, 113]. The basic terminologies are taken from Adamek et al. [2] unless otherwise cited. In this subsection, we recall certain concepts from category theory.

Definition 4.17 A category is a quadruple $\mathbf{A} = (\mathcal{O}, Mor, id, \circ)$ consisting of

1. a class $\mathcal{O}$, whose members are called A-objects
2. for each pair (A, B) of A-objects, a set $Mor(A, B)$(hom-set), whose members are called $A - morphisms$ from A to B
3. for each A-object A, a morphism $A \xrightarrow{id_A} A$ called the A-identity on A,
4. a composition law associating with each A-morphism $A \xrightarrow{f} B$ and each A-morphism $B \xrightarrow{g} C$ an A-morphism $A \xrightarrow{g \circ f} C$ called the composite of f and g, subject to the following conditions:

   (a) composition is associative; i.e., for morphisms $A \xrightarrow{f} B$, $B \xrightarrow{f} C$ and $C \xrightarrow{f} D$, the equation $h \circ (g \circ f) = (h \circ g) \circ f$ holds,
   (b) A-identities act as identities with respect to composition; i.e., for A-morphisms $A \xrightarrow{f} B$ we have $id_B \circ f = f$ and $f \circ id_A = f$,
   (c) the sets $Mor(A, B)$ are pairwise disjoint.

Example 61 The following are examples of category,

(a) **Set** with objects all sets and morphisms all functions between them.
(b) **Vec** with objects all real vector spaces and morphisms all linear transformations between them.

(c) **Grp** with objects all groups and morphisms all homomorphisms between them.
(d) **Rel** with objects all pairs (X, ρ) where X is a set and ρ is a (binary) relation on X. Morphisms $f : (X, \rho) \to (Y, \sigma)$ are relation-preserving maps; i.e., maps $f : X \to Y$ such that $x\rho x'$ implies $f(x)\sigma f(x')$.

Definition 4.18 A morphism $f : A \to B$ in a category is called an isomorphism provided that there exists a morphism $g : B \to A$ with $g \circ f = id_A$ and $f \circ g = id_B$. Such a morphism g is called an inverse of f.

Definition 4.19 Objects A and B in a category are said to be isomorphic provided that there is an isomorphism $f : A \to B$.

Definition 4.20 If A and B are categories, then a functor F from A to B is a function that assigns to each A-object A a B-object $F(A)$, and to each A-morphism $A \xrightarrow{f} A'$ a B-morphism $F(A) \xrightarrow{F(f)} F(A')$, in such a way that

(1) F preserves composition; i.e., $F(f \circ g) = F(f) \circ F(g)$ whenever $f \circ g$ is defined, and
(2) F preserves identity morphisms; i.e., $F(id_A) = id_{F(A)}$ for each A-object A.

Example 62 The following are examples of functor,

(1) For any category **A**, there is the identity functor $id_{\mathbf{A}} : \mathbf{A} \to \mathbf{A}$ defined by

$$id_{\mathbf{A}}(A \xrightarrow{f} B) = A \xrightarrow{f} B$$

(2) For any categories **A** and **B** and any B-object B, there is the constant functor $C_B : \mathbf{A} \to \mathbf{B}$ with value B, defined by

$$C_B(A \xrightarrow{f} A') = B \xrightarrow{id_B} B.$$

(3) For any category **A** mentioned in the Example 61, there is the forgetful functor (or underlying functor) $U : \mathbf{A} \to \mathbf{Set}$, where in each case $U(A)$ is the underlying set of A, and $U(f) = f$ is the underlying function of the morphism f.

Definition 4.21 An object A is said to be an initial object provided that for each object B there is exactly one morphism from A to B.

Definition 4.22 An object A is called a terminal object provided that for each object B there is exactly one morphism from B to A

Definition 4.23 An object A is called a zero object provided that it is both an initial object and a terminal object.

Definition 4.24 An object S is called a separator provided that whenever $A \overset{f}{\underset{g}{\rightrightarrows}} B$ are distinct morphisms, there exists a morphism $S \xrightarrow{h} A$ such that

$$S \xrightarrow{h} A \xrightarrow{f} B \neq S \xrightarrow{h} A \xrightarrow{g} B.$$

Definition 4.25 An object C is a coseparator provided that whenever $B \underset{g}{\overset{f}{\rightrightarrows}} A$ are distinct morphisms, there exists a morphism $A \xrightarrow{h} C$ such that

$$B \xrightarrow{f} A \xrightarrow{h} C \neq B \xrightarrow{g} A \xrightarrow{h} C$$

Definition 4.26 A morphism $A \xrightarrow{f} B$ is called a section provided that there exists some morphism $B \xrightarrow{g} A$ such that $g \circ f = id_A$ (i.e., provided that f has a left-inverse).

Definition 4.27 A morphism $A \xrightarrow{f} B$ is called a retraction provided that there exists some morphism $B \xrightarrow{g} A$ such that $f \circ g = id_B$ (i.e., provided that f has a right-inverse). If there exists such a retraction, then B will be called a retract of A.

Definition 4.28 A morphism $A \xrightarrow{f} B$ is said to be a monomorphism provided that for all pairs $C \underset{k}{\overset{h}{\rightrightarrows}} A$ of morphisms such that $f \circ h = f \circ k$, it follows that $h = k$ (i.e., f is left-cancellable with respect to composition).

Definition 4.29 A morphism $A \xrightarrow{f} B$ is said to be a epimorphism provided that for all pairs $B \underset{k}{\overset{h}{\rightrightarrows}} C$ of morphisms such that $h \circ f = k \circ f$, it follows that $h = k$ (i.e., f is right-cancellable with respect to composition).

Definition 4.30 A morphism is called a bimorphism provided that it is simultaneously a monomorphism and an epimorphism. A category is called balanced provided that each of its bimorphisms is an isomorphism.

Definition 4.31 Let $A \underset{g}{\overset{f}{\rightrightarrows}} B$ be a pair of morphisms. A morphism $E \xrightarrow{e} A$ is called an equalizer of f and g provided that the following conditions hold:

1. $f \circ e = g \circ e$,
2. for any morphism $e' : E' \to A$ with $f \circ e' = g \circ e'$, there exists a unique morphism $\bar{e} : E' \to E$ such that $e' = e \circ \bar{e}$, i.e., such that the triangle commutes (Fig. 4.5).

Definition 4.32 A morphism $E \xrightarrow{e} A$ is called a regular monomorphism provided that it is an equalizer of some pair of morphisms.

Definition 4.33 Let $A \underset{g}{\overset{f}{\rightrightarrows}} B$ be a pair of morphisms. A morphism $B \xrightarrow{c} C$ is called an coequalizer of f and g provided that the following conditions hold:

Fig. 4.5 Equalizer

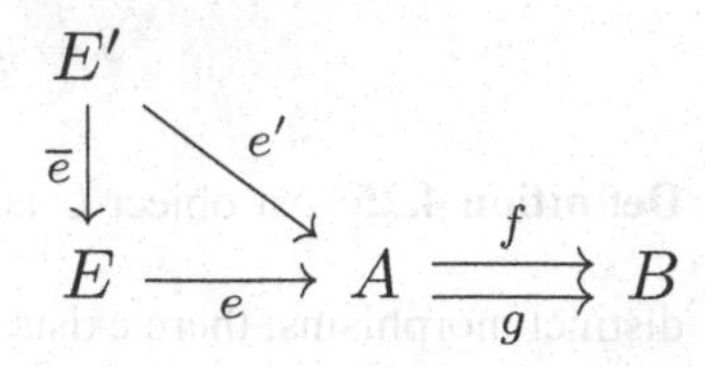

Fig. 4.6 Coequalizer

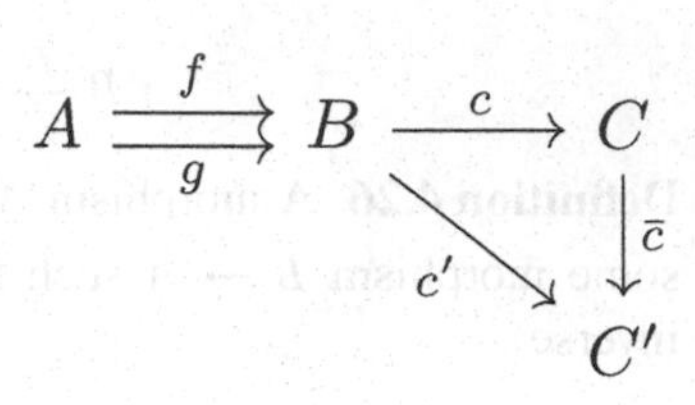

1. $c \circ f = c \circ g$,
2. for any morphism $c' : B \to C'$ with $c' \circ f = c' \circ g$, there exists a unique morphism $\bar{c} : C \to C'$ such that $c' = \bar{c} \circ c$, i.e., such that the triangle commutes (Fig. 4.6).

Definition 4.34 A morphism $B \xrightarrow{c} C$ is called a regular epimorphism provided that it is a coequalizer of some pair of morphisms.

Definition 4.35 Let $\{A_i\}_{i \in I}$ be an indexed family of objects in a category **C**. A product of $\{A_i\}_{i \in I}$ is an object A together with morphisms $f_i : A \to A_i$ with the property that for any morphism $g_i : B \to A_i$, $i \in I$, there exists a unique morphism $h : B \to A$ such that $f_i \circ h = g_i$, for all $i \in I$.

Definition 4.36 Let $\{A_i\}_{i \in I}$ be an indexed family of objects in a category **C**. A coproduct of $\{A_i\}_{i \in I}$ is an object A together with morphisms $f_i : A_i \to A$ with the property that for any morphism $g_i : A_i \to B$, $i \in I$, there exists a unique morphism $h : A \to B$ such that $h \circ f_i = g_i$, for all $i \in I$.

Definition 4.37 Given morphisms $f : A \to B$ and $g : C \to B$, a pullback is an object D with morphisms $p : D \to A$ and $q : D \to C$ such that $f \circ p = g \circ q$ satisfy the universal property i.e., for any object E with morphisms $r : E \to A$ and $s : E \to C$ such that $f \circ r = g \circ s$, then there exists a unique morphism $h : E \to D$ such that $p \circ h = r$ and $q \circ h = s$.

Definition 4.38 Given morphisms $f : B \to A$ and $g : B \to C$, a pushout is an object D with morphisms $p : A \to D$ and $q : C \to D$ such that $p \circ f = q \circ g$ satisfy the universal property i.e., for any object E with morphisms $r : A \to E$ and $s : C \to E$ such that $r \circ f = s \circ g$, then there exists a unique morphism $h : D \to E$ such that $h \circ p = r$ and $h \circ q = s$.

Definition 4.39 A category **A** is called a complete (cocomplete) category, if for each small diagram in **A** there exists a limit (colimit).

$$\begin{array}{ccc} FA & \xrightarrow{\tau_A} & GA \\ {\scriptstyle Ff}\downarrow & & \downarrow{\scriptstyle Gf} \\ FA' & \xrightarrow[\tau_{A'}]{} & GA' \end{array}$$

Fig. 4.7 Natural transformation

Theorem 4.3 *A category* **A** *is complete (cocomplete) if and only if* **A** *has products (coproducts) and equalizers (coequalizer).*

Proposition 4.22 *If* $F : \mathbf{A} \to \mathbf{B}$ *and* $G : \mathbf{B} \to \mathbf{C}$ *are functors, then the composite* $G \circ F : \mathbf{A} \to \mathbf{C}$ *defined by*

$$(G \circ F)(A \xrightarrow{f} A') = G(FA) \xrightarrow{G(Ff)} G(FA')$$

is a functor and is denoted by GF.

Definition 4.40 A functor $F : \mathbf{A} \to \mathbf{B}$ is called an isomorphism provided that there is a functor $G : \mathbf{B} \to \mathbf{A}$ such that $G \circ F = id_{\mathbf{A}}$ and $F \circ G = id_{\mathbf{B}}$.

Definition 4.41 Let $F : \mathbf{A} \to \mathbf{B}$ be a functor, then F is called faithful provided that all the Mor-set restrictions $F : Mor_{\mathbf{A}}(A, A') \to Mor_{\mathbf{B}}(FA, FA')$ are injective.

Definition 4.42 Let **X** be a category. A concrete category over **X** is a pair (**A**,U), where **A** is a category and $U : \mathbf{A} \to \mathbf{X}$ is a faithful functor. A concrete category over **Set** is called a construct.

Definition 4.43 If (**A**,U) and (**B**,V) are concrete category over **X**, then a concrete functor from (**A**,U) to (**B**,V) is a functor $F : \mathbf{A} \to \mathbf{B}$ with $U = V \circ F$.

Proposition 4.23 *Every concrete functor is faithful.*

Definition 4.44 Let $F, G : \mathbf{A} \to \mathbf{B}$ be functors. A natural transformation τ from F to G (denoted by $\tau : F \to G$) is a function that assigns to each **A**-object A a **B**-morphism $\tau_A : FA \to GA$ in such a way that the following naturality condition holds: for each **A**-morphism $A \xrightarrow{f} A'$, the square commutes (Fig. 4.7).

If τ_A is an isomorphism for each A, then τ is called a natural equivalence.

Definition 4.45 If $G, G' : \mathbf{A} \to \mathbf{B}$ are functors and $G \xrightarrow{\tau} G'$ is a natural transformation, then

1. $\tau_F : G \circ F \to G' \circ F$ is defined by $(\tau_F)_C = \tau_{FC}$ is a natural transformation, for each functor $F : \mathbf{C} \to \mathbf{A}$,
2. $H_\tau : H \circ G \to H \circ G'$ is defined by $(H_\tau)_A = H(\tau_A)$ is a natural transformation, for each functor $H : \mathbf{B} \to \mathbf{D}$.

Definition 4.46 Let **A** and **B** be concrete categories over **X**. If $G : \mathbf{A} \to \mathbf{B}$ and $F : \mathbf{B} \to \mathbf{A}$ are concrete functors over **X**, then the pair (F, G) is called a Galois correspondence (between **A** and **B** over **X**) provided that $F \circ G \leq id_A$ and $id_B \leq G \circ F$.

Definition 4.47 ([50]) Suppose **C** and **D** are two categories. Let $F : \mathbf{C} \to \mathbf{D}$ and $G : \mathbf{D} \to \mathbf{C}$ be functors such that there is a natural equivalence of functors, $\eta = \eta_{XY} : \mathbf{D}(FX, Y) \to \mathbf{C}(X, GY)$. We then say that F is left adjoint to G, G is right adjoint to F, and write $\eta : F \dashv G$. We call η the adjugant equivalence or adjugant.

Theorem 4.4 ([50]) *Suppose* ***C*** *and* ***D*** *are two categories. Let* $F : \boldsymbol{C} \to \boldsymbol{D}$ *and* $G : \boldsymbol{D} \to \boldsymbol{C}$ *be functors and let* $\epsilon : Id_C \to GF$, $\delta : FG \to Id_D$ *be natural transformations such that* $\delta_F \circ F_\epsilon = Id_F$, $G_\delta \circ \epsilon_G = Id_G$. *Then the function* $\eta = \eta_{XY} : \boldsymbol{D}(FX, Y) \to \boldsymbol{C}(X, GY)$ *defined by* $\eta(\phi) = G\phi \circ \epsilon_X$, *for* $\phi : FX \to Y$ *is a natural equivalence, so that* $\eta : F \dashv G$. *Moreover,* ϵ, δ *are the unit and counit of the adjugant* η.

4.4.2 Objects in Category Sset(U)

Definition 4.48 ([167]) Let $\mathcal{F}_A$ and $\mathcal{G}_B$ be two soft sets over U. A soft morphism from $\mathcal{F}_A$ to $\mathcal{G}_B$ is a function $\alpha : A \to B$ such that $\mathcal{F}(a) \subseteq (\mathcal{G} \circ \alpha)(a)$, for all $a \in A$.

Remark 4.11 The soft morphism mentioned above is same as the soft function in [167]. In category theory context, it is more convenient to call soft function as soft morphism.

Example 63 Let $U = \{1, 2, \ldots, 10\}$ be a universe set and let $A = \{a_1, a_2, a_3\}$ and $B = \{b_1, b_2\}$ be two parameter sets. Two soft sets $\mathcal{F}_A(U)$ and $\mathcal{G}_B(U)$ are defined by

$$\mathcal{F}(a_1) = \{1, 5\} \quad \mathcal{F}(a_2) = \{4, 5, 8\} \quad \mathcal{F}(a_3) = \{1, 5, 6\} \text{ and}$$
$$\mathcal{G}(b_1) = \{1, 4, 5, 7, 8\} \; \mathcal{G}(b_2) = \{1, 3, 5, 6, 9\}$$

Define $\alpha : A \to B$ by $\alpha(a_1) = b_1, \alpha(a_2) = b_1$ and $\alpha(a_3) = b_2$. Then,
$(\mathcal{G} \circ \alpha)(a_1) = \mathcal{G}(b_1) = \{1, 4, 5, 7, 8\} \supseteq \{1, 5\} = \mathcal{F}(a_1)$
$(\mathcal{G} \circ \alpha)(a_2) = \mathcal{G}(b_1) = \{1, 4, 5, 7, 8\} \supseteq \{4, 5, 8\} = \mathcal{F}(a_2)$
$(\mathcal{G} \circ \alpha)(a_3) = \mathcal{G}(b_2) = \{1, 3, 5, 6, 9\} \supseteq \{1, 5, 6\} = \mathcal{F}(a_3)$
It follows that $\mathcal{F}(a) \subseteq (\mathcal{G} \circ \alpha)(a)$, for all $a \in A$. Hence α is a soft morphism between $\mathcal{F}_A$ and $\mathcal{G}_B$.

Remark 4.12 The soft morphism mentioned in the Definition 4.48 and that in [24] and [163] are different.

Remark 4.13 Let $\alpha : \mathcal{F}_A \to \mathcal{G}_B$ and $\beta : \mathcal{G}_B \to \mathcal{H}_C$ be two soft morphisms, then their composition $\beta \circ \alpha : \mathcal{F}_A \to \mathcal{H}_C$ is again a soft morphism. For this, $[\mathcal{H} \circ (\beta \circ \alpha)](a) = (\mathcal{H} \circ \beta)(\alpha(a)) \supseteq \mathcal{G}(\alpha(a)) \supseteq \mathcal{F}(a)$, for all $a \in A$.

Let **Sset(U)** be the class of all soft sets over U. Each soft morphism $\alpha : \mathcal{F}_A \to \mathcal{G}_B$ in $Mor(\mathcal{F}_A, \mathcal{G}_B)$ is considered as a triple $(\mathcal{F}_A, \alpha, \mathcal{G}_B)$. The composition of soft morphisms is associative and identity function id_A is the identity soft morphism on the soft set $\mathcal{F}_A$. This establishes that **Sset(U)** is a category.

Proposition 4.24 *An object $\mathcal{F}_A \in$ **Sset(U)** is an initial object if and only if $A = \emptyset$.*

Proof Let $\mathcal{F}_A$ be an initial object in the category of soft sets, **Sset(U)**. Then $|Mor(\mathcal{F}_A, \mathcal{G}_B)| = 1$, for every soft set $\mathcal{G}_B$ in **Sset(U)**. If $A \neq \emptyset$, then $|Mor(\mathcal{F}_A, \mathcal{G}_B)| > 1$, for every absolute soft set $\mathcal{G}_B$. It follows that A is a null set.

Conversely, suppose that $A = \emptyset$, then there exists exactly one function from empty set to any set. It follows that $\mathcal{F}_A$ is an initial object in **Sset(U)**. □

Proposition 4.25 *An object $\mathcal{F}_A \in$ **Sset(U)** is a terminal object if and only if $\mathcal{F}_A$ is an absolute soft set with $|A| = 1$.*

Proof Let $\mathcal{F}_A$ be an absolute soft set with $|A| = 1$, then there is exactly one soft morphism (constant function) between any soft set $\mathcal{G}_B$ and $\mathcal{F}_A$.

Conversely, suppose that $|A| > 1$, then $|Mor(\mathcal{G}_B, \mathcal{F}_A)| > 1$ for every null soft set $\mathcal{G}_B$, which is a contradiction. Thus $|A| = 1$. Also, if $\mathcal{F}_A$ is not an absolute soft set, then there is no soft morphism between absolute soft set $\mathcal{G}_B$ and $\mathcal{F}_A$. Thus $\mathcal{F}_A$ must be an absolute soft set with $|A| = 1$. □

Remark 4.14 The category **Sset(U)** has no zero object. Since **Sset(U)** has no object which is both initial and terminal object.

Theorem 4.5 *An object $\mathcal{H}_C$ in **Sset(U)** is a separator if and only if $\mathcal{H}$ is a null soft set with $C \neq \emptyset$.*

Proof Suppose $\mathcal{H}_C$ is a null soft set with $C \neq \emptyset$. Let $\mathcal{F}_A$ and $\mathcal{G}_B$ be two object in **Sset(U)** and $\alpha, \beta : \mathcal{F}_A \to \mathcal{G}_B$ be two distinct soft morphisms, i.e., $\alpha(a) \neq \beta(a)$, for some $a \in A$.

Define $\gamma : C \to A$ by $\gamma(c) = a$, for all $c \in C$. By definition of $\mathcal{H}$, we get $\emptyset = \mathcal{H}(c) \subseteq (\mathcal{F} \circ \gamma)(c)$, for all $c \in C$. It implies that γ is a soft morphism from $\mathcal{H}_C$ to $\mathcal{F}_A$. Also $(\alpha \circ \gamma)(c) = \alpha(a)$ and $(\beta \circ \gamma)(c) = \beta(a)$. Thus $\alpha \circ \gamma \neq \beta \circ \gamma$ and hence $\mathcal{H}_C$ is a separator in **Sset(U)**.

Conversely, assume that $\mathcal{H}_C$ is a separator in **Sset(U)**. If $C = \emptyset$, then we have $\alpha \circ \gamma = \beta \circ \gamma$. Now assume that there exists $c \in C$ such that $\mathcal{H}(c) \neq \emptyset$, then there is no soft morphism between $\mathcal{H}_C$ and null soft set $\mathcal{F}_A$. Hence $\mathcal{H}_C$ is a null soft set with $C \neq \emptyset$. □

Theorem 4.6 *An object $\mathcal{H}_C$ in **Sset(U)** is a co-separator if and only if there exists $c_1, c_2 \in C$, $c_1 \neq c_2$ such that $\mathcal{H}(c_1) = \mathcal{H}(c_2) = U$.*

Proof Suppose $c_1, c_2 \in C, c_1 \neq c_2$ with $\mathcal{H}(c_1) = \mathcal{H}(c_2) = U$. Let $\alpha, \beta : \mathcal{F}_A \to \mathcal{G}_B$ be two distinct soft morphisms, i.e., $\alpha(a_1) \neq \beta(a_1)$, for some $a_1 \in A$.
Define

$$\gamma : B \to C \text{ by } \gamma(b) = \begin{cases} c_1 & \text{if } b = \alpha(a_1); \\ c_2 & \text{otherwise} \end{cases}$$

Then $\gamma : \mathcal{G}_B \to \mathcal{H}_C$ is a soft morphism, since $\mathcal{G}(b) \subseteq U = \mathcal{H}(c_1) = \mathcal{H}(c_2)$. Also, $(\gamma \circ \alpha)(a_1) = c_1 \neq c_2 = (\gamma \circ \beta)(a_1)$. Hence $\mathcal{H}_C$ is a co-separator.

Conversely, suppose $C = \emptyset$, then there is no soft morphism between $\mathcal{G}_B$ and $\mathcal{H}_C$ and if $|C| = 1$, then $\gamma \circ \alpha = \gamma \circ \beta$. Therefore C contains at least two distinct points. Since $\mathcal{H}_C$ is a co-separator, there exists a soft morphism $\gamma : \mathcal{G}_B \to \mathcal{H}_C$ such that $\gamma \circ \alpha \neq \gamma \circ \beta$, i.e., $(\gamma \circ \alpha)(a) = c_1 \neq c_2 = (\gamma \circ \beta)(a)$, for some $a \in A$. Now let $\alpha(a) = b_1$ and $\beta(a) = b_2$ and assume that $\mathcal{G}$ is an absolute soft set, then $\mathcal{H}(c_1) = (\mathcal{H} \circ \gamma)(b_1) = U$ and $\mathcal{H}(c_2) = (\mathcal{H} \circ \gamma)(b_2) = U$. □

4.4.3 Morphisms in Sset(U)

Theorem 4.7 *A soft morphism $\alpha : \mathcal{F}_A \to \mathcal{G}_B$ is a section if and only if*

(a) α is injective
(b) $\mathcal{F}(a) = (\mathcal{G} \circ \alpha)(a)$, for all $a \in A$
(c) for any $b \in B$, there exists $a \in A$ such that $\mathcal{G}(b) \subseteq \mathcal{F}(a)$.

Proof Suppose that $\alpha : \mathcal{F}_A \to \mathcal{G}_B$ is a section, then there is a soft morphism $\beta : \mathcal{G}_B \to \mathcal{F}_A$ with $\beta \circ \alpha = id_{\mathcal{F}_A}$. We have

(a)

$$\begin{aligned} \alpha(a_1) = \alpha(a_2) &\implies (\beta \circ \alpha)(a_1) = (\beta \circ \alpha)(a_2) \\ &\implies id_{\mathcal{F}_A}(a_1) = id_{\mathcal{F}_A}(a_2) \\ &\implies a_1 = a_2 \end{aligned}$$

This shows that α is injective

(b) Since $\alpha : \mathcal{F}_A \to \mathcal{G}_B$ and $\beta : \mathcal{G}_B \to \mathcal{F}_A$ are soft morphisms, then we have $\mathcal{F}(a) \subseteq (\mathcal{G} \circ \alpha)(a)$ and $\mathcal{G}(b) \subseteq (\mathcal{F} \circ \beta)(b)$, for all $a \in A, b \in B$. It implies that $(\mathcal{G} \circ \alpha)(a) \subseteq (\mathcal{F} \circ \beta \circ \alpha)(a)$, for all $a \in A$. Since $\beta \circ \alpha = id_{\mathcal{F}_A}$, we see that $(\mathcal{G} \circ \alpha)(a) \subseteq \mathcal{F}(a)$, for all $a \in A$. Thus $\mathcal{F}(a) = (\mathcal{G} \circ \alpha)(a)$, for all $a \in A$.
(c) Since $\beta : \mathcal{G}_B \to \mathcal{F}_A$ is a soft morphism, we have $\mathcal{G}(b) \subseteq (\mathcal{F} \circ \beta)(b)$, for all $b \in B$. Let $b \in B$, take $a = \beta(b)$ and therefore $\mathcal{G}(b) \subseteq \mathcal{F}(a)$.

Conversely, assume that all conditions of the theorem are satisfied.
Define $\beta : B \to A$ as

$$\beta(b) = \begin{cases} \alpha^{-1}(b) & \text{if } b \in \alpha(A); \\ a_b \in A \text{ with } \mathcal{G}(b) \subseteq \mathcal{F}(a_b) & \text{if b } \in B \setminus \alpha(A) \end{cases}$$

Claim: $\beta : \mathcal{G}_B \to \mathcal{F}_A$ is a soft morphism and $\beta \circ \alpha = id_{\mathcal{F}_A}$.

Case I: If $b \in \alpha(A)$, then $\beta(b) = a$, where $\alpha(a) = b$. By condition (b), it follows that $\mathcal{F}(a) = (\mathcal{G} \circ \alpha)(a) = \mathcal{G}(b)$. Then $(\mathcal{F} \circ \beta)(b) = \mathcal{F}(a) = \mathcal{G}(b)$. Also note that, $(\beta \circ \alpha)(a) = \beta(b) = a = id_{\mathcal{F}_A}(a)$.

Case II: If $b \in B \setminus \alpha(A)$. Using the definition of β, we get $\mathcal{G}(b) \subseteq \mathcal{F}(a_b) = (\mathcal{F} \circ \beta)(b)$, where $\beta(b) = a_b$.

Thus $\beta : \mathcal{G}_B \to \mathcal{F}_A$ is a soft morphism and $(\beta \circ \alpha)(a) = id_{\mathcal{F}_A}(a)$, for all $a \in A$. Hence $\alpha : \mathcal{F}_A \to \mathcal{G}_B$ is a section. □

Theorem 4.8 *A soft morphism $\alpha : \mathcal{F}_A \to \mathcal{G}_B$ is a retraction if and only if for any $b \in B$, there exists $a \in A$ such that $\alpha(a) = b$ and $\mathcal{F}(a) = \mathcal{G}(b)$.*

Proof Let $\alpha : \mathcal{F}_A \to \mathcal{G}_B$ be a retraction, then there exists a soft morphism $\beta : \mathcal{G}_B \to \mathcal{F}_A$ such that $\alpha \circ \beta = id_{\mathcal{G}_B}$. Let $b_0 \in B$, we have

$$\begin{aligned} (\alpha \circ \beta)(b_0) = id_{\mathcal{G}_B}(b_0) &\Longrightarrow \alpha(\beta(b_0)) = b_0 \\ &\Longrightarrow \alpha(a_0) = b_0, \quad \text{where } \beta(b_0) = a_0 \end{aligned}$$

Which implies that there exits $a_0 \in A$ such that $\alpha(a_0) = b_0$.

Since α and β are soft morphisms, then $\mathcal{F}(a) \subseteq (\mathcal{G} \circ \alpha)(a)$ and $\mathcal{G}(b) \subseteq (\mathcal{F} \circ \beta)(b)$, for all $a \in A, b \in B$. This implies that $\mathcal{F}(a_0) \subseteq \mathcal{G}(\alpha(a_0))$ and $\mathcal{G}(b_0) \subseteq \mathcal{F}(\beta(b_0))$, for $a_0 \in A, b_0 \in B$. It follows that $\mathcal{F}(a_0) \subseteq \mathcal{G}(b_0)$ and $\mathcal{G}(b_0) \subseteq \mathcal{F}(a_0)$ and hence $\mathcal{F}(a_0) = \mathcal{G}(b_0)$.

Now conversely suppose that, all the conditions of the theorem are satisfied. Define $\beta : B \to A$ by $\beta(b) = a$, where $a \in \{c \in A : \alpha(c) = b \text{ and } \mathcal{F}(c) = \mathcal{G}(b)\}$. Let $b_0 \in B$ and $\beta(b_0) = a_0$, then $(\mathcal{F} \circ \beta)(b_0) = \mathcal{F}(a_0) = \mathcal{G}(b_0)$. Also, we get $(\alpha \circ \beta)(b_0) = \alpha(a_0) = b_0 = id_{\mathcal{G}_A}(b_0)$.

Thus $\beta : \mathcal{G}_B \to \mathcal{F}_A$ is soft morphism and $(\alpha \circ \beta)(b) = id_{\mathcal{G}_A}(b)$, for all $b \in B$. It follows that $\alpha : \mathcal{F}_A \to \mathcal{G}_B$ is a retraction. □

Theorem 4.9 *A soft morphism $\alpha : \mathcal{F}_A \to \mathcal{G}_B$ is a monomorphism if and only if α is injective.*

Proof Suppose α is a monomorphism. Let $a_1, a_2 \in A$ and let $C = \{c\}$ and $\mathcal{H}(c) = \emptyset$. Define $\beta, \gamma : C \to A$ by $\beta(c) = a_1$ and $\gamma(c) = a_2$. By definition of $\mathcal{H}$, β and γ are soft morphisms from $\mathcal{H}_C$ to $\mathcal{F}_A$. For $a_1, a_2 \in A$,

$$\begin{aligned} \alpha(a_1) = \alpha(a_2) &\Longrightarrow (\alpha \circ \beta)(c) = (\alpha \circ \gamma)(c) \\ &\Longrightarrow \beta(c) = \gamma(c), \quad \text{since } \alpha \text{ is a monomorphism} \\ &\Longrightarrow a_1 = a_2 \end{aligned}$$

Which implies α is injective.

Conversely, suppose α is injective. Let $\beta, \gamma : \mathcal{H}_C \to \mathcal{F}_A$ be soft morphisms with $\alpha \circ \beta = \alpha \circ \gamma$. Now, for any $c \in C$

$$\begin{aligned}(\alpha \circ \beta)(c) = (\alpha \circ \gamma)(c) &\Longrightarrow \alpha(\beta(c)) = \alpha(\gamma(c)) \\ &\Longrightarrow \beta(c) = \gamma(c) \quad \text{Since } \alpha \text{ is injective} \\ &\Longrightarrow \beta = \gamma\end{aligned}$$

Hence α is a monomorphism. □

Remark 4.15 In any category $\mathcal{C}$, every section is a monomorphism [2]. But its converse is not true in general i.e., there exists monomorphisms which are not sections.

Example 64 Let $U = \{1, 2, \ldots, 10\}$ be a universe set and let $A = \{a_1, a_2\}$ and $B = \{b_1, b_2\}$ be the parameter sets. Two soft sets $\mathcal{F}_A$ and $\mathcal{G}_B$ defined by

$$\mathcal{F}(a_1) = \{1, 2\} \quad \mathcal{F}(a_2) = \{2, 3, 5\} \text{ and}$$
$$\mathcal{G}(b_1) = \{1, 2, 4\} \quad \mathcal{G}(b_2) = \{2, 3, 5, 6\}$$

Let us define $\alpha : A \to B$ by $\alpha(a_1) = b_1$ and $\alpha(a_2) = b_2$, then

$$\mathcal{F}(a_1) = \{1, 2\} \subseteq \{1, 2, 4\} = \mathcal{G}(b_1) = (\mathcal{G} \circ \alpha)(a_1) \text{ and}$$
$$\mathcal{F}(a_2) = \{2, 3, 5\} \subseteq \{2, 3, 5, 6\} = \mathcal{G}(b_2) = (\mathcal{G} \circ \alpha)(a_2)$$

Then $\alpha : \mathcal{F}_A \to \mathcal{G}_B$ is soft morphism and also injective, this implies that α is a monomorphism. But $\mathcal{F}(a_1) \neq (\mathcal{G} \circ \alpha)(a_1)$, implies that α is not a section.

Theorem 4.10 *A soft morphism $\alpha : \mathcal{F}_A \to \mathcal{G}_B$ is an epimorphism if and only if α is surjective.*

Proof Assume that $\alpha : \mathcal{F}_A \to \mathcal{G}_B$ is an epimorphism, Let $C = \{0, 1\}$ and $\mathcal{H}_C$ be an absolute soft set. Define $\beta, \gamma : B \to C$ by $\beta(B) = \{0\}$ and

$$\gamma(b) = \begin{cases} 0 \text{ if } b = \alpha(a), \text{ for some } a \in A; \\ 1 \text{ otherwise} \end{cases}$$

Then $(\mathcal{H} \circ \beta)(b) = \mathcal{H}(0) = U \supseteq \mathcal{G}(b)$ and $(\mathcal{H} \circ \gamma)(b) = \mathcal{H}(0 \text{ } or \text{ } 1) = U \supseteq \mathcal{G}(b)$. Thus $\beta, \gamma : \mathcal{G}_B \to \mathcal{H}_C$ are soft morphisms with $\beta \circ \alpha = \gamma \circ \alpha$, which implies $\beta = \gamma$, since α is an epimorphism. From the definition of β and γ, $\beta(B) = \{0\} = \gamma(\alpha(A))$ and hence $\alpha(A) = B$. This shows that α is surjective.

Now conversely assume that α is surjective. Let $\beta, \gamma : \mathcal{G}_B \to \mathcal{H}_C$ be soft morphisms with $\beta \circ \alpha = \gamma \circ \alpha$. Let $b \in B$, then there exists $a \in A$ such that $\alpha(a) = b$. For any $b \in B$, we have

$$\beta(b) = \beta(\alpha(a)) = (\beta \circ \alpha)(a) = (\gamma \circ \alpha)(a) = \gamma(\alpha(a)) = \gamma(b)$$

Which implies $\beta = \gamma$. Hence $\alpha : \mathcal{F}_A \to \mathcal{G}_B$ is an epimorphism. □

Remark 4.16 In any category $\mathcal{C}$, every retraction is an epimorphism [2]. But it converse is not true in general i.e., there exists epimorphisms which are not retractions.

Example 65 In Example 64, the morphism $\alpha : \mathcal{F}_A \to \mathcal{G}_B$ is an epimorphism, since it is surjective. But for $b_1 \in B$, there is no $a_i \in A$ such that $\alpha(a_i) = b_1$ and $\mathcal{F}(a_i) = \mathcal{G}(b_1)$. It implies that α is not a retraction.

Corollary 4.1 *In Sset(U), the bimorphisms are precisely the bijective soft morphisms.*

Proof A bimorphism in a category is a morphism which is both monomorphism and epimorphism. In light of Theorems 4.10 and 4.9, we get bimorphisms in **Sset(U)** are bijective soft morphisms. □

Theorem 4.11 *A soft morphism $\alpha : \mathcal{F}_A \to \mathcal{G}_B$ is an isomorphism if and only if α is bijective and $\mathcal{F}(a) = (\mathcal{G} \circ \alpha)(a)$, for all $a \in A$.*

Proof Suppose $\alpha : \mathcal{F}_A \to \mathcal{G}_B$ is an isomorphism, then there exists a soft morphism $\beta : \mathcal{G}_B \to \mathcal{F}_A$ such that $\beta \circ \alpha = id_{\mathcal{F}_A}$ and $\alpha \circ \beta = id_{\mathcal{G}_B}$, i.e., $\beta \circ \alpha = id_A$ and $\alpha \circ \beta = id_B$ are identity functions. It follows that α is bijective. Let $a \in A$ and $\alpha(a) = b$, then $\beta(b) = a$. Since α and β are soft morphisms, we have

$$\mathcal{F}(a) \subseteq (\mathcal{G} \circ \alpha)(a) = \mathcal{G}(b) \subseteq (\mathcal{F} \circ \beta)(b) = \mathcal{F}(a)$$

Hence $\mathcal{F}(a) = (\mathcal{G} \circ \alpha)(a)$, for all $a \in A$.

Now conversely assume that α is bijective and $\mathcal{F}(a) = (\mathcal{G} \circ \alpha)(a)$, for all $a \in A$. Then there exist a function $\beta : B \to A$ such that $\beta \circ \alpha = id_A$ and $\alpha \circ \beta = id_B$. So $\beta \circ \alpha = id_{\mathcal{F}_A}$ and $\alpha \circ \beta = id_{\mathcal{G}_B}$ are identity soft morphisms. Hence α is an isomorphism. □

Remark 4.17 Two soft sets $\mathcal{F}_A$ and $\mathcal{G}_B$ are said to be isomorphic soft sets, if there exists an isomorphism $\alpha : \mathcal{F}_A \to \mathcal{G}_B$.

Remark 4.18 The category of soft sets, **Sset(U)** is not a balanced category.

Example 66 Consider the Example 65, the soft morphism $\alpha : \mathcal{F}_A \to \mathcal{G}_B$ is both monomorphism and epimorphism. It implies that α is a bimorphism. But $\mathcal{F}(a_1) \neq (\mathcal{G} \circ \alpha)(a_1)$ and $\mathcal{F}(a_2) \neq (\mathcal{G} \circ \alpha)(a_2)$. Hence α is not an isomorphism.

Corollary 4.2 *Let $\mathcal{F}_A$ and $\mathcal{G}_B$ be two isomorphic soft sets, then $|\mathcal{F}^S| = |\mathcal{G}^S|$, for all $S \in \mathcal{P}(U)$.*

Proof Suppose that $\mathcal{F}_A$ and $\mathcal{G}_B$ are two isomorphic soft sets, then there exists an isomorphism $\alpha : \mathcal{F}_A \to \mathcal{G}_B$. By Theorem 4.11, we get α is bijective and $\mathcal{F}(a) = (\mathcal{G} \circ \alpha)(a)$, for all $a \in A$.

The sets $\mathcal{F}^S$ and $\mathcal{G}^S$ are S- cuts of $\mathcal{F}$ and $\mathcal{G}$ respectively. Let us define $\phi : \mathcal{F}^S \to \mathcal{G}^S$ by $\phi = \alpha\big|_{\mathcal{F}^S}$. Since ϕ is a restriction of an injective function α, we get ϕ is injective.

Since $\mathcal{F}(a) = (\mathcal{G} \circ \alpha)(a)$, for all $a \in A$. Consider the S-cut $\mathcal{F}^S$ and $a \in \mathcal{F}^S$, we get $S \subseteq \mathcal{F}(a) = (\mathcal{G} \circ \alpha)(a) = \mathcal{G}(\alpha(a))$, it follows that $\alpha(a) \in \mathcal{G}^S$. Thus $\phi(\mathcal{F}^S) \subseteq \mathcal{G}^S$.

Let $b \in \mathcal{G}^S$, then we have $b \in B$ and there exists $a \in A$ such that $\alpha(a) = b$. Now we have $\mathcal{F}(a) = (\mathcal{G} \circ \alpha)(a) = \mathcal{G}(b) \supseteq S$. This implies $a \in \mathcal{F}^S$ and $\alpha(a) = \alpha(\mathcal{F}^S) = \phi(\mathcal{F}^S)$. Hence $b \in \phi(\mathcal{F}^S)$, which implies $\mathcal{G}^S \subseteq \phi(\mathcal{F}^S)$. Thus ϕ is surjective and hence $|\mathcal{F}^S| = |\mathcal{G}^S|$. □

Theorem 4.12 *The equalizer of two soft morphism $\alpha, \beta \in Mor(\mathcal{F}_A, \mathcal{G}_B)$ in the category of soft sets,* ***Sset(U)*** *is the soft morphism $i \in Mor(\mathcal{F}'_C, \mathcal{F}_A)$ (inclusion map), where $C = \{a \in A : \alpha(a) = \beta(a)\}$ and $\mathcal{F}'$ is the restriction of $\mathcal{F}$ to C.*

Proof Let $\alpha, \beta \in Mor(\mathcal{F}_A, \mathcal{G}_B)$ and $C = \{a \in A : \alpha(a) = \beta(a)\}$, let $i : C \to A$ be the inclusion map. Then $\mathcal{F}'(c) = \mathcal{F}(c) = (\mathcal{F} \circ i)(c)$, for all $c \in C$, which implies that $i : \mathcal{F}'_C \to \mathcal{F}_A$ is a soft morphism. Also note that $\alpha \circ i = \beta \circ i$.
Let $\gamma : \mathcal{H}_D \to \mathcal{F}_A$ be any soft morphism with $\alpha \circ \gamma = \beta \circ \gamma$. Since the image of γ is in C, there exists a map $j : D \to C$ such that $i \circ j = \gamma$. We have $\mathcal{H}(d) \subseteq (\mathcal{F} \circ \gamma)(d)$, for all $d \in D$ and $\mathcal{F} \circ \gamma = (\mathcal{F} \circ i) \circ j = \mathcal{F}' \circ j$. This shows that $\mathcal{H}(d) \subseteq (\mathcal{F}' \circ j)(d)$ for all $d \in D$, it establish that $j \in Mor(\mathcal{H}_D, \mathcal{F}'_C)$
Suppose $j' \in Mor(\mathcal{H}_D, \mathcal{F}'_C)$ satisfy $i \circ j' = \gamma$, then $i \circ j' = i \circ j$. Since i is injective, we have $j = j'$. □

Theorem 4.13 *Every two soft morphisms $\alpha, \beta \in Mor(\mathcal{F}_A, \mathcal{G}_B)$ have a coequalizer.*

Proof Let $\alpha, \beta \in Mor(\mathcal{F}_A, \mathcal{G}_B)$ and let θ be the equivalence relation generated by $\{(\alpha(a), \beta(a)) : a \in A\}$. Then let $C = B/\theta = \{[b] : b \in B\}$, where $[b]$ is the equivalence class generated by b.

Define a soft set $\mathcal{H}_C$ by $\mathcal{H}([b_0]) = \cup\{\mathcal{G}(b) : b \in [b_0]\}$ for all $[b_0] \in C$. Consider the natural map $\gamma : B \to C$ by $\gamma(b) = [b]$ for all $b \in B$. Let $b_1 \in B$, then we get $\mathcal{G}(b_1) \subseteq \cup\{\mathcal{G}(b) : b \in [b_1]\} = \mathcal{H}([b_1]) = (\mathcal{H} \circ \gamma)(b_1)$. It follows that $\gamma \in Mor(\mathcal{G}_B, \mathcal{H}_C)$. Moreover $(\gamma \circ \alpha)(a) = \gamma([\alpha(a)]) = \gamma([\beta(a)]) = (\gamma \circ \beta)(a)$ for all $a \in A$.

Let $\gamma' : \mathcal{G}_B \to \mathcal{I}_D$ be any soft morphism with $\gamma' \circ \alpha = \gamma' \circ \beta$. Then there exists a function $h : C \to D$ by $h([b]) = \gamma'(b)$, it follows that $(h \circ \gamma)(b) = h([b]) = \gamma'(b)$, for all $b \in B$. Let $b \in B$, then $\mathcal{G}(b) \subseteq (\mathcal{I} \circ \gamma')(b) = (\mathcal{I} \circ (h \circ \gamma))(b) = (\mathcal{I} \circ h)([b])$ and $\mathcal{H}([b_0]) = \cup\{\mathcal{G}(b) : b \in [b_0]\} \subseteq (\mathcal{I} \circ h)([b_0])$, for all $[b_0] \in C$. It follows that $h \in Mor(\mathcal{H}_C, \mathcal{I}_D)$ and hence γ is the coequalizer of α and β. □

Theorem 4.14 *A soft morphism $\gamma \in Mor(\mathcal{H}_C, \mathcal{F}_A)$ is a regular monomorphism if and only if γ is injective and $\mathcal{F} \circ \gamma = \mathcal{H}$.*

Proof Suppose $\gamma \in Mor(\mathcal{H}_C, \mathcal{F}_A)$ is a regular monomorphism, then γ is an equalizer of some soft morphisms $\alpha, \beta : \mathcal{F}_A \to \mathcal{G}_B$.

Let $D = \{a \in A : \alpha(a) = \beta(a)\}$, define a soft set $\mathcal{I}$ on D by $\mathcal{I}(d) = \mathcal{F}(d)$. We have $\gamma' : \mathcal{I}_D \to \mathcal{F}_A$ a soft morphism by $\gamma'(d) = d$, for all $d \in D$. Now $(\alpha \circ \gamma')(d) = \alpha(d) = \beta(d) = (\beta \circ \gamma')(d)$ and γ is an equalizer of α and β, then there exists isomorphism (by uniqueness of equalizer) $\eta : \mathcal{I}_D \to \mathcal{H}_C$ such that $\gamma \circ \eta = \gamma'$. Since γ' is injective, it implies that γ is injective.

Since $\gamma \in Mor(\mathcal{H}_C, \mathcal{F}_A)$, we have $\mathcal{H}(c) \subseteq (\mathcal{F} \circ \gamma)(c)$, for all $c \in C$. Let $c \in C$, there exists $d \in D$ such that $\eta(d) = c$ and

$$\begin{aligned}(\mathcal{F} \circ \gamma)(c) &= (\mathcal{F} \circ \gamma)(\eta(d)) \\ &= \mathcal{F}((\gamma \circ \eta)(d)) = (\mathcal{F} \circ \gamma')(d) = \mathcal{F}(d) \\ &= \mathcal{I}(d) \\ &\subseteq (\mathcal{H} \circ \eta)(d) = \mathcal{H}(c)\end{aligned}$$

Thus, we can conclude that $\mathcal{H}(c) = (\mathcal{F} \circ \gamma)(c)$, for all $c \in C$.

Conversely, assume that the conditions hold. Let $\mathcal{G}_B$ be the absolute soft set with $B = \{b_0, b_1\}$, then any functions α and β from A to B are soft morphisms from $\mathcal{F}_A$ to $\mathcal{G}_B$. Thus we can take $\beta(a) = b_0$, for all $a \in A$, but $\alpha(a) = b_0$, $a \in \gamma(C)$ and $\alpha(a) = b_1$, $a \in A \setminus \gamma(C)$. It follows that $\alpha \circ \gamma = \beta \circ \gamma$. If there exists an object $\mathcal{J}_E$ and a soft morphism $\eta \in Mor(\mathcal{J}_E, \mathcal{F}_A)$ such that $\alpha \circ \eta = \beta \circ \eta$. Then there exists a unique map $\eta' : E \to C$ such that $\gamma \circ \eta' = \eta$. Since $(\mathcal{H} \circ \eta')(e) = (\mathcal{H} \circ \gamma^{-1} \circ \eta)(e) = (\mathcal{F} \circ \gamma \circ \gamma^{-1} \circ \eta)(e) = (\mathcal{F} \circ \eta)(e) \supseteq \mathcal{J}(e)$. This implies that η' is a soft morphism. Thus γ is an equalizer of α and β and hence γ is a regular monomorphism. □

Theorem 4.15 *A soft morphism $\gamma \in Mor(\mathcal{G}_B, \mathcal{H}_C)$ is a regular epimorphism if and only if γ is surjective and $\mathcal{H}(c) = \cup\{\mathcal{G}(b) : \gamma(b) = c\}$.*

Proof Suppose $\gamma \in Mor(\mathcal{G}_B, \mathcal{H}_C)$ is a regular epimorphism, then γ is a coequalizer of some soft morphisms $f, g : \mathcal{F}_A \to \mathcal{G}_B$. Thus γ is an epimorphism, it follows that γ is surjective.

Let $D = B/\theta = \{[b] : b \in B\}$, where θ is the equivalence relation generated by the set $\{(b_1, b_2) \in B \times B : \gamma(b_1) = \gamma(b_2)\}$. Define a soft set $\mathcal{I}$ on D by $\mathcal{I}([b_0]) = \cup\{\mathcal{G}(b) : b \in [b_0]\}$. We have an induced map $\gamma' : B \to D$ by $\gamma'(b) = [b]$ and $\gamma' \in Mor(\mathcal{G}_B, \mathcal{I}_D)$.

Let $b_0 \in B$, $\mathcal{G}(b_0) \subseteq \cup\{\mathcal{G}(b) : b \in [b_0]\} = \mathcal{I}([b_0]) = (\mathcal{I} \circ \gamma')(b_0)$. Since γ is a coequalizer, there exists a soft morphism $\pi : \mathcal{H}_C \to \mathcal{I}_D$ with $\gamma' = \pi \circ \gamma$.

Let $c \in C$, there exist $b \in B$ with $\gamma(b) = c$. Then $\mathcal{H}(c) = (\mathcal{H} \circ \gamma)(b) \supseteq \mathcal{G}(b)$, it follows $\mathcal{H}(c) \supseteq \cup\{\mathcal{G}(b_0) : \gamma(b_0) = c\}$.

Since $\pi \in Mor(\mathcal{H}_C, \mathcal{I}_D)$, we have

$$\begin{aligned}\mathcal{H}(c) &\subseteq (\mathcal{I} \circ \pi)(c) \\ &= (\mathcal{I} \circ \pi \circ \gamma)(b) \\ &= (\mathcal{I} \circ \gamma')(b) = \mathcal{I}([b]) \\ &= \cup\{\mathcal{G}(b) : b \in [b_0]\} \\ &= \cup\{\mathcal{G}(b) : \gamma(b_0) = \gamma(b) = c\}\end{aligned}$$

Thus we get, $\mathcal{H}(c) = \cup\{\mathcal{G}(b_0) : \gamma(b_0) = c\}$.

Conversely, assume that the conditions hold. Let $\mathcal{F}_A$ be the null soft set, then any functions $f, g : A \to B$ with $\gamma \circ f = \gamma \circ g$ are soft morphisms from $\mathcal{F}_A$ to $\mathcal{G}_B$. The proof for γ is a coequalizer of f and g is same as in the Theorem 4.13. Hence γ is a regular epimorphism. □

4.4.4 General Properties of the Category of Soft Sets

In [107], Preuss describes the concepts of initial structure, final structure and topological construct in the category theory. In [167], M. Zhou et al. proved that the category of soft sets has a unique initial lift (initial structure).

Lemma 4.1 *The category soft sets has an initial structure.*

Theorem 4.16 *The category of soft sets is a topological construct.*

Proof By lemma 4.1, **Sset(U)** has a unique initial structure. Moreover, for any set A, the class $\{\mathcal{G}_B \in \textbf{Sset(U)} : B = A\}$ is a set. Since there is only one set of different functions from A to $\mathcal{P}(U)$, it follows that **Sset(U)** is a topological construct. □

Theorem 4.17 *The category Sset(U) has a final structure.*

Proof Let A be a set. Any family $(\mathcal{F}_{i_{A_i}})_{i \in I}$ of **Sset(U)** objects indexed by some class I and any family $(f_i : A_i \to A)_{i \in I}$ of functions, there exists unique **Sset(U)**- object $\mathcal{F}$ on A (i.e., $\mathcal{F} : A \to \mathcal{P}(U)$ is a soft set, final structure) such that for any soft set $\mathcal{G}_B$, a function $\alpha : \mathcal{F}_A \to \mathcal{G}_B$ is a soft morphism if and only if $\alpha \circ f_i : \mathcal{F}_{i_{A_i}} \to \mathcal{G}_B$ is a soft morphism, for all $i \in I$.

Define a soft set $\mathcal{F}$ on A by $\mathcal{F}(a) = \bigcup_{i \in I}(\cup \mathcal{F}_i \circ f_i^{-1}(a))$.

For $j \in I$ and $a_j \in A_j$, we have

$$\begin{aligned}\mathcal{F}_j(a_j) &\subseteq \cup \mathcal{F}_j \circ f_j^{-1} \circ f_j(a_j)\\ &\subseteq \bigcup_{i \in I}(\cup \mathcal{F}_i \circ f_i^{-1} \circ f_j(a_j))\\ &= (\mathcal{F} \circ f_j)(a_j)\end{aligned}$$

It implies that $f_j \in Mor(\mathcal{F}_j, \mathcal{F})$.

If $\alpha \in Mor(\mathcal{F}, \mathcal{G})$, then clearly $\alpha \circ f_i \in Mor(\mathcal{F}_i, \mathcal{G})$, for all $i \in I$.

Conversely, suppose that for each $i \in I$, $\alpha \circ f_i \in Mor(\mathcal{F}_i, \mathcal{G})$ and let $a \in A$.
Case I: If $f_i^{-1}(a) = \emptyset$, for all $i \in I$. Then $\mathcal{F}(a) = \emptyset \subseteq (\mathcal{G} \circ \alpha)(a)$, for all $a \in A$.
Case II: There exits $i_0 \in I$ such that $f_{i_0}^{-1}(a) \neq \emptyset$. Let $b \in f_{i_0}^{-1}(a)$, we have

$$\begin{aligned}\mathcal{F}_{i_0}(b) &\subseteq (\mathcal{G} \circ \alpha \circ f_{i_0})(b)\\ &= (\mathcal{G} \circ \alpha)(a)\\ \text{i.e.,}\quad \cup \mathcal{F}_{i_0} \circ f_{i_0}^{-1}(a) &\subseteq (\mathcal{G} \circ \alpha)(a)\\ \bigcup_{i \in I}(\cup \mathcal{F}_i \circ f_i^{-1}(a)) &\subseteq (\mathcal{G} \circ \alpha)(a)\\ \mathcal{F}(a) &\subseteq (\mathcal{G} \circ \alpha)(a)\end{aligned}$$

Hence $\alpha \in Mor(\mathcal{F}, \mathcal{G})$.

Now we have to demonstrate the uniqueness of final structure. Consider a final structure $\mathcal{F}' : A \to \mathcal{P}(U)$ which is different from $\mathcal{F}$.

Since $id_A \in Mor(\mathcal{F}, \mathcal{F}')$, we have $f_i \in Mor(\mathcal{F}_i, \mathcal{F}')$ (take $\mathcal{G} = \mathcal{F}'$ and $\alpha = id_A$).
Let $a \in A$, we have
Case I: If $f_i^{-1}(a) = \emptyset$, for all $i \in I$. Then $\mathcal{F}(a) = \emptyset \subseteq \mathcal{F}'(a)$, for all $a \in A$.
Case II: There exits $i_0 \in I$ such that $f_{i_0}^{-1}(a) \neq \emptyset$. Let $b \in f_{i_0}^{-1}(a)$

$$\begin{aligned}\mathcal{F}'(a) &= (\mathcal{F}' \circ f_{i_0})(b)\\ &\supseteq \mathcal{F}_{i_0}(b)\\ &\supseteq \cup \mathcal{F}_{i_0} \circ f_{i_0}^{-1}(a), \text{ for } i_0 \in I\end{aligned}$$

$$\text{It follows that } \mathcal{F}'(a) \supseteq \bigcup_{i \in I} (\cup \mathcal{F}_i \circ f_i^{-1}(a)) = \mathcal{F}(a)$$

Let $\mathcal{G} = \mathcal{F}$ and $\alpha = id_A$. Since $\alpha \circ f_i = f_i \in Mor(\mathcal{F}_i, \mathcal{F})$ for all $i \in I$, then α is soft morphism. It follows that $\mathcal{F}(a) = (\mathcal{F} \circ \alpha)(a) \supseteq \mathcal{F}'(a)$, for all $a \in A$. Thus $\mathcal{F}(a) = \mathcal{F}'(a)$, for all $a \in A$. Therefore the final structure is unique. □

Theorem 4.18 *The category* ***Sset(U)*** *has products.*

Proof Let $(\mathcal{F}_{i_{A_i}})_{i \in I}$ be a family of soft sets in **Sset(U)** indexed by a set I. Let $A = \prod_{i \in I} A_i$, define a soft set $\prod_{i \in I} \mathcal{F}_{i_{A_i}} = \mathcal{F}_A$ by $\mathcal{F}((a_i)_{i \in I}) = \bigcap_{i \in I} \mathcal{F}_{i_{A_i}}(a_i)$, for all $(a_i)_{i \in I} \in A$.
Then the essential projection maps $P_{\mathcal{F}_j} : \mathcal{F}_A \to \mathcal{F}_{j_{A_j}}$ are given by $P_{\mathcal{F}_j}((a_i)_{i \in I}) = a_j$.

For any soft morphisms $f_i : \mathcal{H}_C \to \mathcal{F}_{i_{A_i}}$ for $i \in I$, there exists a mapping $h : \mathcal{H}_C \to \mathcal{F}_A$ by $h(c) = (f_i(c))_{i \in I}$.

Let us verify the following,

(a) h and $P_{\mathcal{F}_j}$, for all $j \in I$ are soft morphisms.
(b) $P_{\mathcal{F}_j} \circ h = f_j$, for all $j \in I$ and h is unique.

(a)$(\mathcal{F}_{j_{A_j}} \circ P_{\mathcal{F}_j})((a_i)_{i \in I}) = \mathcal{F}_{j_{A_j}}(a_j) \supseteq \mathcal{F}(a)$. It implies that for $j \in I$, $P_{\mathcal{F}_j}$ is a soft morphism.
Now, $(\mathcal{F}_A \circ h)(c) = \mathcal{F}_A((f_i(c))_{i \in I}) = \bigcap_{i \in I} \mathcal{F}_{i_{A_i}}(f_i(c)) \supseteq \mathcal{H}(c)$, since $f_i : \mathcal{H}_C \to \mathcal{F}_{i_{A_i}}$ is a soft morphisms. Thus h is a soft morphism.
(b) $(P_{\mathcal{F}_j} \circ h)(c) = P_{\mathcal{F}_j}((f_i(c))_{i \in I}) = f_j(c)$ and hence $P_{\mathcal{F}_j} \circ h = f_j$, for all $j \in I$.
Let $h' : \mathcal{H}_C \to \mathcal{F}_A$ be a soft morphism satisfying $P_{\mathcal{F}_j} \circ h' = f_j$, for all $j \in I$.
To prove that $h = h'$. For $c \in C$, we have $h'(c) = (a'_j)_{j \in I} \in A$. Then,
$f_j(c) = (P_{\mathcal{F}_j} \circ h')(c) = P_{\mathcal{F}_j}((a'_j)_{j \in I}) = a'_j$.
Now, $h'(c) = (a'_j)_{j \in I} = (f_j(c))_{j \in J} = h(c)$, for all $c \in C$. Thus $h = h'$ and hence h is unique. This completes the proof. □

Remark 4.19 For $I = \{1, 2\}$, consider the diagram given in Fig. 4.8. Here one can observe the following two parts of the product,

(1) Object part: $\mathcal{F}_{1_{A_1}} \times \mathcal{F}_{2_{A_2}}$ is the product of the soft sets $\mathcal{F}_{1_{A_1}}$ and $\mathcal{F}_{2_{A_2}}$
(2) Morphism part: $h = (f_1, f_2)$ is the product of the morphisms f_1 and f_2.

Theorem 4.19 ([113]) *If a category C has finite products and equalizers, then C has pullbacks.*

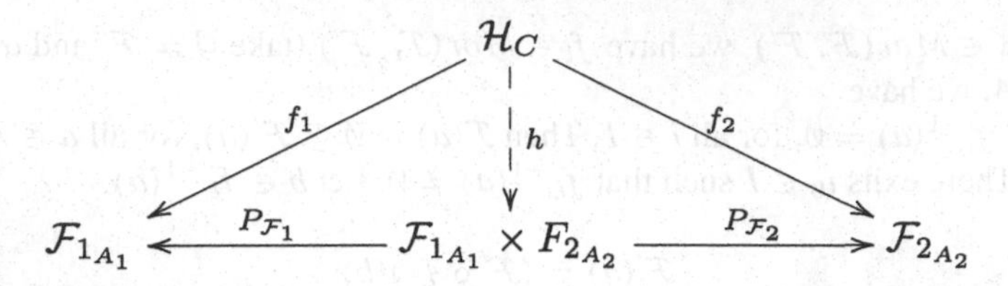

Fig. 4.8 Product of two soft sets

Fig. 4.9 Pullback of α and β

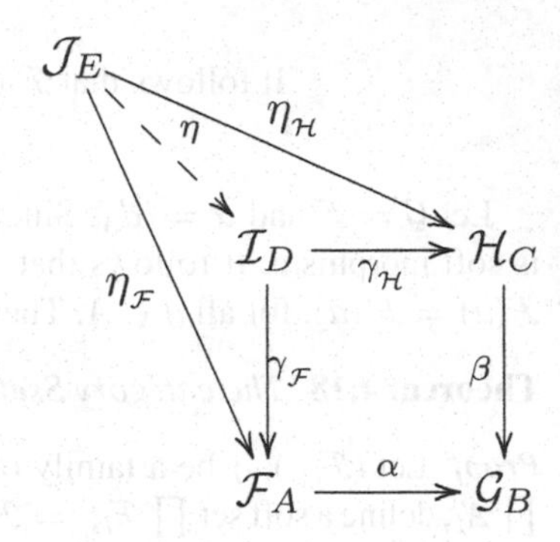

Corollary 4.3 *The category **Sset(U)** has pullbacks.*

Proof By Theorem 4.18 and Proposition 4.12, **Sset(U)** has finite products and equalizers. Then **Sset(U)** has pullbacks by Theorem 4.19. □

Now we see that how pullbacks appear in **Sset(U)**.

Let $\mathcal{F}_A$, $\mathcal{G}_B$ and $\mathcal{H}_C$ be three objects in **Sset(U)** and let $\alpha \in Mor(\mathcal{F}_A, \mathcal{G}_B)$ and $\beta \in Mor(\mathcal{H}_C, \mathcal{G}_B)$ be two soft morphisms. We have to construct an object $\mathcal{I}_D$ in **Sset(U)** and two soft morphisms $\gamma_{\mathcal{F}} \in Mor(\mathcal{I}_D, \mathcal{F}_A)$ and $\gamma_{\mathcal{H}} \in Mor(\mathcal{I}_D, \mathcal{H}_C)$ with $\alpha \circ \gamma_{\mathcal{F}} = \beta \circ \gamma_{\mathcal{H}}$ that for each object $\mathcal{J}_E$ and any two soft morphisms $\eta_{\mathcal{F}} \in Mor(\mathcal{J}_E, \mathcal{F}_A)$ and $\eta_{\mathcal{H}} \in Mor(\mathcal{J}_E, \mathcal{H}_C)$ with $\alpha \circ \eta_{\mathcal{F}} = \beta \circ \eta_{\mathcal{H}}$, there exists a unique soft morphism $\eta \in Mor(\mathcal{J}_E, \mathcal{I}_D)$ such that the diagram commutes (Fig. 4.9).

Let $D = \{(a, c) \in A \times C : \alpha(a) = \beta(c)\}$. We proceed with two cases,
Case I: Suppose $D \neq \emptyset$.
Define a soft set $\mathcal{I}_D$ as $\mathcal{I}(a, c) = \mathcal{F}(a) \cap \mathcal{H}(c)$, for all $(a, c) \in D$, then the essential projection maps as follows $\gamma_{\mathcal{F}}(a, c) = a$ and $\gamma_{\mathcal{H}}(a, c) = c$, for all $(a, c) \in D$.
For any two soft morphisms $\eta_{\mathcal{F}} \in Mor(\mathcal{J}_E, \mathcal{F}_A)$ and $\eta_{\mathcal{H}} \in Mor(\mathcal{J}_E, \mathcal{H}_C)$, define a map $\eta : E \to D$ by $\eta(e) = (\eta_{\mathcal{F}}(e), \eta_{\mathcal{H}}(e))$, for all $e \in E$.

Claim I: $\gamma_{\mathcal{F}}$ and $\gamma_{\mathcal{H}}$ are soft morphisms and $\alpha \circ \gamma_{\mathcal{F}} = \beta \circ \gamma_{\mathcal{H}}$.

$(\mathcal{F} \circ \gamma_{\mathcal{F}})(a, c) = \mathcal{F}(a) \supseteq \mathcal{F}(a) \cap \mathcal{H}(c) = \mathcal{I}(a, c)$, for all $(a, c) \in D$. It implies that $\gamma_{\mathcal{F}} \in Mor(\mathcal{I}_D, \mathcal{F}_A)$. Similarly we can prove $\gamma_{\mathcal{H}} \in Mor(\mathcal{I}_D, \mathcal{H}_C)$.

Also, for any $(a, c) \in D$, we have $(\alpha \circ \gamma_{\mathcal{F}})(a, c) = \alpha(a) = \beta(c) = (\beta \circ \gamma_{\mathcal{H}})(a, c)$.

Claim II: η is a soft morphism.

Since $(\alpha \circ \eta_{\mathcal{F}})(e) = (\alpha \circ \eta_{\mathcal{H}})(e)$, we have $\eta(e) = (\eta_{\mathcal{F}}(e), \eta_{\mathcal{H}}(e)) \in D$, for all $e \in E$. Then $(\mathcal{I} \circ \eta)(e) = \mathcal{F}(\eta_{\mathcal{F}}(e)) \cap \mathcal{H}(\eta_{\mathcal{H}}(e))$. Since $\eta_{\mathcal{F}} \in Mor(\mathcal{J}_E, \mathcal{F}_A)$ and $\eta_{\mathcal{H}} \in Mor(\mathcal{J}_E, \mathcal{H}_C)$, we get $(\mathcal{F} \circ \eta_{\mathcal{F}})(e) \supseteq \mathcal{J}(e)$ and $(\mathcal{H} \circ \eta_{\mathcal{H}})(e) \supseteq \mathcal{J}(e)$. Thus we can conclude that $(\mathcal{I} \circ \eta)(e) \supseteq \mathcal{J}(e)$. Therefore $\eta \in Mor(\mathcal{J}_E, \mathcal{I}_D)$.

Case II: Suppose $D = \emptyset$.

If $E \neq \emptyset$, then $(\alpha \circ \eta_{\mathcal{F}})(e) = (\alpha \circ \eta_{\mathcal{H}})(e)$, which implies that $(\eta_{\mathcal{F}}(e), \eta_{\mathcal{H}}(e)) \in D$. Thus $E = \emptyset$.

Since products are unique, we get $\eta : \mathcal{J}_E \to \mathcal{I}_D$ is unique.

To prove: the diagram commutes, i.e., $\eta_{\mathcal{F}} = \gamma_{\mathcal{F}} \circ \eta$ and $\eta_{\mathcal{H}} = \gamma_{\mathcal{H}} \circ \eta$. Let $e \in E$,

$$(\gamma_{\mathcal{F}} \circ \eta)(e) = \gamma_{\mathcal{F}}(\eta_{\mathcal{F}}(e), \eta_{\mathcal{H}}(e)) = \eta_{\mathcal{F}}(e) \text{ and}$$

$$(\gamma_{\mathcal{H}} \circ \eta)(e) = \gamma_{\mathcal{H}}(\eta_{\mathcal{F}}(e), \eta_{\mathcal{H}}(e)) = \eta_{\mathcal{H}}(e)$$

Therefore $\eta : \mathcal{J}_E \to \mathcal{I}_D$ is a unique pullback of α and β.

Theorem 4.20 *The category **Sset(U)** has coproducts.*

Proof Let $(\mathcal{F}_{i_{A_i}})_{i \in I}$ be a family of soft sets in **Sset(U)** indexed by a set I. Let $A = \bigcup_{i \in I}(A_i \times \{i\})$, define a soft set $\mathcal{F}_A$ by $\mathcal{F}(a_i, i) = \mathcal{F}_{i_{A_i}}(a_i)$, for all $(a_i, i) \in A$ and the functions $\Psi_{\mathcal{F}_j} : \mathcal{F}_{j_{A_j}} \to \mathcal{F}_A$ by $\Psi_{\mathcal{F}_j}(a_j) = (a_j, j)$, for $j \in I$.

Let $\mathcal{G}_B$ be an object in **Sset(U)** and let $f_i : \mathcal{F}_{i_{A_i}} \to \mathcal{G}_B$ be soft morphisms, for $i \in I$, then there exists a mapping $h : \mathcal{F}_A \to \mathcal{G}_B$ as $h(a_i, i) = f_i(a_i)$.

Let us verify the following,

(a) h and $\Psi_{\mathcal{F}_j}$, for all $j \in I$ are soft morphisms.
(b) $h \circ \Psi_{\mathcal{F}_j} = f_j$, for all $j \in I$ and h is unique.

(a) $(\mathcal{G} \circ h)(a_i, i) = \mathcal{G}(f_i(a_i)) \supseteq \mathcal{F}_{i_{A_i}}(a_i) = \mathcal{F}(a_i, i)$, since f_i is a soft morphism for all $i \in I$. Thus $h \in Mor(\mathcal{F}_A, \mathcal{G}_B)$.

Now $(\mathcal{F} \circ \Psi_{\mathcal{F}_j})(a_j) = \mathcal{F}(a_j, j) = \mathcal{F}_{j_{A_j}}(a_j)$. It implies that $\Psi_{\mathcal{F}_j} \in Mor(\mathcal{F}_{j_{A_j}}, \mathcal{F}_A)$.

(b) $(h \circ \Psi_{\mathcal{F}_j})(a_j) = h(a_j, j) = f_j(a_j)$, for all $j \in I$.

Let $h' : \mathcal{F}_A \to \mathcal{G}_B$ be a soft morphism satisfying $h' \circ \Psi_{\mathcal{F}_j} = f_j$, for all $j \in I$.

To prove that $h = h'$. For $(a_j, j) \in A$, we have $h'(a_j, j) = b_j \in B$. Then, $f_j(a_j) = (h' \circ \Psi_{\mathcal{F}_j})(a_j) = h'(a_j, j) = b_j$.

Now, $h'(a_j, j) = b_j = f_j(a_j) = h(a_j, j)$, for all $(a_j, j) \in A$. Thus $h = h'$ and hence h is unique. Hence **Sset(U)** has coproducts. □

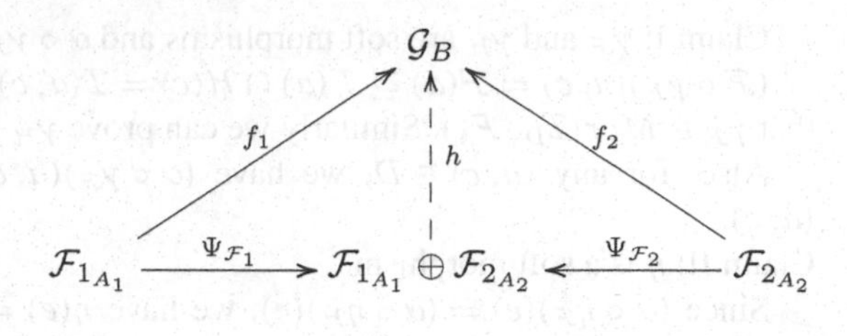

Fig. 4.10 Coproduct of two soft sets

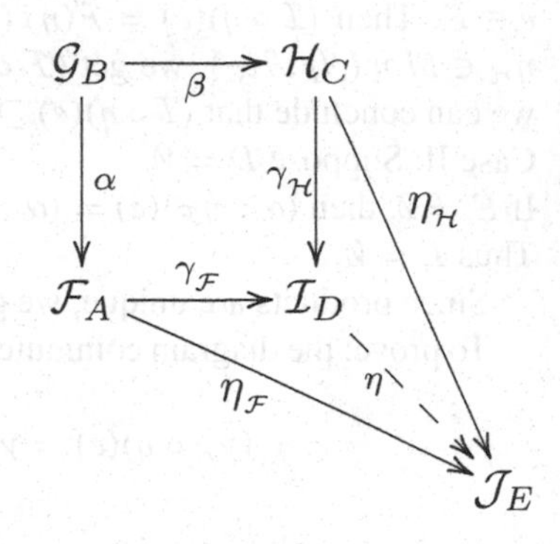

Fig. 4.11 Pushout of α and β

Remark 4.20 For $I = \{1, 2\}$, we have the diagram given as Fig. 4.10, where $\oplus$ denotes the coproduct.

Here one can observe the following two parts of the coproduct,

(1) Object part: $\mathcal{F}_{1_{A_1}} \oplus \mathcal{F}_{2_{A_2}}$ is the coproduct of the soft sets $\mathcal{F}_{1_{A_1}}$ and $\mathcal{F}_{2_{A_2}}$
(2) Morphism part: $h(a) = \begin{cases} f_1(a) \text{ if } a \in A_1; \\ f_2(a) \text{ if } a \in A_2 \end{cases}$ is the coproduct of the morphisms f_1 and f_2, i.e., $h = f_1 \oplus f_2$.

Theorem 4.21 *The category **Sset(U)** has pushouts.*

Proof Let $\mathcal{F}_A, \mathcal{G}_B$ and $\mathcal{H}_C$ be three objects in **Sset(U)** and let $\alpha \in Mor(\mathcal{G}_B, \mathcal{F}_A)$ and $\beta \in Mor(\mathcal{G}_B, \mathcal{H}_C)$ be two soft morphisms.

We have to construct an object $\mathcal{I}_D$ in the category **Sset(U)** and two soft morphisms $\gamma_{\mathcal{F}} \in Mor(\mathcal{F}_A, \mathcal{I}_D)$ and $\gamma_{\mathcal{H}} \in Mor(\mathcal{H}_C, \mathcal{I}_D)$ with $\gamma_{\mathcal{F}} \circ \alpha = \gamma_{\mathcal{H}} \circ \beta$ that for each object $\mathcal{J}_E$ and any two soft morphisms $\eta_{\mathcal{F}} \in Mor(\mathcal{F}_A, \mathcal{J}_E)$ and $\eta_{\mathcal{H}} \in Mor(\mathcal{H}_C, \mathcal{J}_E)$ with $\eta_{\mathcal{F}} \circ \alpha = \eta_{\mathcal{H}} \circ \beta$, there exists a unique soft morphism $\eta \in Mor(\mathcal{I}_D, \mathcal{J}_E)$ such that the diagram (Fig. 4.11) commutes.

Assume that A, B and C are disjoint. Define a relation R on $A \cup C$ by aRc if and only if there exists $b \in B$ such that $\alpha(b) = a$ and $\beta(b) = c$. Let θ be the equivalence relation generated by R.

Let $D = (A \cup B)/\theta$, quotient of $A \cup B$ with respect to θ.

Define a soft set $\mathcal{I}_D$ by

$$\mathcal{I}(d) = \begin{cases} \mathcal{F}(d) & \text{if } d = [a]_\theta = \{a\}; \\ \mathcal{H}(d) & \text{if } d = [c]_\theta = \{c\}; \\ (\bigcup\limits_{x \in [a]_\theta \cap A} \mathcal{F}(x)) \cup (\bigcup\limits_{y \in [c]_\theta \cap C} \mathcal{H}(y)) & \text{if } d = [a]_\theta = [c]_\theta \end{cases}$$

Define two maps $\gamma_{\mathcal{F}} : \mathcal{F}_A \to \mathcal{I}_D$ and $\gamma_{\mathcal{H}} : \mathcal{H}_C \to \mathcal{I}_D$ as $\gamma_{\mathcal{F}}(a) = [a]_\theta$ and $\gamma_{\mathcal{H}}(c) = [c]_\theta$.

For any two soft morphisms $\eta_{\mathcal{F}} \in Mor(\mathcal{F}_A, \mathcal{J}_E)$ and $\eta_{\mathcal{H}} \in Mor(\mathcal{H}_C, \mathcal{J}_E)$ with $\eta_{\mathcal{F}} \circ \alpha = \eta_{\mathcal{H}} \circ \beta$, define a map $\eta : D \to E$ as

$$\eta(d) = \begin{cases} \eta_{\mathcal{F}}(d) & \text{if } d = [a]_\theta = \{a\}; \\ \eta_{\mathcal{H}}(d) & \text{if } d = [c]_\theta = \{c\}; \\ \eta_{\mathcal{F}}(a) = \eta_{\mathcal{H}}(c) & \text{if } d = [a]_\theta = [c]_\theta \end{cases}$$

Claim I: $\gamma_{\mathcal{F}}$ and $\gamma_{\mathcal{H}}$ are soft morphisms with $\gamma_{\mathcal{F}} \circ \alpha = \gamma_{\mathcal{H}} \circ \beta$.
If $a \in A$ and $[a]_\theta = \{a\}$, then $(\mathcal{I} \circ \gamma_{\mathcal{F}})(a) = \mathcal{I}(\gamma_{\mathcal{F}}(a)) = \mathcal{I}([a]_\theta) = \mathcal{F}(a)$.
If $a \in A$ and $[a]_\theta = [c]_\theta$, then

$$\begin{aligned} (\mathcal{I} \circ \gamma_{\mathcal{F}})(a) &= \mathcal{I}(\gamma_{\mathcal{F}}(a)) \\ &= \mathcal{I}([a]_\theta) \\ &= (\bigcup_{x \in [a]_\theta \cap A} \mathcal{F}(x)) \cup (\bigcup_{y \in [c]_\theta \cap C} \mathcal{H}(y)) \\ &\supseteq \mathcal{F}(a) \end{aligned}$$

Then $\gamma_{\mathcal{F}}$ is a soft morphism. Similarly, we can prove $\gamma_{\mathcal{H}}$ is a soft morphism. Also, if $\alpha(b) = a$, then there exists $c \in C$ such that $\alpha(b) = c$, which implies that $[a]_\theta = [c]_\theta$.
Therefore $(\gamma_{\mathcal{F}} \circ \alpha)(b) = \gamma_{\mathcal{F}}(\alpha(b)) = [a]_\theta = [c]_\theta = \gamma_{\mathcal{H}}(\beta(b)) = (\gamma_{\mathcal{H}} \circ \beta)(b)$.
Hence the claim I.
Claim II: η is a unique soft morphism, for which the diagram commutes.
If $d = [a]_\theta = \{a\}$, then $(\mathcal{J} \circ \eta)(d) = \mathcal{J}(\eta_{\mathcal{F}})(d) \supseteq \mathcal{F}(d) = \mathcal{I}(d)$.
If $d = [c]_\theta = \{c\}$, then $(\mathcal{J} \circ \eta)(d) = \mathcal{J}(\eta_{\mathcal{H}})(d) \supseteq \mathcal{H}(d) = \mathcal{I}(d)$.
If $d = [a]_\theta = [c]_\theta$, then $(\mathcal{J} \circ \eta)(d) = \mathcal{J}(\eta_{\mathcal{F}}(a)) \supseteq \mathcal{F}(a) = \mathcal{I}(d)$. Hence η is a soft morphism. Since coproducts are unique, we get η is unique.

The only point remaining to show is that the diagram commutes:
If $a \in A$ and $[a]_\theta = \{a\}$, then $(\eta \circ \gamma_{\mathcal{F}})(a) = \eta(a) = \eta_{\mathcal{F}}(a)$.
If $a \in A$ and $[a]_\theta = [c]_\theta$, then $(\eta \circ \gamma_{\mathcal{F}})(a) = \eta([a]_\theta) = \eta_{\mathcal{F}}(a)$, this gives $\eta \circ \gamma_{\mathcal{F}} = \eta_{\mathcal{F}}$, Similarly, we can shows that $\eta \circ \gamma_{\mathcal{H}} = \eta_{\mathcal{H}}$. Hence the theorem. □

We conclude this section with showing a Galois correspondence between the categories **Fset** and **Sset**([0,1]).

Theorem 4.22 *The category of fuzzy sets,* ***Fset*** *is a concrete category over* ***Set****.*

Proof Let A be any set, then $pt(A) = \bigcup_{a \in A} \{\{a\} \times [0, 1]\}$. Let f_A be any fuzzy set and $f_A = \{(a, f(a)) : a \in A\} \subseteq pt(A)$. Define Ψ : **Fset** $\to$ **Set** by $\Psi(f_A) = \bigcup_{a \in A} (\{a\} \times [0, f(a)])$ and is denoted by $f_A{}^*$. Let $h \in Mor(f_A, g_B)$, define $h^* : f_A{}^* \to g_B{}^*$ by

$$h^*(a,t) = \begin{cases} (h(a), 0) & \text{if } t = 0; \\ (h(a), r_{fg}^{h(a)}(t)) & \text{if } 0 < t \leq f(a) \end{cases}$$

where $r_{fg}^{h(a)}(t) = \frac{(g \circ h)(a)}{f(a)} t$ and note that if $h(a) = b$, then

$$r_{fg}^{h(a)}(t) = \frac{(g \circ h)(a)}{f(a)} t = g(b)(\frac{t}{f(a)}) \leq g(b)$$

.

Thus $h^* \in Mor(f_A{}^*, g_B{}^*)$ and take $\Psi(h) = h^*$.
Now we have to show that,

(i) Ψ is a covariant functor.
(ii) Ψ is a faithful functor.

(i) It enough to prove that (a). Ψ preserves identity morphism and (b). Ψ preserves composition.

(a) Let $id_{f_A} : f_A \to f_A$ is the identity morphism. Then

$$\begin{aligned} \Psi(id_{f_A})(a,t) &= (id_{f_A})^*(a,t) \\ &= \begin{cases} (id_{f_A}(a), 0) & \text{if } t = 0; \\ (id_{f_A}(a), r_{ff}^{id_{f_A}(a)}(t)) & \text{if } 0 < t \leq f(a) \end{cases} \\ &= \begin{cases} (a, 0) & \text{if } t = 0; \\ (a, t) & \text{if } 0 < t \leq f(a) \end{cases} \\ &= id_{f_A{}^*}(a,t), \quad \text{for all } (a,t) \in f_A{}^* \end{aligned}$$

Therefore $\Psi(id_{f_A}) = id_{f_A{}^*}$.

(b) Let $h_1 \in Mor(f_A, g_B)$ and $h_2 \in Mor(g_B, j_C)$ be two fuzzy morphisms, then $h_2 \circ h_1 \in Mor(f_A, j_C)$.

Case I: If $t = 0$,

$$\begin{aligned} (\Psi(h_2) \circ \Psi(h_1))(a, 0) &= (h_2^* \circ h_1^*)(a, 0) \\ &= h_2^*(h_1(a), 0) \\ &= ((h_2 \circ h_1)(a), 0) \\ &= (h_2 \circ h_1)^*(a, 0) = \Psi(h_2 \circ h_1)(a, 0) \end{aligned}$$

Case II: If $0 < t \leq f(a)$, then $g(h_1(a)) \neq 0$ since h_1 is a fuzzy morphism.

$$\begin{aligned}(\Psi(h_2) \circ \Psi(h_1))(a,t) &= (h_2^* \circ h_1^*)(a,t) \\ &= h_2^*(h_1(a), r_{fg}^{h_1(a)}(t)) \\ &= (h_2(h_1(a)), r_{gj}^{h_2(h_1(a))}(r_{fg}^{h_1(a)}(t))) \\ &= ((h_2 \circ h_1)(a), r_{fj}^{(h_2 \circ h_1)(a)}(t)) \\ &= (h_2 \circ h_1)^*(a,t) = \Psi(h_2 \circ h_1)(a,t)\end{aligned}$$

This shows that $\Psi(h_2 \circ h_1) = \Psi(h_1) \circ \Psi(h_2)$ and hence Ψ is a covariant functor.

(ii) From the definition of h^*, we get $\Psi : Mor(f_A, g_B) \to Mor(f_A^*, g_B^*)$ is injective and hence it is faithful.
It follows that Ψ : **Fset** $\to$ **Set** is faithful functor. Thus **Fset** is a concrete category over **Set**.

□

Theorem 4.23 *The category of soft sets over* [0, 1], ***Sset([0,1])*** *is a concrete category over* ***Set***.

Proof Let $\mathcal{F}_A \in Obj$(**Sset([0,1])**). Define $\mathcal{F}_A^* = \bigcup_{a \in A} \{\{a\} \times [0, \sup \mathcal{F}(a)]\}$, then $\mathcal{F}_A^* \in Obj$(**Set**).
Now define Φ : **Sset([0,1])** $\to$ **Set** by $\Phi(\mathcal{F}_A) = \mathcal{F}_A^*$.

Let $k \in Mor(\mathcal{F}_A, \mathcal{G}_B)$, define $k^* : \mathcal{F}_A{}^* \to \mathcal{G}_B{}^*$ by

$$k^*(a,t) = \begin{cases} (k(a), 0) & \text{if } t = 0; \\ (k(a), s_{\mathcal{F}\mathcal{G}}^{k(a)}(t)) & \text{if } 0 < t \leq \sup \mathcal{F}(a) \end{cases}$$

where $s_{\mathcal{F}\mathcal{G}}^{k(a)}(t) = \frac{\sup(\mathcal{G} \circ k)(a)}{\sup \mathcal{F}(a)} t$ and if $k(a) = b$, then $s_{\mathcal{F}\mathcal{G}}^{k(a)}(t) = \frac{\sup \mathcal{G}(b)}{\sup \mathcal{F}(a)} t \leq \sup \mathcal{G}(b)$.

Thus $k^* \in Mor(\mathcal{F}_A{}^*, \mathcal{G}_B{}^*)$. Define $\Phi(k) = k^*$.
Now we have to show that (i). Φ is a covariant functor and
(ii). Φ is a faithful functor.

(i) To prove: $\Phi(id_{\mathcal{F}_A}) = id_{\mathcal{F}_A{}^*}$, where $id_{\mathcal{F}_A} : \mathcal{F}_A \to \mathcal{F}_A$ is the identity morphism.

$$(id_{\mathcal{F}_A})^*(a,t) = \begin{cases} (id_{\mathcal{F}_A}(a), 0) & \text{if } t = 0 \\ (id_{\mathcal{F}_A}(a), r_{\mathcal{F}\mathcal{F}}^{a}(t)) & \text{if } 0 < t \leq \sup \mathcal{F}(a) \end{cases}$$

$$= \begin{cases} (a, 0) & \text{if } t = 0 \\ (a, t) & \text{if } 0 < t \leq \sup \mathcal{F}(a) \end{cases}$$

$$= id_{\mathcal{F}_A{}^*}(a,t) \qquad \text{for all } (a,t) \in \mathcal{F}_A{}^*$$

It implies that $\Phi(id_{\mathcal{F}_A}) = (id_{\mathcal{F}_A})^* = id_{\mathcal{F}_A^*}$
Let $k_1 : \mathcal{F}_A \to \mathcal{G}_B$ and $k_2 : \mathcal{G}_B \to \mathcal{J}_C$ be two soft morphisms.
Case I: If $t = 0$,

$$\begin{aligned}(k_2^* \circ k_1^*)(a, 0) &= k_2^*(k_1(a), 0)\\ &= ((k_2 \circ k_1)(a), 0)\\ &= (k_2 \circ k_1)^*(a, 0)\end{aligned}$$

Case II: If $0 < t \leq \sup \mathcal{F}(a)$, then $\mathcal{G}(k_1(a)) \neq 0$

$$\begin{aligned}(k_2^* \circ k_1^*)(a, t) &= k_2^*(k_1(a), s_{\mathcal{FG}}^{k_1(a)}(t))\\ &= ((k_2 \circ k_1)(a), s_{\mathcal{GJ}}^{k_2(a)}(s_{\mathcal{FG}}^{k_1(a)}(t)))\\ &= ((k_2 \circ k_1)(a), s_{\mathcal{FJ}}^{(k_2 \circ k_1)(a)}(t))\\ &= (k_2 \circ k_1)^*(a, t)\end{aligned}$$

Thus $\Phi(k_2 \circ k_1) = (k_2 \circ k_1)^* = k_2^* \circ k_1^* = \Phi(k_2) \circ \Phi(k_1)$ and hence Φ is a covariant functor.

(ii) From the definition of k^*, we get $\Phi : Mor(\mathcal{F}_A, \mathcal{G}_B) \to Mor(\mathcal{F}_A^*, \mathcal{G}_B^*)$ is injective and hence it is faithful.

This shows that Φ : **Sset([0,1])** $\to$ **Set** is a faithful functor and hence **Sset([0,1])** is a concrete category over **Set**. □

Theorem 4.24 *Let **Fset** and **Sset([0,1])** be constructs with the faithful functors defined as in theorems 4.22 and 4.23, then*

*(a) H : **Fset** $\to$ **Sset([0,1])** by $H(f_A) = \mathcal{F}_A$, where $\mathcal{F}_A(a) = [0, f(a)]$, for all $f_A \in Obj($**Fset**$)$ and $H(h) = h$ for all $h \in Mor(f_A, g_B)$ is a concrete functor.*
*(b) K : **Sset([0,1])** $\to$ **Fset** by $K(\mathcal{F}_A) = f_A$, where $f_A(a) = \sup \mathcal{F}(a)$, for all $\mathcal{F}_A \in Obj($**Sset([0,1])**$)$ and $K(k) = k$ for all $k \in Mor(\mathcal{F}_A, \mathcal{G}_B)$ is a concrete functor.*

Proof (a) Let $h \in Mor(f_A, g_B)$, we have $f(a) \leq (g \circ h)(a)$, for all $a \in A$.

$$\begin{aligned}\mathcal{G}(h(a)) &= [0, g(h(a))]\\ &\supseteq [0, f(a)]\\ &= \mathcal{F}(a)\end{aligned}$$

So, $h \in Mor(\mathcal{F}_A, \mathcal{G}_B) = Mor(H(f_A), H(g_B))$. Thus H is well defined and trivially H is a covariant functor.
Now,

$$\begin{aligned}(\Phi \circ H)(f_A) &= \Phi(\mathcal{F}_A)\\ &= \mathcal{F}_A^*\\ &= \bigcup_{a \in A}\{\{a\} \times [0, \sup \mathcal{F}(a)]\}\end{aligned}$$

$$= \bigcup_{a\in A}\{\{a\} \times [0, f(a)]\}$$
$$= f_A^* = \Psi(f_A)$$

Therefore $\Phi \circ H = \Psi$. Hence H is a concrete functor.

(b) Let $k \in Mor(\mathcal{F}_A, \mathcal{G}_B)$, we have $\mathcal{F}(a) \subseteq (\mathcal{G} \circ k)(a)$, for all $a \in A$.

$$\begin{aligned} g(k(a)) &= \sup \mathcal{G}(k(a)) \\ &\geq \sup \mathcal{F}(a) \\ &= f(a) \end{aligned}$$

It implies that $k \in Mor(f_A, g_B) = Mor(K(\mathcal{F}_A), K(\mathcal{G}_B))$. Thus K is well defined and trivially K is a covariant functor.
Now,

$$\begin{aligned} (\Psi \circ K)(\mathcal{F}_A) &= \Psi(f_A) \\ &= f_A^* \\ &= \bigcup_{a\in A}\{\{a\} \times [0, \sup \mathcal{F}(a)]\} \\ &= \mathcal{F}_A^* = \Phi(\mathcal{F}_A) \end{aligned}$$

Therefore $\Psi \circ K = \Phi$. Hence K is a concrete functor.

□

Theorem 4.25 ([2]) *Composite of concrete functors over a category* **X** *is a concrete functor over* **X**.

Theorem 4.26 *If K and H are the concrete functors defined as in Theorem 4.24, then the pair (K, H) is a Galois correspondence between* **Fset** *and* **Sset([0,1])**.

Proof Let us consider the concrete functors $K \circ H$ and id_{Fset} from **Fset** to **Fset**. For $f_A \in Obj(\textbf{Fset})$ and $\mathcal{F}_A(a) = [0, f(a)]$ we have,

$$\begin{aligned} (K \circ H)(f_A) &= K(F_A) \\ &= f_A \end{aligned}$$

Thus $K \circ H \leq id_{Fset}$.

Now consider the concrete functors $H \circ K$ and $id_{Sset([0,1])}$ from **Sset([0,1])** to **Sset([0,1])**.
For $\mathcal{F}_A \in Obj(\textbf{Sset([0,1])})$ and $f_A(a) = \sup(\mathcal{F}_A(a))$ for all $a \in A$, we have,

$$\begin{aligned} (H \circ K)(\mathcal{F}_A) &= H(f_A) \\ &= \mathcal{F}'_A, \quad \text{where } \mathcal{F}'_A(a) = [0, \sup \mathcal{F}_A(a)] \end{aligned}$$

Clearly $\mathcal{F}_A(a) \subseteq \mathcal{F}'_A(a)$ for all $a \in A$ and hence $id_{\mathcal{F}_A} \in Mor(\mathcal{F}_A, \mathcal{F}'_A)$. Also, $\Phi(\mathcal{F}_A) = \Phi(\mathcal{F}'_A) = \mathcal{F}^*_A$ and $\Phi(id_{\mathcal{F}_A}) = id_{\mathcal{F}^*_A}$. It follows that $id_{Sset([0,1])} \leq (H \circ K)$. Thus the pair (K, H) is a Galois correspondence between **Fset** and **Sset([0,1])**. □

4.5 Category of Soft Graphs

This section introduces the concept of the category of soft graphs. First subsection explains the initial object, terminal object, equalizer, coequalizer, product, co product etc. and the existence of a left adjoint of a functor between **Sset(U)** and **SGr(U)** is proved in the next subsection.

4.5.1 *General Properties of the Category of Soft Graphs, SGr(U)*

Definition 4.49 Let $G = (\mathcal{F}, \mathcal{G})_A$ and $G' = (\mathcal{F}', \mathcal{G}')_B$ be two soft graphs over U. A soft graph morphism from $(\mathcal{F}, \mathcal{G})_A$ to $(\mathcal{F}', \mathcal{G}')_B$ is a function $\alpha : A \to B$ such that

1. $\mathcal{F}(a) \subseteq (\mathcal{F}' \circ \alpha)(a)$, for each $a \in A$.
2. $\mathcal{G}(a_1, a_2) \subseteq (\mathcal{G}' \circ \alpha)(a_1, a_2)$, for each $a_1, a_2 \in A$.

Example 67 Let $U = \{1, 2, \ldots, 10\}$ be a universe set and let $A = \{a_1, a_2, a_3\}$ and $B = \{b_1, b_2, b_3\}$ be two parameter sets. Two soft graphs $(\mathcal{F}, \mathcal{G})_A$ and $(\mathcal{F}', \mathcal{G}')_B$ are given in Fig. 4.12

Here we defined $\alpha : A \to B$ as follows, $\alpha(a_1) = b_1$ and $\alpha(a_2) = b_3$. Then,

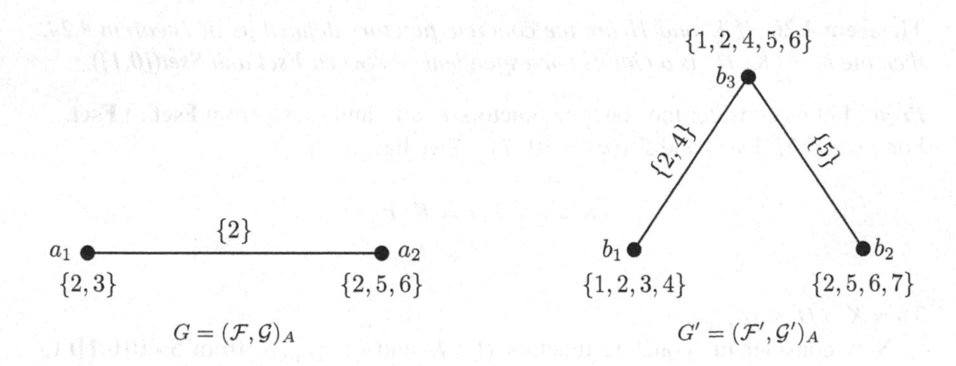

Fig. 4.12 Soft graph morphism between G and G'

$$\mathcal{F}(a_1) = \{2, 3\} \subseteq \{1, 2, 3, 4\} = (\mathcal{F}' \circ \alpha)(a_1)$$
$$\mathcal{F}(a_2) = \{2, 5, 6\} \subseteq \{1, 2, 4, 5.6\} = (\mathcal{F}' \circ \alpha)(a_2)$$
$$\mathcal{G}(a_1, a_2) = \{2\} \subseteq \{2, 4\} = (\mathcal{G} \circ \alpha)(a_1, a_2).$$

Thus $\alpha : G \to G'$ is a soft graph morphism.

Remark 4.21 The composition of two soft graph morphisms is again a soft graph morphism. The composition of soft graph morphisms is associative and identity function is the identity soft graph morphism. The collection of all soft graphs over U is denoted by **SGr(U)** and is a category.

Proposition 4.26 *An object $(\mathcal{F}, \mathcal{G})_A \in$ **SGr(U)** is an initial object if and only if $A = \emptyset$ and is a terminal object if and only if $(\mathcal{F}, \mathcal{G})_A$ is an absolute soft graph with $|A| = 1$.*

Proof Let $(\mathcal{F}, \mathcal{G})_A$ be an initial object in the category of soft graphs, **SGr(U)**. Then $|Mor((\mathcal{F}, \mathcal{G})_A, (\mathcal{F}', \mathcal{G}')_B)| = 1$, for every soft graph $(\mathcal{F}', \mathcal{G}')_B$ in **SGr(U)**. If $A \neq \emptyset$, then $|Mor((\mathcal{F}, \mathcal{G})_A, (\mathcal{F}', \mathcal{G}')_B)| > 1$, for every absolute soft graph $(\mathcal{F}', \mathcal{G}')_B$. So A is a null set. Conversely assume $A = \emptyset$, there is exactly one function from empty set to any set. So $(\mathcal{F}, \mathcal{G})_A$ is an initial object in **SGr(U)**.

Let $(\mathcal{F}, \mathcal{G})_A$ be an absolute soft graph with $|A| = 1$, then there is exactly one soft morphism (constant function) between any soft graph $(\mathcal{F}', \mathcal{G}')_B$ and $(\mathcal{F}, \mathcal{G})_A$. Conversely, suppose $|A| > 1$. Then $|Mor((\mathcal{F}, \mathcal{G}')_A, (\mathcal{F}, \mathcal{G})_A| > 1$, for the soft graph $(\mathcal{F}, \mathcal{G}')_A$ with $\mathcal{G}'(a, b) = \emptyset$, for each $(a, b) \in A \times A$. This is not possible and hence $|A| = 1$. Also, if $(\mathcal{F}, \mathcal{G})_A$ is not an absolute soft graph, then there is no soft morphism between absolute soft graph $(\mathcal{F}', \mathcal{G}')_B$ and $(\mathcal{F}, \mathcal{G})_A$. So $(\mathcal{F}, \mathcal{G})_A$ must be an absolute soft graph with $|A| = 1$. □

Theorem 4.27 *Let $G_1 = (\mathcal{F}_1, \mathcal{G}_1)_A$ and $G_2 = (\mathcal{F}_2, \mathcal{G}_2)_B$ be two soft graphs over U. The equalizer of two soft graph morphism $\alpha, \beta \in Mor(G_1, G_2)$ in the category* **SGr(U)** *is the soft graph morphism $i : G' = (\mathcal{F}', \mathcal{G}')_C \to (\mathcal{F}_1, \mathcal{G}_1)_A$, where $C = \{a \in A : \alpha(a) = \beta(a)\}$ and G' is the restriction of G to C.*

Proof Let $\alpha, \beta \in Mor(G_1, G_2)$ and $C = \{a \in A : \alpha(a) = \beta(a)\}$ and let $i : C \to A$ be the inclusion map. Then,

1. $\mathcal{F}'(c) = \mathcal{F}(c) = (\mathcal{F} \circ i)(c)$, for all $c \in C$
2. $\mathcal{G}'(c_1, c_2) = \mathcal{G}(c_1, c_2) = (\mathcal{G} \circ i)(c_1, c_2)$, for all $c_1, c_2 \in C$

It follows that $i \in Mor(G', G)$. Also note that $\alpha \circ i = \beta \circ i$.

Let $G_3 = (\mathcal{F}_3, \mathcal{G}_3)_D$ be a soft graph and let $\gamma : G_3 \to G_1$ be any soft graph morphism with $\alpha \circ \gamma = \beta \circ \gamma$. Since the image of γ is in C, there exists a map $j : D \to C$ such that $i \circ j = \gamma$. Then we have,

1. $\mathcal{F}_1 \circ \gamma = (\mathcal{F} \circ i) \circ j = \mathcal{F}' \circ j$
2. $\mathcal{G}_1 \circ \gamma = (\mathcal{G} \circ i) \circ j = \mathcal{G}' \circ j$

Since γ is a soft graph morphism, we get

1. $\mathcal{F}_3(d) \subseteq (\mathcal{F}_1 \circ \gamma)(d)$, for all $d \in D$
2. $\mathcal{G}_3(d_1, d_2) \subseteq (\mathcal{G}_1 \circ \gamma)(d_1, d_2)$, for all $d_1, d_2 \in D$

Therefore,

1. $\mathcal{F}_3(d) \subseteq (\mathcal{F}' \circ j)(d)$ for all $d \in D$
2. $\mathcal{G}_3(d_1, d_2) \subseteq (\mathcal{G}' \circ j)(d_1, d_2)$ for all $d_1, d_2 \in D$

It establish that $j : G_3 \to G'$ is a soft graph morphism. Suppose $j' \in Mor(G_3, G')$ satisfy $i \circ j' = \gamma$, then $i \circ j' = i \circ j$. Since i is injective, $j = j'$. Hence $i : G' \to G_1$ is the equalizer of α and β. □

Theorem 4.28 *Let $G_1 = (\mathcal{F}_1, \mathcal{G}_1)_A$ and $G_2 = (\mathcal{F}_2, \mathcal{G}_2)_B$ be two soft graphs over U. Then any two morphisms $\alpha, \beta \in Mor(G_1, G_2)$ have a coequalizer.*

Proof Let $\alpha, \beta \in Mor(G_1, G_2)$ and let θ is the equivalence relation generated by $\{(\alpha(a), \beta(a)) : a \in A\}$. Then let $C = B/\theta = \{[b] : b \in B\}$, where $[b]$ is the equivalence class generated by b. Define a soft graph $G' = (\mathcal{F}', \mathcal{G}')_C$ by

1. $\mathcal{F}'([b_1]) = \cup\{\mathcal{F}_2(b) : b \in [b_1]\}$ for all $[b_1] \in C$ and
2. $\mathcal{G}'([b_1], [b_2])) = \cup\{\mathcal{G}_2(b_i, b_j)\} : b_i \in [b_1], b_j \in [b_2]$, for all $[b_1], [b_2] \in C$.

Consider the natural map $\gamma : B \to C$ by $\gamma(b) = [b]$ for all $b \in B$. Let $b_1 \in B$, then

1. $\mathcal{F}_2(b_1) \subseteq \cup\{\mathcal{F}_2(b) : b \in [b_1]\} = \mathcal{F}'([b_1]) = (\mathcal{F}' \circ \gamma)(b_1)$ and
2. $\mathcal{G}_2(b_1, b_2) \subseteq \cup\{\mathcal{G}_2(b_i, b_j) : b_i \in [b_1], b_j \in [b_2]\} = \mathcal{G}'([b_1], [b_2]) = (\mathcal{G}' \circ \gamma)(b_1, b_2)$.

It follows that $\gamma : G_2 \to G'$ is a soft graph morphism. Moreover, for all $a \in A$ we have, $(\gamma \circ \alpha)(a) = \gamma([\alpha(a)]) = \gamma([\beta(a)]) = (\gamma \circ \beta)(a)$. i.e., $\gamma \circ \alpha = \gamma \circ \beta$.

Let $\gamma' : G_2 \to G_3 = (\mathcal{F}_3, \mathcal{G}_3)_D$ be any soft graph morphism with $\gamma' \circ \alpha = \gamma' \circ \beta$. Then there exists a function $h : C \to D$ by $h([b]) = \gamma'(b)$, this implies that $(h \circ \gamma)(b) = h([b]) = \gamma'(b)$ for all $b \in B$.

Let $b \in B$. Then,

$$\begin{aligned} \mathcal{F}_2(b) &\subseteq (\mathcal{F}_3 \circ \gamma')(b) \\ &= (\mathcal{F}_3 \circ (h \circ \gamma))(b) \\ &= (\mathcal{F}_3 \circ h)([b]), \quad \text{which implies that} \\ \mathcal{F}'([b_0]) &= \cup\{\mathcal{F}_2(b) : b \in [b_0]\} \\ &\subseteq (\mathcal{F}_3 \circ h)([b_0]), \quad \text{for all } [b_0] \in C. \end{aligned}$$

Also, for all $b_1, b_2 \in B$

$$\begin{aligned} \mathcal{G}_2(b_1, b_2) &\subseteq (\mathcal{G}_3 \circ \gamma')(b_1, b_2) \\ &= (\mathcal{F}_3 \circ (h \circ \gamma))(b_1, b_2) \\ &= (\mathcal{F}_3 \circ h)([b_1], [b_2]), \quad \text{which implies that} \\ \mathcal{G}'([b_1], [b_2]) &= \cup\{\mathcal{G}_2(b_i, b_j) : b_i \in [b_1], b_j \in [b_2]\} \\ &\subseteq (\mathcal{G}_3 \circ h)([b_1], [b_2]) \quad \text{for all } [b_1], [b_2] \in C. \end{aligned}$$

Thus $h : G' \to G_3$ is a soft graph morphism and hence h is the coequalizer of α and β. □

Theorem 4.29 *The category* ***SGr(U)*** *has products*

Proof Let $G_{i_{A_i}} = (\mathcal{F}_i, \mathcal{G}_i)_{A_i}, i \in I$ be an indexed family of soft graphs over U. Let $A = \prod_{i\in I} A_i$. Then by Theorem 4.18, there exists a soft set $\mathcal{F}_A = \prod_{i\in I} \mathcal{F}_{i_{A_i}}$ (*i.e.*, $\mathcal{F}_A(a) = \bigcap_{i\in I} \mathcal{F}_i(a_i)$, for all $a \in A$) and a soft relation $\mathcal{G}_{A\times A} = \prod_{i\in I} \mathcal{G}_{i_{A_i\times A_i}}$ (*i.e.*, $\mathcal{G}_{A\times A}(a, b) = \bigcap_{i\in I} \mathcal{G}_i(a_i, b_i)$, for all $a, b \in A$).

Since each $G_{i_{A_i}} = (\mathcal{F}_i, \mathcal{G}_i)_{A_i}, i \in I$ is a soft graphs, we obtain

$$\begin{aligned}\mathcal{G}(a,b) &= \bigcap_{i\in I} \mathcal{G}_i(a_i, b_i)\\ &\subseteq \bigcap_{i\in I} (\mathcal{F}_i(a_i) \cap \mathcal{F}_i(b_i))\\ &= \bigcap_{i\in I} \mathcal{F}_i(a_i) \cap \bigcap_{i\in I} \mathcal{F}_i(b_i)\\ &= \mathcal{F}(a) \cap \mathcal{F}(b) \quad \text{for all } a, b \in A.\end{aligned}$$

This implies that $\mathcal{G}_{A\times A}$ is a soft relation on a soft set $\mathcal{F}_A$ and also it is symmetric. Hence $G_A = (\mathcal{F}, \mathcal{G})_A$ is a soft graph.

Then the essential projection maps are $P_{G_j} : G_A \to G_{j_{A_j}}$ by $P_{G_j}(a) = a_j$, for all $a \in A$. For every soft graph morphism $f_i : G'_C = (\mathcal{F}', \mathcal{G}')_C \to G_{i_{A_i}}$, there exists a mapping $h : G'_C \to G_A$ by $h(c) = (f_i(c))$.

Let us verify the following,

(a) P_{G_j} and h are soft graph morphisms.
(b) $(P_{G_j} \circ h) = f_j$, for all $j \in I$ and h is unique.

(a) For all $a, b \in A$ and $j \in I$, we have

1. $(\mathcal{F}_j \circ P_{G_j})(a) = \mathcal{F}_j(a_j) \supseteq \bigcap_{i\in I} \mathcal{F}_i(a_i) = \mathcal{F}(a)$
2. $(\mathcal{G}_j \circ P_{G_j})(a, b) = \mathcal{G}_j(a_j, b_j) \supseteq \bigcap_{i\in I} \mathcal{G}_i(a_i, b_i) = \mathcal{G}(a, b)$.

It implies that P_{G_j} is a soft graph morphism, for all $j \in I$. Now,

1. $(\mathcal{F} \circ h)(c) = \mathcal{F}(f_i(c)) = \bigcap_{i\in I}(\mathcal{F}_i(f_i(c)) \supseteq \mathcal{F}'(c)$.
2. $(\mathcal{G} \circ h)(c, d) = \mathcal{G}(f_i(c), f_i(d)) = \bigcap_{i\in I}(\mathcal{G}_i(f_i(c), f_i(d)) \supseteq \mathcal{G}'(c, d)$.

Therefore h is a soft graph morphism.

(b) For $c \in C$ and $j \in I$, we get

$(P_{G_j} \circ h)(c) = P_{G_j}(f_j(c)) = f_j(c)$. This implies that $P_{G_j} \circ h = f_j$, for all $j \in I$. Uniqueness of h follows from the definition of P_{G_j}.

Hence the soft graph $G_A = (\mathcal{F}, \mathcal{G})_A$ is the product of an indexed family of soft graphs $G_{i_{A_i}} = (\mathcal{F}_i, \mathcal{G}_i)_{A_i}, i \in I$. □

Theorem 4.30 *The category* ***SGr(U)*** *has coproducts*

Proof Let $G_{i_{A_i}} = (\mathcal{F}_i, \mathcal{G}_i)_{A_i}$, $i \in I$ be an indexed family of soft graphs over U. Let $A = \bigcup_{i \in I}(A_i \times \{i\})$, then by Theorem 4.20, there exists a soft set $\mathcal{F}_A$ defined by $\mathcal{F}_A(a_i, i) = \mathcal{F}_i(a_i)$, for all $a = (a_i, i) \in A$ and a soft relation $\mathcal{G}_{A\times A}$ defined by $\mathcal{G}_{A\times A}((a_i, i), (b_j, j)) = \begin{cases} \mathcal{G}_i(a_i, b_j) & \text{if } i = j; \\ \emptyset & \text{if } i \neq j \end{cases}$

Since each $G_{i_{A_i}} = (\mathcal{F}_i, \mathcal{G}_i)_{A_i}$, $i \in I$ are soft graphs, we get

$$\begin{aligned} \mathcal{G}((a_i, i), (b_j, j)) &= \mathcal{G}_i(a_i, b_j) \cap \mathcal{G}_j(b_j, b_j) \\ &\subseteq \mathcal{F}_i(a_i) \cap \mathcal{F}_j(b_j) \\ &= \mathcal{F}(a_i, i) \cap \mathcal{F}(b_j, j), \text{ for all } a = (a_i, i), b = (b_j, j) \in A \text{ and } i = j. \end{aligned}$$

It follows that $\mathcal{G}_{A\times A}$ is a soft relation on a soft set $\mathcal{F}_A$ and also it is symmetric. Hence $G_A = (\mathcal{F}, \mathcal{G})_A$ is a soft graph.

Now we define functions $\Psi_{G_j} : G_{j_{A_j}} \to G_A$ by $\Psi_{G_j}(a_j) = (a_j, j)$, for all $j \in I$. For every soft graph morphism $f_i : G_{i_{A_i}} \to G'_C = (\mathcal{F}', \mathcal{G}')_C$, there exists a mapping $h : G_A \to G'_C$ by $h(a) = (f_i(a_i))$, for all $a = (a_i, i) \in A$.

Let us verify following,

(a) Ψ_{G_j} and h are soft graph morphisms
(b) $h \circ \Psi_j = f_j$ and h is unique.

(a) For $a_j, b_j \in A_j$ and $j \in I$, we have

1. $(\mathcal{F} \circ \Psi_{G_j})(a_j) = \mathcal{F}(a_j, j) = \mathcal{F}_j(a_j)$
2. $(\mathcal{G} \circ \Psi_{G_j})(a_j, b_j) = \mathcal{G}((a_j, j), (b_j, j)) = \mathcal{G}_{,}(a_j, b_j)$

It implies that Ψ_{G_j} are soft graph morphisms, for $j \in I$.

1. $(\mathcal{F}' \circ h)(a_i, i) = \mathcal{F}'(f_i(a_i)) \supseteq \mathcal{F}_i(a_i) = \mathcal{F}(a_i, i)$
2. $(\mathcal{G}' \circ h)((a_j, j), (b_j, j)) = \mathcal{G}'(f_i(a_i), f_j(b_j)) = (\mathcal{G}' \circ f_i)(a_i, b_j) = \mathcal{G}_i)(a_i, b_j)$ $= \mathcal{G}((a_j, j), (b_j, j))$, for $i = j$. If $i \neq j$, it is obvious.

It implies that h is a soft graph morphism.

(b) For $a_j \in A_j$ and $j \in I$, we get $(h \circ \Psi_{G_j})(a_j) = h(a_j, j) = f_j(a_j)$. This implies that $h \circ \Psi_{G_j} = f_j$, for all $j \in I$. Uniqueness of h followed from the definition of Ψ_{G_j}.

Hence the soft graph $G_A = (\mathcal{F}, \mathcal{G})_A$ is the coproduct of an indexed family of soft graphs $G_{i_{A_i}} = (\mathcal{F}_i, \mathcal{G}_i)_{A_i}$, $i \in I$. □

Theorem 4.31 *The category* ***SGr(U)*** *is a complete category.*

Proof Theorems 4.27, 4.29 and 4.3 establish that **SGr(U)** is a complete category. □

Theorem 4.32 *The category* ***SGr(U)*** *is a cocomplete category.*

Proof Theorems 4.28, 4.30 and 4.3 establish that **SGr(U)** is a cocomplete category. □

4.5.2 Existence of an Adjoint

Theorem 4.33 *Let* **Sset(U)** *and* **SGr(U)** *be category of soft sets and category of soft graphs over U respectively. Then*

*(a) there exists a functor G : **Sset(U)** $\rightarrow$ **SGr(U)** which maps a soft set $\mathcal{F}_A$ to the soft graph $(\mathcal{F}, \mathcal{G}^*)_A$, where $\mathcal{G}^*(a_1, a_2) = \mathcal{F}(a_1) \cap \mathcal{F}(a_2), \forall a_1, a_2 \in A$.*

*(b) there exists a functor F : **SGr(U)** $\rightarrow$ **Sset(U)** which maps a soft graph $(\mathcal{F}, \mathcal{G})_A$ to the soft set $\mathcal{F}_A$.*

Proof (a) Let $h : \mathcal{F}_A \rightarrow \mathcal{F}'_B$ be a soft morphism in **Sset(U)**.
To prove that $G(h) : (\mathcal{F}, \mathcal{G}^*)_A \rightarrow (\mathcal{F}', \mathcal{G}'^*)_B$ by $G(h) = h$ is soft graph morphism. For convenience, we denote $G(h)$ by Gh.

1. Since h is a soft morphism, we have $\mathcal{F}(a) \subseteq (\mathcal{F}' \circ h)(a)$, for all $a \in A$.
2. Let $a_1, a_2 \in A$, then we have

$$\begin{aligned}(\mathcal{G}'^* \circ Gh)(a_1, a_2) &= \mathcal{G}'^*(Gh(a_1, a_2))\\ &= \mathcal{G}'^*(Gh(a_1), Gh(a_2))\\ &= \mathcal{G}'^*(h(a_1,), h(a_2))\\ &= \mathcal{F}'(h(a_1)) \cap \mathcal{F}'(h(a_2))\\ &\supseteq \mathcal{F}(a_1) \cap \mathcal{F}(a_2) = \mathcal{G}^*(a_1, a_2)\end{aligned}$$

   It follows that Gh is a soft graph morphism.
3. $Gid_{\mathcal{F}_A} : (\mathcal{F}, \mathcal{G}^*)_A \rightarrow (\mathcal{F}, \mathcal{G}^*)_A$ is the identity soft morphism.
4. G preserves composition
   Suppose $h : \mathcal{F}_A \rightarrow \mathcal{F}'_B$ and $h' : \mathcal{F}'_B \rightarrow \mathcal{F}''_C$ are soft morphisms. We have $h' \circ h : \mathcal{F}_A \rightarrow \mathcal{F}''_C$ is a soft morphism, then

   4a. For $a \in A$, we have

$$\begin{aligned}[\mathcal{F}'' \circ G(h' \circ h)](a) &= [\mathcal{F}'' \circ (h' \circ h)](a)\\ &= (\mathcal{F}'' \circ h')(h((a)))\\ &\supseteq \mathcal{F}'(h(a))\\ &\supseteq \mathcal{F}(a)\end{aligned}$$

   4b. For $a_1, a_2 \in A$, we have

$$\begin{aligned}[\mathcal{G}''^* \circ G(h' \circ h)](a_1, a_2) &= [\mathcal{G}''^* \circ (h' \circ h)](a_1, a_2)\\ &= (\mathcal{G}''^* \circ h')(h(a_1), h(a_2))\\ &\supseteq \mathcal{G}'^*(h(a_1), h(a_2))\\ &\supseteq \mathcal{G}^*(a_1, a_2)\end{aligned}$$

Fig. 4.13 ϵ is a natural transformation

$$\begin{array}{ccc} (\mathcal{F},\mathcal{G})_A & \xrightarrow{\epsilon_{(\mathcal{F},\mathcal{G})_A}} & GF(\mathcal{F},\mathcal{G})_A \\ h\downarrow & & \downarrow GF(h) \\ (\mathcal{F}',\mathcal{G}')_B & \xrightarrow{\epsilon_{(\mathcal{F}',\mathcal{G}')_B}} & GF(\mathcal{F}',\mathcal{G}')_B \end{array}$$

Thus $G(h' \circ h) : (\mathcal{F}, \mathcal{G}^*) \to (\mathcal{F}, \mathcal{G}''^*)$ is a soft graph morphism in **SGr(U)** and $G(h' \circ h) = h' \circ h = Gh' \circ Gh$. Hence G : **Sset(U)** $\to$ **SGr(U)** is a functor.

(b) Let $k : (\mathcal{F}, \mathcal{G})_A \to (\mathcal{F}', \mathcal{G}')_B$ be a soft graph morphism. By the definition of soft graph morphism, $k : \mathcal{F}_A \to \mathcal{F}'_B$ is a soft morphism. It implies that $Fk : \mathcal{F}_A \to \mathcal{F}'_B$ by $Fk = k$ is a soft morphism and F : **SGr(U)** $\to$ **Sset(U)** is a functor. □

Remark 4.22 Suppose F and G are functors in the Theorem 4.33, then the compositions $F \circ G$: **Sset(U)** $\to$ **Sset(U)** and $G \circ F$: **SGr(U)** $\to$ **SGr(U)** are functors and denoted by FG and GF respectively.

Now we turn to construct two natural transformations ϵ and δ as follows.

Lemma 4.2 *Let F, G be the functors in the Theorem 4.33, then the function $\epsilon : Id_{\mathbf{SGr(U)}} \to GF$ defined by $\epsilon_{(\mathcal{F},\mathcal{G})_A} : (\mathcal{F}, \mathcal{G})_A \to (\mathcal{F}, \mathcal{G}^*)_A$ is a natural transformation.*

Proof From the definition of $\epsilon_{(\mathcal{F},\mathcal{G})_A}$, it is evident that $\epsilon_{(\mathcal{F},\mathcal{G})_A}$ maps every element of A to itself. We proceed to show that $\epsilon_{(\mathcal{F},\mathcal{G})_A}$ is a soft graph morphism in **SGr(U)**.

Let $a, a_1, a_2 \in A$, we have

1. $(\mathcal{F} \circ \epsilon_{(\mathcal{F},\mathcal{G})_A})(a) = \mathcal{F}(a)$
2. Since $(\mathcal{F}, \mathcal{G})_A$ is a soft graph, we have
$(\mathcal{G}^* \circ \epsilon_{(\mathcal{F},\mathcal{G})_A})(a_1, a_2) = \mathcal{G}^*(a_1, a_2) = \mathcal{F}(a_1) \cap \mathcal{F}(a_2) \supseteq \mathcal{G}(a_1, a_2)$.

It implies that $\epsilon_{(\mathcal{F},\mathcal{G})_A}$ is a soft graph morphism. This means ϵ maps an object $(\mathcal{F}, \mathcal{G})_A$ in **SGr(U)** to a soft graph morphism $\epsilon_{(\mathcal{F},\mathcal{G})_A}$.
Let $h \in Mor((\mathcal{F}, \mathcal{G})_A, (\mathcal{F}', \mathcal{G}')_B)$, then the diagram (Fig. 4.13) commutes.

Let $a \in A$. Then we obtain $(GF(h) \circ \epsilon_{(\mathcal{F},\mathcal{G})_A})(a) = GF(h)(a) = h(a)$ and $(\epsilon_{(\mathcal{F}',\mathcal{G}')_B} \circ h)(a) = h(a)$. Thus ϵ is a natural transformation. □

Lemma 4.3 *Let F, G be the functors in the Theorem 4.33, then the function $\delta : FG \to Id_{\mathbf{Sset(U)}}$ defined by $\delta_{\mathcal{F}_A} : \mathcal{F}_A \to \mathcal{F}_A$ is a natural transformation.*

Proof From the definition, $\delta_{\mathcal{F}_A}$ maps every element of A to itself. So it is soft morphism in **Sset(U)**. Let $k \in Mor(\mathcal{F}_A, \mathcal{F}_B)$, then the diagram (Fig. 4.14) commutes.

Let $a \in A$, we have
$(k \circ \delta_{\mathcal{F}_A})(a) = k(a)$ and $(\delta_{\mathcal{F}_A} \circ FG(k))(a) = \delta_{\mathcal{F}_A}(k(a)) = k(a)$.
It follows that δ is a natural transformation. □

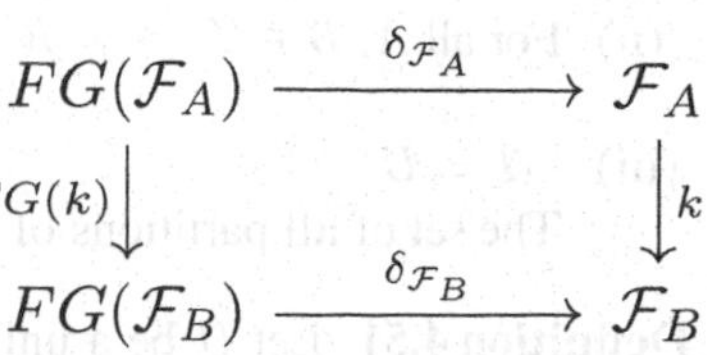

Fig. 4.14 δ is a natural transformation

$$\begin{array}{ccc} FG(\mathcal{F}_A) & \xrightarrow{\delta_{\mathcal{F}_A}} & \mathcal{F}_A \\ {\scriptstyle FG(k)}\downarrow & & \downarrow{\scriptstyle k} \\ FG(\mathcal{F}_B) & \xrightarrow{\delta_{\mathcal{F}_B}} & \mathcal{F}_B \end{array}$$

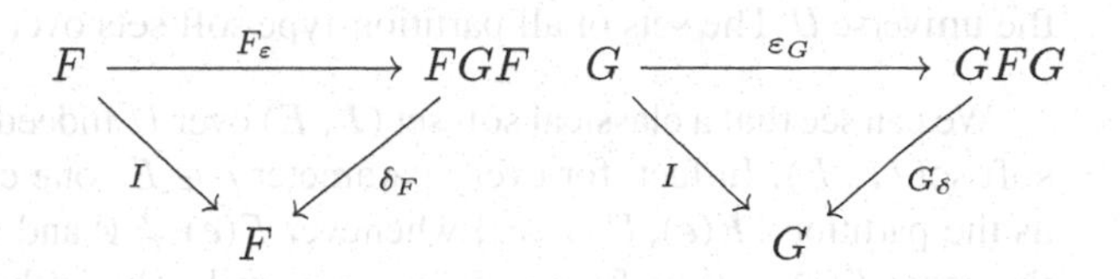

Fig. 4.15 F is a left adjoint of G

$$F \xrightarrow{F_\epsilon} FGF \xrightarrow{\delta_F} F$$

$$G \xrightarrow{\epsilon_G} GFG \xrightarrow{G_\delta} G$$

Theorem 4.34 *Let F and G be the functors in the Theorem 4.33, Then F is a left adjoint of G.*

Proof From Lemmas 4.2 and 4.3 and Theorem 4.4, it remains to show that the diagrams (Fig. 4.15) commutes.
$FGF((\mathcal{F}, \mathcal{G})_A) = FG(\mathcal{F}_A) = F((\mathcal{F}, \mathcal{G}^*)_A) = \mathcal{F}_A = F((\mathcal{F}, \mathcal{G})_A)$ and
$GFG(\mathcal{F}_A) = GF((\mathcal{F}, \mathcal{G}^*)_A)) = G(\mathcal{F}_A)$.
Therefore $\delta_F \circ F_\epsilon = Id_F$ and $G_\delta \circ \epsilon_G = Id_G$. It implies that F is a left adjoint of G. □

4.6 Relationships of Soft Sets with Information Systems

In this section we discuss the relationship between soft sets and classical information systems. A soft set can be identified with a simple information system in which the attributes only take a finite number of values. Most of the results given in this section is taken from D. Pei and D. Miao [102].

Definition 4.50 Let $\mathcal{T} \subseteq \mathcal{P}(U)$. $\mathcal{T}$ is called a partition of the universe U if the following conditions hold:

(i) $\emptyset \neq \mathcal{T}$

(ii) For all $A, B \in \mathcal{T}, A \neq B \Longrightarrow A \cap B = \emptyset$

(iii) $\cup \boldsymbol{T} = U$
The set of all partitions of the universe U is denoted by $\text{Par}(U)$

Definition 4.51 Let U be a universe. An ordered pair (F, E) is called a partition-type soft set over U if F is a mapping from E to the set $\text{Par}(U)$ of all partitions of the universe U The sets of all partition-type soft sets over U are denoted by $PS(U)$

We can see that a classical soft set (F, E) over U indeed is a special partition-type soft set (F, E). In fact, for every parameter $e \in E$, one can define the image $F(e)$ as the partition $\{F(e), U \backslash F(e)\}$ whenever $F(e) \neq \emptyset$ and $F(e) \neq U$. In addition, in the cases $F(e) = \emptyset$ or $F(e) = U$, we can easily obtain the corresponding partitions of U (they only contain an element U).

By making use of the fact that a partition of some universe must be a covering of the universe, we may more generally consider the situations in which the image of every parameter under the mapping is a covering of the universe.

Definition 4.52 Let $\mathcal{C} \subseteq \mathcal{P}(U)$. $\mathcal{C}$ is called a covering of the universe U, if the following conditions hold: (i) $\emptyset \notin \mathcal{C}$ (ii) $\cup \mathcal{C} = U$. The set of all coverings of the universe U is denoted by $\text{Cov}(U)$

Definition 4.53 Let U be a universe. An ordered pair (F, E) is called a covering-type soft set over U if F is a mapping from E to the set $Cov(U)$ of all coverings of the universe U. The sets of all covering-type soft sets over U are denoted by $CS(U)$

Definition 4.54 The quadruple (U, A, F, V) is called an information system, or a database system, or an information table, where $U = \{x_1, \ldots, x_n\}$ is a universe containing all interested objects, $A = \{a_1, \ldots, a_m\}$ is a set of attributes, $V = \cup_{i=1}^{m} V_i$ where V_j is the value set of the attribute a_j, and $F = \{f_1, \ldots, f_m\}$ where $f_j : U \to V_j$

Usually, one assumes that every V_j only contains finite elements (these elements may be and may not be numbers) for every $j \leq m$. Such information systems are called classical information systems. However, if $V_j = [0, 1]$ for every $j \leq m$ then the corresponding information systems are called fuzzy information systems. Furthermore, if $f_j : \quad U \to \mathcal{P}\left(V_j\right)$ is a mapping from U to the power set of V_j for all $j \leq m$, then the corresponding information systems are called set-valued information systems. Similarly, many other types information systems are applied to information processing and data analysis.

Theorem 4.35 *If (F, E) is a soft set over the universe U with the parameter set $E = \{e_1, \ldots, e_m\}$ then (F, E) is an information system.*

Proof In fact, the given soft set (F, E) can be seen as an information system (U, G, A, V) according to the following manner:

$G = \{g_1, \ldots, g_m\}, g_i : U \to V_i, g_i(x) = \begin{cases} 1, & x \in F(e_i) \\ 0, & \text{otherwise} \end{cases}$
$A = E, \quad V = \bigcup_{i=1}^{m} V_i, \quad V_i = \{0, 1\}, 1 \leq i \leq m$ □

Remark 4.23 If R is the indiscernibility relation induced by the attribute set $A = \{a_1, \ldots, a_m\}$ as: $R = \{(x, y) \in U \times U | g_i(x) = g_i(y), 1 \leq i \leq m\}$, then information system (U, G, A, V) can be naturally translated into an approximation space (U, R).

Corollary 4.4 *Let (F, E) is a soft set over the universe U with the parameter set $E = \{e_1, \ldots, e_m\}$ then (U, R) is an approximation space where R is defined as: $R = \{(x, y) \in U \times U | g_i(x) = g_i(y), 1 \leq i \leq m\}$.*

Theorem 4.36 *There is a one-one correspondence between PS(U) and Set of all classical information systems.*

Proof If $(F, E) \in PS(U)$ is a partition-type soft set over U with the parameter set $E = \{e_1, \ldots, e_m\}$ and the mapping $F : E \to \text{Par}(U)$, then (F, E) is an information system with the form (U, A, G, V) according to the following manner: $G = \{g_1, \ldots, g_m\}, g_i : U \to V_i, g_j(x) = v_j^{(i)}, x \in [x_j]_i, j \leq n_i$

$$A = E, \quad V = \bigcup_{i=1}^{m} V_i, \quad V_i = \left\{v_1^{(i)}, \ldots, v_{n_i}^{(i)}\right\}, i \leq m$$

where $[x_j]_i$ is the block of the partition $F(e_i)$ which containing x_j and $|F(e_i)| = n_i$ for all $i \leq m$.

Conversely, if (U, A, G, V) is an information system with the attribute set $A = \{a_1, \ldots, a_m\}$ and the set of mappings $G = \{g_1, \ldots, g_m\}$ with $g_i : \quad U \to V_i$ then (U, A, G, V) is a partition-type soft set (F, E) with the parameter set $E = A$ and the mapping $F : E \to Par(U)$ with $F(e_i) = U/R$ where R is the indiscernibility relation induced by the attribute set A, and U/R is the partition of U consisting of all equivalence classes with respect to R (also called the quotient set of U with respect to R). □

According to the original definition given by Pawlak [97, 98], the relation R under consideration in rough set theory is an equivalence relation on the universe U. Such approximation spaces are called classical approximation spaces, or Pawlak approximation spaces. We use the notation PAS(U) to represent the set of all Pawlak approximation spaces

Corollary 4.5 *The sets $PS(U)$ of all partition-type soft sets and the set $PAS(U)$ of all Pawlak approximation spaces are the same, or there exists a bijection between $PS(U)$ and $PAS(U)$*

Recently, many researchers generalized Pawlak approximation spaces to more general cases. For example, Yao [159] extended Pawlak approximation spaces to the cases where the equivalence relations are replaced by arbitrary binary relations on

U which are called generalized approximation spaces. A detailed further discussion regarding these is provided in section titled "Parameter reduction of soft sets by means of attribute reductions in information systems" in Chap. 6.

Chapter 5
Hybrid Structures Involving Soft Sets

Hybridization of existing structures always yielded better results in decision making and other applications of uncertainty modelling structures. Soft sets are also not an exception and many hybrid structures involving soft sets have already been evolved. In this chapter we will have a quick look into these structures. They include hybridizations including Fuzzy sets, Intuitionistic fuzzy sets, Hesitant fuzzy sets, Rough sets etc.

5.1 Fuzzy Soft Sets and Soft Fuzzy Sets

Two different types of hybrid structures involving soft sets and fuzzy sets, namely fuzzy soft sets and soft fuzzy sets will be considered in this section.

5.1.1 Fuzzy Soft Sets

The concept of fuzzy soft sets (fs-sets) was introduced by Maji et al. [78] by merging the ideas of fuzzy sets and soft sets. Later Som [130] defined soft relation and fuzzy soft relation on the theory of soft sets. The operations of soft sets and fs-sets defined by Maji et al. and Roy et al. [79, 116] are used in many applications. But, Chen et al. [32], Pei and Miao [102], and many other researchers pointed out that these works have some weak points. Therefore, Cagman et al. [26] redefined the fs-sets and their operations. This subsection contains fuzzy soft set theory results given in this sense and mainly adapted from [26].

Let us recall the definition of a fuzzy set. Let U be a universe. A fuzzy set X over U is a set defined by a function μ_X representing a mapping

$$\mu_X : U \to [0, 1]$$

S. J. John, *Soft Sets*, Studies in Fuzziness and Soft Computing 400,
https://doi.org/10.1007/978-3-030-57654-7_5

μ_X is called the membership function of X, and the value $\mu_X(u)$ is called the grade of membership of $u \in U$. The value represents the degree of u belonging to the fuzzy set X. Thus, a fuzzy set X over U can be represented as follows:

$$X = \{(\mu_X(u)/u) : u \in U, \mu_X(x) \in [0, 1]\}.$$

Note that the set of all the fuzzy sets over U will be denoted by $F(U)$.

First we define fuzzy soft sets (fs-sets) and their operations. In classical soft sets, the parameter sets and the approximate functions are crisp. But in the fs-sets, while the parameters sets are crisp, the approximate functions are fuzzy subsets of U. We use $\Gamma_A, \Gamma_B, \Gamma_C, \ldots$, etc. for fs -sets and $\gamma_A, \gamma_B, \gamma_C, \ldots$, etc. for their fuzzy approximate functions, respectively.

Definition 5.1 An fs-set Γ_A over U is a set defined by a function γ_A representing a mapping

$$\gamma_A : E \to F(U) \text{ such that } \gamma_A(x) = \emptyset \text{ if } x \notin A.$$

Here, γ_A is called fuzzy approximate function of the $fs-$ set Γ_A, and the value $\gamma_A(x)$ is a set called x -element of the fs-set for all $x \in E$. Thus, an fs -set Γ_A over U can be represented by the set of ordered pairs

$$\Gamma_A = \{(x, \gamma_A(x)) : x \in E, \gamma_A(x) \in F(U)\}.$$

Note that the set of all $fs-$ sets over U will be denoted by $FS(U)$.

Example 68 Let $U = \{u_l, u_2, u_3, u_4, u_5\}$ be a universal set and $E = \{x_1, x_2, x_3, x_4\}$ be a set of parameters. If $A = \{x_1, x_2, x_4\} \subseteq E$, $\gamma_A(x_1) = \{0.9/u_2, 0.5/u_4\}$, $\gamma_A(x_2) = U$, and $\gamma_A(x_4) = \{0.2/u_1, 0.4/u_3, 0.8/u_5\}$, then the soft set F_A is written by

$$F_A = \{(x_1, \{0.9/u_2, 0.5/u_4\}), (x_2, U), (x_4, \{0.2/u_1, 0.4/u_3, 0.8/u_5\})\}.$$

Definition 5.2 Let $\Gamma_A \in FS(U)$. If $\gamma_A(x) = \emptyset$ for all $x \in E$, then Γ_A is called an empty fs -set, denoted by Γ_Φ.

Definition 5.3 Let $\Gamma_A \in FS(U)$. If $\gamma_A(x) = U$ for all $x \in A$, then Γ_A is called A-universal f s-set, denoted by $\Gamma_{\tilde{A}}$.

If $A = E$, then the A -universal $fs-$ set is called universal $fs-$ set, denoted by $\Gamma_{\tilde{E}}$.

Example 69 Assume that $U = \{u_1, u_2, u_3, u_4, u_5\}$ is a universal set and $E = \{x_1, x_2, x_3, x_4\}$ is a set of all parameters.

If $A = \{x_2, x_3, x_4\}$, $\gamma_A(x_2) = \{0.5/u_2, 0.9/u_4\}$, $\gamma_A(x_3) = \emptyset$ and $\gamma_A(x_4) = U$ then the fs-set Γ_A is written by $\Gamma_A = \{(x_2, \{0.5/u_2, 0.9/u_4\}), (x_4, U)\}$.

If $B = \{x_1, x_3\}$, and $\gamma_B(x_1) = \emptyset$, $\gamma_B(x_3) = \emptyset$, then the fs-set Γ_B is an empty fs-set, i.e., $\Gamma_B = \Gamma_\Phi$.

If $C = \{x_1, x_2\}$, $\gamma_C(x_1) = U$, and $\gamma_C(x_2) = U$, then the f s-set Γ_C is a $C-$ universal $fs_{-\text{Set}}$, i.e., , $\Gamma_C = \Gamma_{\tilde{C}}$.

If $D = E$, and $\gamma_D(x_i) = U$ for all $x_i \in E$, where $i = 1, 2, 3, 4$, then the $fs-$ set Γ_D is a universal fs -set, i.e., $\Gamma_D = \Gamma_{\tilde{E}}$.

Definition 5.4 Let $\Gamma_A, \Gamma_B \in FS(U)$. Then, Γ_A is an fs-subset of Γ_B, denoted by $\Gamma_A \tilde{\subseteq} \Gamma_B$, if $\gamma_A(x) \subseteq \gamma_B(x)$ for all $x \in E$.

Remark $\Gamma_A \subseteq \Gamma_B$ does not imply that every element of Γ_A is an element of Γ_B as in the definition of the classical subset.

For example, assume that $U = \{u_1, u_2, u_3, u_4\}$ is a universal set of objects and $E = \{x_1, x_2, x_3\}$ is a set of all the parameters. If $A = \{x_1\}$, $B = \{x_1, x_3\}$, $\Gamma_A = \{(x_1, \{0.2/u_2\})\}$ and $\Gamma_B = \{(x_1, \{0.9/u_2, 0.3/u_3, 0.5/u_4\})\ (x_3, \{0.2/u_1, 0.7/u_5\})\}$, then for all $x \in E, \gamma_A(x) \subseteq \gamma_B(x)$ is valid. Hence, $\Gamma_A \tilde{\subseteq} \Gamma_B$. It is clear that $(x_1, \{0.2/u_2\}) \in \Gamma_A$, but $(x_1, \{0.2/u_2\}) \notin \Gamma_B$.

Proposition 5.1 *Let* $\Gamma_A, \Gamma_B \in FS(U)$. *Then,*

(1) $\Gamma_A \tilde{\subseteq} \Gamma_{\tilde{E}}$
(2) $\Gamma_\Phi \tilde{\subseteq} \Gamma_A$
(3) $\Gamma_A \tilde{\subseteq} \Gamma_A$
(4) $\Gamma_A \tilde{\subseteq} \Gamma_B$ *and* $\Gamma_B \tilde{\subseteq} \Gamma_C \Rightarrow \Gamma_A \tilde{\subseteq} \Gamma_C$.

Proof Clear from definitions. □

Definition 5.5 Let $\Gamma_A, \Gamma_B \in FS(U)$. Then, Γ_A and Γ_B are fs-equal, written as $\Gamma_A = \Gamma_B$, if and only if $\gamma_A(x) = \gamma_B(x)$ for all $x \in E$.

Proposition 5.2 *Let* $\Gamma_A, \Gamma_B, \Gamma_C \in FS(U)$. *Then*

(1) $\Gamma_A = \Gamma_B$ *and* $\Gamma_B = \Gamma_C \Rightarrow \Gamma_A = \Gamma_C$
(2) $\Gamma_A \tilde{\subseteq} \Gamma_B$ *and* $\Gamma_B \tilde{\subseteq} \Gamma_A \Leftrightarrow \Gamma_A = \Gamma_B$.

Proof Clear from definitions. □

Definition 5.6 Let $\Gamma_A \in FS(U)$. Then, the complement $\Gamma_A^{\tilde{c}}$ of Γ_A is an fs-set such that

$$\gamma_{A^{\tilde{c}}}(x) = \gamma_A^c(x), \text{ for all } x \in E$$

where $\gamma_A^c(x)$ is complement of the set $\gamma_A(x)$.

Proposition 5.3 *Let* $\Gamma_A \in FS(U)$. *Then,*

(1) $\left(\Gamma_A^{\tilde{c}}\right)^{\tilde{c}} = \Gamma_A$
(2) $\Gamma_\Phi^{\tilde{c}} = \Gamma_{\tilde{E}}$.

Proof By using the fuzzy approximate functions of the fs-sets, the proofs are straight forward. □

Definition 5.7 Let $\Gamma_A, \Gamma_B \in FS(U)$. Then, the union of Γ_A and Γ_B, denoted by $\Gamma_A \widetilde{\cup} \Gamma_B$, is defined by its fuzzy approximate function

$$\gamma_{A\widetilde{\cup}B}(x) = \gamma_A(x) \cup \gamma_B(x) \text{ for all } x \in E\,.$$

Proposition 5.4 *Let $\Gamma_A, \Gamma_B, \Gamma_C \in FS(U)$. Then,*
(1) $\Gamma_A \widetilde{\cup} \Gamma_A = \Gamma_A$
(2) $\Gamma_A \widetilde{\cup} \Gamma_\Phi = \Gamma_A$
(3) $\Gamma_A \widetilde{\cup} \Gamma_{\tilde{E}} = \Gamma_{\tilde{E}}$
(4) $\Gamma_A \widetilde{\cup} \Gamma_B = \Gamma_B \widetilde{\cup} \Gamma_A$
(5) $\left(\Gamma_A \widetilde{\cup} \Gamma_B\right) \widetilde{\cup} \Gamma_C = \Gamma_A \widetilde{\cup} \left(\Gamma_B \widetilde{\cup} \Gamma_C\right)$.

Proof Clear from definitions. □

Definition 5.8 Let $\Gamma_A, \Gamma_B \in FS(U)$. Then, the intersection of Γ_A and Γ_B, denoted by $\Gamma_A \widetilde{\cap} \Gamma_B$, is defined by its fuzzy approximate function

$$\gamma_{A\widetilde{\cap}B}(x) = \gamma_A(x) \cap \gamma_B(x) \text{ for all } x \in E\,.$$

Proposition 5.5 *Let $\Gamma_A, \Gamma_B, \Gamma_C \in FS(U)$. Then,*
(1) $\Gamma_A \widetilde{\cap} \Gamma_A = \Gamma_A$
(2) $\Gamma_A \widetilde{\cap} \Gamma_\Phi = \Gamma_\Phi$
(3) $\Gamma_A \widetilde{\cap} \Gamma_{\tilde{E}} = \Gamma_{\tilde{A}}$
(4) $\Gamma_A \widetilde{\cap} \Gamma_B = \Gamma_B \widetilde{\cap} \Gamma_A$
(5) $\left(\Gamma_A \widetilde{\cap} \Gamma_B\right) \widetilde{\cap} \Gamma_C = \Gamma_A \widetilde{\cap} \left(\Gamma_B \widetilde{\cap} \Gamma_C\right)$.

Proof Clear from definitions. □

Remark Let $\Gamma_A \in FS(U)$. If $\Gamma_A \neq \Gamma_\Phi$ and $\Gamma_A \neq \Gamma_{\tilde{E}}$, then $\Gamma_A \widetilde{\cup} \Gamma_A^{\tilde{c}} \neq \Gamma_{\tilde{E}}$ and $\Gamma_A \widetilde{\cap} \Gamma_A^{\tilde{c}} \neq \Gamma_\Phi$.

Theorem 5.1 *(DeMorgan Laws) Let $\Gamma_A, \Gamma_B \in FS(U)$, then*
(1) $\left(\Gamma_A \widetilde{\cup} \Gamma_B\right)^{\tilde{c}} = \Gamma_A^{\tilde{c}} \widetilde{\cap} \Gamma_B^{\tilde{c}}$.
(2) $\left(\Gamma_A \widetilde{\cap} \Gamma_B\right)^{\tilde{c}} = \Gamma_A^{\tilde{c}} \widetilde{\cup} \Gamma_B^{\tilde{c}}$.

Proof The proofs can be obtained by using the respective approximate functions.
(1) For all $x \in E$,

$$\begin{aligned}
\gamma_{(A\widetilde{\cup}B)^{\tilde{c}}(x)} &= \gamma^c_{A\widetilde{\cup}B}(x) \\
&= (\gamma_A(x) \cup \gamma_B(x))^c \\
&= (\gamma_A(x))^c \cap (\gamma_B(x))^c \\
&= \gamma^c_A(x) \cap \gamma^c_B(x) \\
&= \gamma_{A^{\tilde{c}}}(x) \cap \gamma_{B^{\tilde{c}}}(x) \\
&= \gamma_{A^{\tilde{c}}\widetilde{\cap}B^{\tilde{c}}}(x).
\end{aligned}$$

The proof of (2) is similar. □

Proposition 5.6 *Let $\Gamma_A, \Gamma_B, \Gamma_C \in FS(U)$. Then,*
(1) $\Gamma_A \tilde{\cup} \left(\Gamma_B \tilde{\cap} \Gamma_C\right) = \left(\Gamma_A \tilde{\cup} \Gamma_B\right) \tilde{\cap} \left(\Gamma_A \tilde{\cup} \Gamma_C\right)$
(2) $\Gamma_A \tilde{\cap} \left(\Gamma_B \tilde{\cup} \Gamma_C\right) = \left(\Gamma_A \tilde{\cap} \Gamma_B\right) \tilde{\cup} \left(\Gamma_A \tilde{\cap} \Gamma_C\right)$.

Proof For all $x \in E$

$$
\begin{aligned}
(1)\gamma_{A\tilde{\cup}(B\tilde{\cap}C)}(x) &= \gamma_A(x) \cup \gamma_{B\tilde{\cap}C}(x) \\
&= \gamma_A(x) \cup (\gamma_B(x) \cap \gamma_C(x)) \\
&= (\gamma_A(x) \cup \gamma_B(x)) \cap (\gamma_A(x) \cup \gamma_C(x)) \\
&= \gamma_A \tilde{\cup}_B(x) \cap \gamma_{A\tilde{\cup}C}(x) \\
&= \gamma_{(A\tilde{\cup}B)\tilde{\cap}(A\tilde{\cup}C)}(x).
\end{aligned}
$$

Likewise, the proof of (2) can be made in a similar way. □

5.1.2 Soft Fuzzy Sets

A fuzzy soft set is a parameterized collection of fuzzy subsets of the universe U but soft fuzzy set discusses theories based on fuzzy relationship between an initial universe set and a set of parameters. Results and examples given are taken from Bing-xue Yao et al. [156].

Let X and Y be initial universes, a relation R from X to Y is a subset of $X \times Y$. A fuzzy relation from X to Y is a fuzzy subset of $X \times Y$.

Definition 5.9 Let $\tilde{R}$ be a fuzzy subset of $X \times Y$, and $\tilde{R}$ is defined as a fuzzy relationship from X to Y. We write $X \xrightarrow{\tilde{R}} Y$, and $\tilde{R}(x, y)$ denotes the degree of corresponding between x and y based on the relationship $\tilde{R}$.

Let $F(X \times Y)$ denotes the family of fuzzy relationships from X to Y.

The set $\tilde{R}_\alpha = \{(x, y) \in X \times Y | \tilde{R}(x, y) \geq \alpha\} \subset X \times Y$ is defined as cut-set if $\tilde{R} \in F(X \times Y)$ for $\alpha \in [0, 1]$. Then $\tilde{R}_\alpha$ denote a relationship from X to Y and $\tilde{R}_\alpha$ is called $\alpha-$ cut relationship.

Definition 5.10 Let U be an initial universe set and E be a set of parameters. Let $P(U)$ denotes the power set of U. Let $A \subset E$. A pair (F, A) is called a soft fuzzy set over U, where F is a mapping given by $F : A \to P(U)$ and $F(x) = \{y \in U | \tilde{R}(x, y) \geq \alpha \; x \in A, y \in U, \alpha \in [0, 1]\}$ (Here $\tilde{R}$ denotes fuzzy relationship between E and U).

Definition 5.11 Let U be an initial universe set and E be a set of parameters. Let $P(U)$ denotes the power set of U. Let $A \subset E$. A pair (F, A) is called a soft fuzzy set over U, where F is a mapping given by $F : A \to P(U)$ and $F(x) = \left\{y \in U | (x, y) \in \tilde{R}_\alpha \; x \in A, y \in U, \alpha \in [0, 1]\right\}$ (Here $\tilde{R}_\alpha$ denotes α -cut relationship between E and U).

Clearly both definitions given above are equivalent.

Example 70 Suppose that U is the set of houses under consideration and E is the set of parameters where each parameter is a fuzzy word or a sentence involving fuzzy words, $E = \{$ expensive (e_1), beautiful (e_2), wooden (e_3), cheap (e_4), in the green surroundings $(e_5)\}$. In the case, a fuzzy soft set means pointing out expensive houses, beautiful houses, and so on. The fuzzy soft set (F, E) describes the "attractiveness of the houses" which Mr. X thinks.

Suppose that fuzzy relationship $\widetilde{R} = 0.6/(h_1, e_1) + 0.3/(h_1, e_2) + \cdots 0.3/(h_1, e_5) + 0.8/(h_2, e_1) + \cdots +0.7/(h_2, e_5) + 0.7/(h_3, e_1) + \cdots .0.3/(h_3, e_5) + 0.8/(h_4, e_1) + \cdots 0.7/(h_4, e_5) + 0.6/(h_5, e_1) + \cdots +0.9/(h_5, e_5) + 0.2/(h_6, e_1) + \cdots 0.7/(h_6, e_5)$.
$F(e_1) = \{(h_1, 0.6), (h_2, 0.8), (h_3, 0.7), (h_4, 0.8), (h_5, 0.6)\ (h_6, 0.2)\}$,
$F(e_2) = \{\cdots, \cdots\}$, $F(e_3) = \{\cdots, \cdots\}$
$F(e_4) = \{\cdots, \cdots\}$, $F(e_5) = \{(h_1, 0.3), (h_2, 0.7), (h_3, 0.3), (h_4, 0.7), (h_5, 0.9), (h_6, 0.7)\}$

The soft fuzzy set (F, E) is a parameterized family $\{F(e_i), i = 1, 2, \ldots, 8\}$ and gives us a collection of approximate description of an object. The mapping F here is "house ()" where () is to be filled up by a parameter $e \in E$. Therefore $F(e_1)$ means "house (expensive)" whose functional-value is the fuzzy set $\{(h_1, 0.6), (h_2, 0.8), (h_3, 0.7), (h_4, 0.8), (h_5, 0.6)\ (h_6, 0.2)\}$. Thus, we can view the soft fuzzy set (F, E) as a collection of fuzzy approximations (which are under the fuzzy relationship $\widetilde{R}$) as below:
$(F, E) = \{$expensive houses$= \{(h_1, 0.6), (h_2, 0.8), (h_3, 0.7), (h_4, 0.8)\ (h_5, 0.6), (h_6, 0.2)\}$, beautiful houses $= F(e_2)$, wooden houses $= F(e_3)$, in the green surroundings houses $= F(e_4)$, modern houses $= \{(h_1, 0.3), (h_2, 0.7)\ (h_3, 0.3), (h_4, 0.7), (h_5, 0.9), (h_6, 0.7)\}$.

Remark 5.1 Just like soft sets, soft fuzzy sets also can be represented in a tabular form and the difference is that $h_{ij} = \theta_{ij}, \theta_{ij} \in [0, 1]$ (membership value of h_i in $F(e_j)$).

Definition 5.12 For two soft fuzzy sets (F, A) and (H, B) over a common universe U, we say that (F, A) is a soft fuzzy subset of (H, B) if
i. $A \subset B$
ii. $\forall \varepsilon \in A$, $F(\varepsilon) \subset H(\varepsilon)$.

We write $(F, A) \hat{\subset} (H, B)$.

Definition 5.13 (*Equality of two soft fuzzy sets*) Two soft fuzzy sets (F, A) and (H, B) over a common universe U are said to be soft fuzzy equal if (F, A) is a soft fuzzy subset of (H, B) and (H, B) is a soft fuzzy subset of (F, A).

Definition 5.14 (*Complement of soft fuzzy set*) The complement of a soft fuzzy set (F, A) is denoted by $(F, A)^C$ and is defined by $(F, A)^C = (F^C, \neg A)$ where $F^C : \neg A \to P(U)$ is a mapping given by $F^C(a) = fuzzy\ complement\ of$ $F(\neg a)$, $\forall a \in \neg A$. Let us call F^C to be the soft complement function of F.

Clearly, $\left((F, A)^C\right)^C = (F, A)$.

Definition 5.15 (*Null soft fuzzy set*) A soft fuzzy set (F, A) over U is said to be Null soft fuzzy set denoted by Φ, if $\forall \varepsilon \in A$, $F(\varepsilon) =$ null fuzzy set of U.

Definition 5.16 (*Absolute soft fuzzy set*) A soft fuzzy set (F, A) over U is said to be Absolute soft fuzzy set denoted by $\hat{A}$, if $\forall \varepsilon \in A$, $F(\varepsilon) = U$.

Clearly, $\hat{A}^C = \Phi$, $\Phi^C = \hat{A}^C$.

Definition 5.17 (*AND operation on two soft fuzzy sets*) If (F, A) and (H, B) be two soft fuzzy sets then "(F, A) AND (H, B)" is a soft fuzzy set denoted by $(F, A)\hat{\wedge}(H, B) = (P, A \times B)$, where $P(\alpha, \beta) = F(\alpha) \cap H(\beta), \forall(\alpha, \beta) \in A \times B$.

Definition 5.18 (*OR operation on two soft fuzzy sets*) If (F, A) and (H, B) be two soft fuzzy sets then "$(F, A)\mathrm{OR}(H, B)$" is a soft fuzzy set denoted by$(F, A)\hat{\vee}(H, B) = (O, A \times B)$ where $O(\alpha, \beta) = F(\alpha) \cup H(\beta), \forall(\alpha, \beta) \in A \times B$.

Also the following De Morgan's type results hold:

(i) $((F, A)\hat{\vee}(G, B))^C = (F, A)^C\hat{\wedge}(G, B)^C$
(ii) $((F, A)\hat{\wedge}(G, B))^C = (F, A)^C\hat{\vee}(G, B)^C$.

Definition 5.19 (*Union operation*) Union of two soft fuzzy sets of (F, A) and (H, B) over a common universe U is the soft fuzzy set (P, C), where $C = A \cup B$ and $\forall e \in C$,

$$P(e) = \begin{cases} F(e) & e \in A - B \\ H(e) & e \in B - A \\ F(e) \cup H(e) & e \in A \cap B \end{cases}$$

We write $(F, A)\hat{\cup}(H, B) = (P, C)$.

Definition 5.20 (*Intersection operation*) Intersection of two soft fuzzy sets of (F, A) and (H, B) over a common universe U is the soft fuzzy set (O, C) where $C = A \cap B$ and $\forall e \in C$, $O(e) = F(e) \cap H(e)$.

We write $(F, A)\hat{\cap}(H, B) = (O, C)$.

Proposition 5.7 *Let (F, A) be a soft fuzzy set over a universe U. Then clearly we have*
(i) $(F, A)\hat{\cup}(F, A) = (F, A)$
(ii) $(F, A)\hat{\cap}(F, A) = (F, A)$
(iii) $(F, A)\hat{\cup}\Phi = \Phi$ *where Φ is the null soft fuzzy set.*
(iv) $(F, A)\hat{\cap}\Phi = \Phi$
(v) $(F, A)\hat{\cup}\hat{A} = \hat{A}$ *where $\hat{A}$ is the absolute soft fuzzy set.*
(vi) $(F, A)\hat{\cap}\hat{A} = \hat{A}$.

5.2 Intuitionistic Fuzzy Soft Sets and Soft Intuitionistic Fuzzy Sets

One of the major generalizations of Zadeh's fuzzy sets is the concept of intuitionistic fuzzy sets by Atanassov [11]. In this section we consider two types of amalgamations of soft sets with Atanassov intuitionistic fuzzy sets namely, intuitionistic fuzzy soft sets and soft intuitionistic fuzzy sets. Results and examples given here are taken from Yong-jie Xu et al. [154] and Alhazaymeh et al. [6].

We recall some fundamental concepts related to intuitionistic fuzzy sets as given by Atanassov [11].

Definition 5.21 Let a set E be fixed. An intuitionistic fuzzy set or IFS in E is an object having the form $\mathrm{A} = \{< \mathrm{x}, \mu_{\mathrm{A}}(\mathrm{x}), \nu_{\mathrm{A}}(\mathrm{x}) > |\mathrm{x} \in \mathrm{E}\}$ where, the function $\mu_{\mathrm{A}} : \mathrm{E} \to [0, 1]$ and $\nu_{\mathrm{A}} : \mathrm{E} \to [0, 1]$ define the degree of membership and the de- gree of non-membership respectively of the element $x (\in E)$ to the set A. For any $x \in \mathrm{E}$, $0 \le \mu_{\mathrm{A}}(\mathrm{x}) + \nu_{\mathrm{A}}(\mathrm{x}) \le 1$. The indeterministic part for x denoted by $\pi_{\mathrm{A}}(\mathrm{x})$, where $\pi_{\mathrm{A}}(\mathrm{x}) = 1 - \mu_{\mathrm{A}}(\mathrm{x}) - \gamma_{\mathrm{A}}(\mathrm{x})\cdot$ Clearly , $0 \le \pi_{\mathrm{A}}(\mathrm{x}) \le 1$.

Definition 5.22 If A and B are two IFSs of the set E, then
$A = B$ iff $\forall x \in E, \mu_A(x) = \mu_B(x)$ and $\gamma_A(x) = \gamma_B(x)$
$\mathrm{A^c} = \{< \mathrm{x}, \nu_{\mathrm{A}}(\mathrm{x}), \mu_{\mathrm{A}}(\mathrm{x}) > |\mathrm{x} \in \mathrm{E}\}$
$\mathrm{A} \cup \mathrm{B} = \{< \mathrm{x}, \max(\mu_{\mathrm{A}}(\mathrm{x}), \mu_{\mathrm{B}}(\mathrm{x})), \min(\nu_{\mathrm{A}}(\mathrm{x}) \quad \nu_{\mathrm{B}}(\mathrm{x})) > |\mathrm{x} \in \mathrm{E}\}$
$\mathrm{A} \cap \mathrm{B} = \{< \mathrm{x}, \min(\mu_{\mathrm{A}}(\mathrm{x}), \mu_{\mathrm{B}}(\mathrm{x})), \max(\nu_{\mathrm{A}}(\mathrm{x}), \nu_{\mathrm{B}}(\mathrm{x})) > |\mathrm{x} \in \mathrm{E}\}$
$\mathrm{A} \cdot \mathrm{B} = \{< \mathrm{x}, \mu_{\mathrm{A}}(\mathrm{x}) \cdot \mu_{\mathrm{B}}(\mathrm{x}), \quad \nu_{\mathrm{A}}(\mathrm{x}) + \nu_{\mathrm{B}}(\mathrm{x}) - \nu_{\mathrm{A}}(\mathrm{x}) \cdot \nu_{\mathrm{B}}(\mathrm{x}) > |\mathrm{x} \in \mathrm{E}\}$.

5.2.1 *Intuitionistic Fuzzy Soft Sets*

Definition 5.23 Let U be an initial universe E be a set of parameters and $IFS(U)$ denotes the intuitionistic fuzzy power set of U and $A \subset E$. A pair $(\tilde{F}, A)$ is called an intuitionistic fuzzy soft set over U, where F is a mapping given by

$$\tilde{F} : A \to IFS(U).$$

An intuitionistic fuzzy soft set is a parameterized family of intuitionistic fuzzy subsets of U, a fuzzy soft set is a special case of an intuitionistic fuzzy soft set, because when all the intuitionistic fuzzy subset of U degenerates into fuzzy subsets, the corresponding intuitionistic fuzzy soft set degenerates into a fuzzy soft set.

In general, $\forall \varepsilon \in A \subseteq E$, $\tilde{F}(\varepsilon)$ is an intuitionistic fuzzy set on U, which is called the intuitionistic fuzzy set of parameter ε. The intuitionistic fuzzy value $< \mu_{\tilde{F}(\varepsilon)}(x), \nu_{\tilde{F}(\varepsilon)}(x) >$ denotes the degree that object $x \in U$ holds parameter ε. $\tilde{F}(\varepsilon)$ can be written as: $\tilde{F}(\varepsilon) = \left\{< x, \mu_{\tilde{F}(\varepsilon)}(x), \nu_{\tilde{F}(\varepsilon)}(x) > |x \in U\right\}$.

If $\forall x \in U, \mu_{\tilde{F}(\varepsilon)}(x) + \nu_{\tilde{F}(\varepsilon)}(x) = 1$, then $\tilde{F}(\varepsilon)$ degenerates into a fuzzy set; if $\forall x \in U$ and $\forall \varepsilon \in A \subseteq E$, $\mu_{\tilde{F}(\varepsilon)}(x) + \nu_{\tilde{F}(\varepsilon)}(x) = 1$, then intuitionistic fuzzy soft set $(\tilde{F}, A)$ degenerates into a fuzzy soft set.

Definition 5.24 Let U be an initial universe set, E be a set of parameters. Suppose that $A, B \subset E$, $(\tilde{F}, A)$ and $(\tilde{G}, B)$ are two intuitionistic fuzzy soft sets. We say that $(\tilde{F}, A)$ is an intuitionistic fuzzy soft subset of $(\tilde{G}, B)$ if and only if

1) $A \subset B$;
2) $\forall \varepsilon \in A, \tilde{F}(\varepsilon) \subset \tilde{G}(\varepsilon)$; which is denoted by $(\tilde{F}, A)\tilde{\subset}(\tilde{G}, B)$.

Definition 5.25 Let U be an initial universe set, E be a set of parameters. Suppose that $A, B \subset E$, $(\tilde{F}, A)$ and $(\tilde{G}, B)$ are two intuitionistic fuzzy soft sets. We say that $(\tilde{F}, A)$ and $(\tilde{G}, B)$ are intuitionistic fuzzy soft equal if and only if
(1) $(\tilde{F}, A)\tilde{\subset}(\tilde{G}, B)$;
(2) $(\tilde{G}, B)\tilde{\subset}(\tilde{F}, A)$; which is denoted by $(\tilde{F}, A) \cong (\tilde{G}, B)$.

Definition 5.26 The complement of an intuitionistic fuzzy soft set $(\tilde{F}, A)$ is denoted by $(\tilde{F}, A)^C$ and is defined by

$$(\tilde{F}, A)^C = \left(\tilde{F}^C, \neg A\right)$$

where

$$\tilde{F}^c : \neg A \to IFS(U)$$

is the mapping given by $F^C(\beta) = (F(\neg\beta))^C, \forall \beta \in \neg A$.

Definition 5.27 Let $(\tilde{F}, A)$ and $(\tilde{G}, B)$ are two intuitionistic fuzzy soft sets on the same universe set U, the "OR" operation is

$$(\tilde{F}, A) \vee (\tilde{G}, B) = (\tilde{H}, A \times B)$$

where $\tilde{H}(\alpha, \beta) = \tilde{F}(\alpha) \cup G(\beta), \forall(\alpha, \beta) \in A \times B$.

Definition 5.28 Let $(\tilde{F}, A)$ and $(\tilde{G}, B)$ are two intuitionistic fuzzy soft sets on the same universe set U, the "AND" operation is

$$(\tilde{F}, A) \wedge (\tilde{G}, B) = (\tilde{H}, A \times B)$$

where $\tilde{H}(\alpha, \beta) = \tilde{F}(\alpha) \cap G(\beta), \forall(\alpha, \beta) \in A \times B$.

Definition 5.29 Let $(\tilde{F}, A)$ and $(\tilde{G}, B)$ are two intuitionistic fuzzy soft sets on the same universe set U, the "union" operation is $(\tilde{F}, A)\tilde{\cup}(\tilde{G}, B) = (\tilde{H}, C)$, where $C = A \cup B, \forall \varepsilon \in C$

$$\tilde{H}(\varepsilon) = \begin{cases} \tilde{F}(\varepsilon) & \varepsilon \in A - B \\ \tilde{G}(\varepsilon) & \varepsilon \in B - A \\ \tilde{F}(\varepsilon) \cup \tilde{G}(\varepsilon) & \varepsilon \in A \cap B \end{cases}$$

Definition 5.30 Let $(\tilde{F}, A)$ and $(\tilde{G}, B)$ are two intuitionistic fuzzy soft sets on the same universe set U, the "intersection" operation is

$$(\tilde{F}, A)\tilde{\cap}(\tilde{G}, B) = (\tilde{H}, C)$$

where $C = A \cap B, \forall \varepsilon \in C, \tilde{H}(\varepsilon) = \tilde{F}(\varepsilon) \cap \tilde{G}(\varepsilon)$.

We state the following theorems without proof.

Theorem 5.2 *The complement of an intuitionistic fuzzy soft set satisfies:*

$$\left((\tilde{F}, A)^C\right)^C = (\tilde{F}, A).$$

Theorem 5.3 *Let $(\tilde{F}, A)$, $(\tilde{G}, B)$ and $(\tilde{H}, C)$ be three intuitionistic fuzzy soft sets, we have*
(1) $(\tilde{F}, A) \wedge ((\tilde{G}, B) \wedge (\tilde{H}, C)) = ((\tilde{F}, A) \wedge (\tilde{G}, B)) \wedge (\tilde{H}, C)$
(2) $(\tilde{F}, A) \vee ((\tilde{G}, B) \vee (\tilde{H}, C)) = ((\tilde{F}, A) \vee (\tilde{G}, B)) \vee (\tilde{H}, C)$
(3) $(\tilde{F}, A) \wedge ((\tilde{G}, B) \vee (\tilde{H}, C)) = ((\tilde{F}, A) \wedge (\tilde{G}, B)) \vee ((\tilde{F}, A) \wedge (\tilde{H}, C))$
(4) $(\tilde{F}, A) \vee ((\tilde{G}, B) \wedge (\tilde{H}, C)) = ((\tilde{F}, A) \vee (\tilde{G}, B)) \wedge ((\tilde{F}, A) \vee (\tilde{H}, C))$.

Theorem 5.4 *Let $(\tilde{F}, A)$, $(\tilde{G}, B)$ and $(\tilde{H}, C)$ be three intuitionistic fuzzy soft sets, we have*
(1) $(\tilde{F}, A)\tilde{\cup}(\tilde{F}, A) = (\tilde{F}, A)$; $(\tilde{F}, A)\tilde{\cap}(\tilde{F}, A) = (\tilde{F}, A)$
(2) $(\tilde{F}, A)\tilde{\cup}(\tilde{G}, B) = (\tilde{G}, B)\tilde{\cup}(\tilde{F}, A)$; $(\tilde{F}, A)\tilde{\cap}(\tilde{G}, B) = (\tilde{G}, B)\tilde{\cap}(\tilde{F}, A)$
(3) $(\tilde{F}, A)\tilde{\cap}((\tilde{G}, B)\tilde{\cap}(\tilde{H}, C)) = ((\tilde{F}, A)\tilde{\cap}(\tilde{G}, B))\tilde{\cap}(\tilde{H}, C)$
(4) $(\tilde{F}, A) \cup ((\tilde{G}, B) \cup (\tilde{H}, C)) = ((\tilde{F}, A) \cup (\tilde{G}, B)) \cup (\tilde{H}, C)$
(5) $(\tilde{F}, A) \cap (\tilde{G}, B) \cup (\tilde{H}, C)) = ((\tilde{F}, A) \cap (\tilde{G}, B)) \cup ((\tilde{F}, A) \cap (\tilde{H}, C))$
(6) $(\tilde{F}, A) \cup (\tilde{G}, B) \cap (\tilde{H}, C)) = ((\tilde{F}, A) \cup (\tilde{G}, B)) \cap ((\tilde{F}, A) \cup (\tilde{H}, C))$.

5.2.2 *Soft Intuitionistic Fuzzy Sets*

Definition 5.31 Let $\tilde{R}_\alpha =< \tilde{R}_{\mu_\alpha}, R_{\nu_\alpha} >$ be an intuitionistic fuzzy subset of $X \times Y$, and $\tilde{R}_\alpha$ is defined as intuitionistic fuzzy relationship from X to Y. We write $X \xrightarrow{\tilde{R}} Y$ and $\tilde{R}(x, y)$ denotes the degree of correspondence between x and y based on the relationship $\tilde{R}$. Let $F(X \times Y)$ denotes the family of an intuitionistic fuzzy relationship on X to Y. The set $\tilde{R}_\alpha = \left\{(x, y) \in X \times Y : \tilde{R}_{\mu_\alpha}(x, y) \geq \alpha \text{ and } \tilde{R}_{\nu_\alpha}(x, y) \leq \alpha\right\} \subset X \times Y$ is defined as $\alpha-$ cut set if $\tilde{R} \in F(X \times Y)$ for $\alpha \in [0, 1]$.

Definition 5.32 Let U be an initial set and E be a set of parameters. Let $P(U)$ denotes the power set of U. Let $A \subset E$. A pair (F, A) is called a soft intuitionistic fuzzy set over U, where F is a mapping given by $F : A \to P(U)$ and

$$F(x) = \{y \in U : (x, y) \in R_\alpha, x \in A, y \in U, \alpha \in [0, 1]\}.$$

Definition 5.33 For two intuitionistic fuzzy sets $(F, A)_{\tilde{R}}$ and $(G, B)_{\tilde{R}}$ over common universe U, we say that $(F, A)_{\tilde{R}}$ is a soft intuistiontic fuzzy subset of $(G, B)_{\tilde{R}}$ if: (i) $A \subset B$ and (ii) $\forall \varepsilon \in A$, $F(\varepsilon)$ is an intuitionistic fuzzy subset of $G(\varepsilon)$, denoted by $(F, A)_{\tilde{R}} \subset (G, B)_{\tilde{R}}$.

Definition 5.34 Two soft intuitionistic fuzzy sets (F, A), and (G,B), over common universe U are said to be soft intuitionistic fuzzy sets equal if $(F, A)_{\tilde{R}}$ is a soft intuitionistic subset of $(G, B)_{\tilde{R}}$ and $(G, B)_{\tilde{R}}$ is a soft intuitionistic subset of $(F, A)_{\tilde{R}}$.

Definition 5.35 The complement of soft intuitionistic fuzzy set $(F, A)_{\tilde{R}}$ is denoted by $(F, A)^c_{\tilde{R}}$ and defined by $(F, A)^c_{\tilde{R}} = (F^c, \neg A)_{\tilde{R}}$ where $F^c : \neg A \to P(U)$ is a mapping given by $F^c(A)$ = intuitionistic fuzzy complement of $F(\neg e)$, $\forall e \in \neg A$.

Definition 5.36 Let $(F, A)_{\tilde{R}}$ a soft intuitionistic fuzzy set over U is said to be a null soft intuitionistic fuzzy set denoted by Φ, if $\forall \varepsilon \in A$, $F(\varepsilon)$ is null intuitionistic fuzzy subset of U.

Definition 5.37 The union of two soft intuitionistic fuzzy sets $(F, A)_{\tilde{R}}$ and $(G, B)_{\tilde{R}}$ over a common universe U is the soft intuitionistic fuzzy set $(P, C)_{\tilde{R}}$ where $C = A \cup B$ and $\forall e \in C$

$$P(e) = \begin{cases} F(e) & \text{if } e \in A - B \\ G(e) & \text{if } e \in B - A \\ F(e) \cup G(e) & \text{if } e \in A \cap B \end{cases}$$

we write $(F, A)_{\tilde{R}} \cup (G, B)_{\tilde{R}} = (P, C)_{\tilde{R}}$.

Definition 5.38 The intersection of two soft intuitionistic fuzzy sets $(F, A)_{\tilde{R}}$ and $(G, B)_{\tilde{R}}$ over a common universe U is the soft intuitionistic fuzzy set $(H, C)_{\tilde{R}}$ where $C = A \cup B$ and $\forall e \in C$

$$H(e) = \begin{cases} F(e) & \text{if } e \in A - B \\ G(e) & \text{if } e \in B - A \\ F(e) \cap G(e) & \text{if } e \in A \cap B \end{cases}$$

we write $(F, A)_{\tilde{R}} \cap_E (G, B)_{\tilde{R}} = (H, C)_{\tilde{R}}$.

Definition 5.39 (*AND operator*) If $(F, A)_{\tilde{R}}$ and $(G, B)_{\tilde{R}}$ be two soft intuitionistic fuzzy sets then, "(F, A) $_{\tilde{R}}$AND$(G, B)_{\tilde{R}}$" is the soft intuitionistic fuzzy set $((H, A \times B)_{\tilde{R}}$, defined as $H(\alpha, \beta) = F(\alpha) \cap G(\beta)$, $\forall (\alpha, \beta) \in A \times B$.

Definition 5.40 (*OR operator*) If $(F, A)_{\tilde{R}}$ and $(G, B)_{\tilde{R}}$ be two soft intuitionistic fuzzy sets then, "(F, A) $_{\tilde{R}}$OR$(G, B)_{\tilde{R}}$" is the soft intuitionistic fuzzy set $((O, A \times B)_{\tilde{R}}$, defined as $O(\alpha, \beta) = F(\alpha) \cup G(\beta)$, $\forall (\alpha, \beta) \in A \times B$.

Results similar and parallel to soft fuzzy sets can be obtained for soft intuitionistic fuzzy sets also.

5.3 Hesitant Fuzzy Soft Sets

As an attempt to tackle the problem of hesitancy in assigning the membership values in fuzzy sets, Torra [143], introduced the concept of hesitant fuzzy sets. Hesitant fuzzy set can more accurately reflect the people's hesitancy in stating their preferences over objects, compared to the fuzzy set and its many classical extensions. In this section we discuss soft set model extended to hesitant fuzzy set. For fundamental concepts related to hesitant fuzzy sets we refer [143, 144, 151]. Definitions, results and examples given in this section is taken from Wang et al. [147].

Definition 5.41 A hesitant fuzzy set (HFS) on U is in terms of a function that when applied to U returns a subset of [0, 1], which can be represented as the following mathematical symbol:

$$\widetilde{A} = \{\langle u, h_{\widetilde{A}}(u)\rangle \, | u \in U\}$$

where $h_{\widetilde{A}}(u)$ is a set of values in [0, 1], denoting the possible membership degrees of the element $u \in U$ to the set $\widetilde{A}$. For convenience, we call $h_{\widetilde{A}}(u)$ a hesitant fuzzy element (HFE) and H the set of all HFEs.

Definition 5.42 Given hesitant fuzzy set $\widetilde{A}$, if $h(u) = \{0\}$ for all u in U, then $\widetilde{A}$ is called the null hesitant fuzzy set, denoted by $\widetilde{\phi}$. If $h(u) = \{1\}$ for all u in U, then $\widetilde{A}$ is called the full hesitant fuzzy set, denoted by $\widetilde{1}$.

Definition 5.43 For an HFE h, $s(h) = (1/l(h)) \sum_{\gamma \in h} \gamma$ is called the score function of h, where $l(h)$ is the number of the values in h. For two HFEs h_1 and h_2, if $s(h_1) > s(h_2)$, then $h_1 > h_2$; if $s(h_1) = s(h_2)$, then $h_1 = h_2$.

Remark 5.2 Let h_1 and h_2 be two HFEs. It is noted that the number of values in different HFEs h_1 and h_2 are commonly different; that is, $l(h_1) \neq l(h_2)$. For convenience, let $l = \max\{l(h_1), l(h_2)\}$. To operate correctly, we should extend the shorter one until both of them have the same length when we compare them. To extend the shorter one, the best way is to add the same value several times in it. In fact, we can extend the shorter one by adding any value in it. The selection of this value mainly depends on the decision makers' risk preferences. Optimists anticipate desirable outcomes and may add the maximum value, while pessimists expect unfavorable outcomes and may add the minimum value. For example, let $h_1 = \{0.1, 0.2, 0.3\}$, let $h_2 = \{0.4, 0.5\}$, and let $l(h_1) > l(h_2)$. To operate correctly, we should extend h_2 to $h_2' = \{0.4, 0.4, 0.5\}$ until it has the same length of h_1, the optimist may extend h_2 as $h_2' = \{0.4, 0.5, 0.5\}$ and the pessimist may extend it as $h_2' = \{0.4, 0.4, 0.5\}$. Although the results may be different if we extend the shorter one by adding different values, this is reasonable because the decision makers' risk preferences can directly influence the final decision. We assume that the decision makers are all pessimistic (other situations can be studied similarly).

We arrange the elements in $h_{\widetilde{A}}(u)$ in decreasing order, and let $h_{\widetilde{A}}^{\sigma(j)}(u)$ be the jth largest value in $h_{\widetilde{A}}(u)$.

Definition 5.44 Given two hesitant fuzzy sets $\widetilde{M} = \{\langle u, h_{\widetilde{M}}(u)\rangle \mid u \in U\}$ and $\widetilde{N} = \{\langle u, h_{\widetilde{N}}(u)\rangle \mid u \in U\}$, $\widetilde{M}$ is called the fuzzy subset of $\widetilde{N}$ if and only if $h_M^{\sigma(j)}(u) \leq h_N^{\sigma(j)}(u)$, for $\forall u \in U$ and $j = 1, 2, \ldots, l$, which can be denoted by $\widetilde{M} \widetilde{\subseteq} \widetilde{N}$.

Definition 5.45 $\widetilde{M}$ and $\widetilde{N}$ are two hesitant fuzzy sets, we call $\bar{M}$ and $\widetilde{N}$ is hesitant fuzzy equal if and only if
(1) $\widetilde{M} \widetilde{\subseteq} \widetilde{N}$
(2) $\widetilde{M} \supseteq \widetilde{N}$
which can be denoted by $\widetilde{M} = \widetilde{N}$.

Further we have

Definition 5.46 (1) $h^c = \cup_{\gamma \in h} \{1 - \gamma\}$
(2) $h_1 \cup h_2 = \{h \in (h_1 \cup h_2) \mid h \geq \max(h_1^-, h_2^-)\}$
(3) $h_1 \cap h_2 = \{h \in (h_1 \cup h_2) \mid h \leq \min(h_1^+, h_2^+)\}$
where $h^-(u) = \min h(u)$ and $h^+(u) = \max h(u)$ are the lower bound and upper bound of the given hesitant fuzzy elements, respectively.

Definition 5.47 Given two hesitant fuzzy sets $\widetilde{A}$ and $\widetilde{B}$, we can define the following operations:

(1) $\widetilde{A}^c = \left\{\left\langle u, h^c_{\widetilde{A}}(u)\right\rangle \mid u \in U\right\}$
(2) $\widetilde{A} \cup \widetilde{B} = \{\langle u, h_{\widetilde{A}}(u) \cup h_{\widetilde{B}}(u)\rangle \mid u \in U\}$
(3) $\widetilde{A} \cap \widetilde{B} = \{\langle u, h_{\widetilde{A}}(u) \cap h_{\widetilde{B}}(u)\rangle \mid u \in U\}$.

Definition 5.48 For the aggregation of hesitant fuzzy information, the following new operations can be defined on HFEs h, h_1, and h_2:

(1) $h^\lambda = \cup_{\gamma \in h} \{\gamma^\lambda\}$
(2) $\lambda h = \cup_{\gamma \in h} \{1 - (1 - \gamma)^\lambda\}$
(3) $h_1 \oplus h_2 = \cup_{\gamma_1 \in h_1, \gamma_2 \in h_2} \{\gamma_1 + \gamma_2 - \gamma_1\gamma_2\}$
(4) $h_1 \otimes h_2 = \cup_{\gamma_1 \in h_1, \gamma_2 \in h_2} \{\gamma_1\gamma_2\}$.

Definition 5.49 Let $\widetilde{H}(U)$ be the set of all hesitant fuzzy sets in U; a pair $(\widetilde{F}, A)$ is called a hesitant fuzzy soft set over U, where $\widetilde{F}$ is a mapping given by

$$\widetilde{F} : A \longrightarrow \widetilde{H}(U).$$

A hesitant fuzzy soft set is a mapping from parameters to $\widetilde{H}(U)$. It is a parameterized family of hesitant fuzzy subsets of U. For $e \in A$, $\widetilde{F}(e)$ may be considered as the set of $e-$ approximate elements of the hesitant fuzzy soft set $(\widetilde{F}, A)$.

Definition 5.50 Let $A, B \in E$. $(\widetilde{F}, A)$ and $(\widetilde{G}, B)$ are two hesitant fuzzy soft sets over U. $(\widetilde{F}, A)$ is said to be a hesitant fuzzy soft subset of $(\widetilde{G}, B)$ if

(1) $A \subseteq B$

(2) For all $e \in A, \widetilde{F}(e) \cong \widetilde{G}(e)$.

In this case, we write $(\widetilde{F}, A)\widetilde{\subseteq}(\widetilde{G}, B)$.

Definition 5.51 Two hesitant fuzzy soft sets $(\widetilde{F}, A)$ and $(\widetilde{G}, B)$ are said to be hesitant fuzzy soft equal if $(\widetilde{F}, A)$ is a hesitant fuzzy soft subset of $(\widetilde{G}, B)$ and $(\widetilde{G}, B)$ is a hesitant fuzzy soft subset of $(\widetilde{F}, A)$. In this case, we write $(\widetilde{F}, A) \cong (\widetilde{G}, B)$.

Definition 5.52 A hesitant fuzzy soft set $(\widetilde{F}, A)$ is said to be empty hesitant fuzzy soft set, denoted by $\widetilde{\Phi}_A$, if $\widetilde{F}(e) = \widetilde{\phi}$ for all $e \in A$.

Definition 5.53 A hesitant fuzzy soft set $(\widetilde{F}, A)$ is said to be full hesitant fuzzy soft set, denoted by $\widetilde{U}_A$, if $\widetilde{F}(e) = \widetilde{1}$ for all $e \in A$.

Definition 5.54 The complement of a hesitant fuzzy soft set $(\widetilde{F}, A)$ is denoted by $(\widetilde{F}, A)^c$ and is defined by

$$(\widetilde{F}, A)^c = \left(\widetilde{F}^c, A\right)$$

where $\widetilde{F}^c : A \to \widetilde{H}(U)$ is a mapping given by $\widetilde{F}^c(e) = (\widetilde{F}(e))^c$ for all $e \in A$.

Clearly, $\left(\widetilde{F}^c\right)^c$ is the same as $\widetilde{F}$ and $\left((\widetilde{F}, A)^c\right)^c = (\widetilde{F}, A)$.

It may be noted that in the above definition of complement, the parameter set of the complement $(\widetilde{F}, A)^c$ is still the original parameter set A, instead of $\neg A$.

Definition 5.55 The AND operation on two hesitant fuzzy soft sets$(\widetilde{F}, A)$ and $(\widetilde{G}, B)$ which is denoted by $(\widetilde{F}, A) \wedge (\widetilde{G}, B)$ is defined by $(\widetilde{F}, A) \wedge (\widetilde{G}, B) = (\widetilde{J}, A \times B)$, where $\widetilde{J}(\alpha, \beta) = \widetilde{F}(\alpha) \cap \widetilde{G}(\beta)$, for all $(\alpha, \beta) \in A \times B$.

Definition 5.56 The OR operation on the two hesitant fuzzy soft sets $(\widetilde{F}, A)$ and $(\widetilde{G}, B)$ which is denoted by $(\widetilde{F}, A) \vee (\widetilde{G}, B)$ is defined by $(\widetilde{F}, A) \vee (\widetilde{G}, B) = (\widetilde{O}, A \times B)$, where $\widetilde{O}(\alpha, \beta) = \widetilde{F}(\alpha) \cup \widetilde{G}(\beta)$, for all $(\alpha, \beta) \in A \times B$.

Definition 5.57 Union of two hesitant fuzzy soft sets $(\widetilde{F}, A)$ and $(\widetilde{G}, B)$ over U is the hesitant fuzzy soft set $(\widetilde{J}, C)$, where $C = A \cup B$, and for all $e \in C$

$$\begin{aligned}
\widetilde{J}(e) &= \widetilde{F}(e), \quad \text{if } e \in A - B \\
&= \widetilde{G}(e), \quad \text{if } e \in B - A \\
&= \widetilde{F}(e) \cup \widetilde{G}(e), \quad \text{if } e \in A \cap B.
\end{aligned}$$

We write $(\widetilde{F}, A)\widetilde{\cup}(\widetilde{G}, B) = (\widetilde{J}, C)$.

Definition 5.58 Intersection of two hesitant fuzzy soft sets $(\widetilde{F}, A)$ and $(\widetilde{G}, B)$ with $A \cap B \neq \phi$ over U is the hesitant fuzzy soft set $(\widetilde{J}, C)$, where $C = A \cap B$, and for all $e \in C$, $\widetilde{J}(e) = \widetilde{F}(e) \cap \widetilde{G}(e)$.

We write $(\widetilde{F}, A)\widetilde{\cap}(\widetilde{G}, B) = (\widetilde{J}, C)$.

Theorem 5.5 *(De Morgan's laws of hesitant fuzzy soft sets). Let* $(\widetilde{F}, A)$ *and* $(\widetilde{G}, B)$ *be two hesitant fuzzy soft sets over* U*; we have*
(1) $((\widetilde{F}, A) \wedge (\widetilde{G}, B))^c = (\widetilde{F}, A)^c \vee (\widetilde{G}, B)^c$
(2) $((\widetilde{F}, A) \vee (\widetilde{G}, B))^c = (\widetilde{F}, A)^c \wedge (\widetilde{G}, B)^c$.

Proof Suppose that $(\widetilde{F}, A) \wedge (\widetilde{G}, B) = (\widetilde{J}, A \times B)$. Therefore $((\widetilde{F}, A) \wedge (\widetilde{G}, B))^c = (\widetilde{J}, A \times B)^c = \left(\widetilde{J}^c, A \times B\right)$. Similarly $(\widetilde{F}, A)^c \vee (\widetilde{G}, B)^c = \left((\widetilde{F}^c, A) \vee (\widetilde{G}^c, B\right) = \widetilde{O}, A \times B)$. Now take $(\alpha, \beta) \in A \times B$; therefore

$$\begin{aligned}
&\widetilde{J}^c(\alpha, \beta)\\
&= (\widetilde{J}(\alpha, \beta))^c = (F(\alpha) \cap G(\beta))^c\\
&= \left\{\left\langle u, h_{\bar{A}}(u) \cap h_{\bar{B}}(u)\right\rangle | u \in U\right\}^c\\
&= \left\{\left\langle u, \left\{h \in \left(h_{\bar{A}} \cup h_{\bar{B}}\right) | h \leq \min\left(h^{+}_{\bar{A}}, h^{+}_{\bar{B}}\right)\right\}\right\rangle | u \in U\right\}^c\\
&= \left\{\left\langle u, \left\{1 - h | h \in \left(h_{\bar{A}} \cup h_{\bar{B}}\right) \wedge h \leq \min\left(h^{\dagger}_{\bar{A}}, h^{+}_{\bar{B}}\right)\right\}\right\rangle | u \in U\right\}\\
&= \left\{\left\langle u, \left\{1 - h | (1 - h) \in \left(1 - h_{\bar{A}}\right) \cup \left(1 - h_{\bar{B}}\right) \wedge (1 - h) \geq \max\left(\left(1 - h_{\bar{A}}\right)^{-} \left(1 - h_{\bar{B}}\right)^{-}\right)\right\}\right\rangle | u \in U\right\}\\
&= \left\{\left\langle u, \left\{(1 - h) \in \left(1 - h_{\bar{A}}\right) \cup \left(1 - h_{\bar{B}}\right) | (1 - h) \geq \max\left(\left(1 - h_{\bar{A}}\right)^{-}, \left(1 - h_{\bar{B}}\right)^{-}\right)\right\}\right\rangle | u \in U\right\}
\end{aligned}$$

$$\begin{aligned}
&\widetilde{O}(\alpha, \beta)\\
&= F^c(\alpha) \cup G^c(\beta)\\
&= \left\{\left\langle u, h_{\bar{A}}(u)\right\rangle | u \in U\right\}^c \cup \left\{\left\langle u, h_{\bar{B}}(u)\right\rangle | u \in U\right\}^c\\
&= \left\{\left\langle u, 1 - h_{\bar{A}}(u)\right\rangle | u \in U\right\} \cup \left\{\left\langle u, 1 - h_{\bar{B}}(u)\right\rangle | u \in U\right\}\\
&= \left\{\left\langle u, 1 - h_{\bar{A}}(u)\right\rangle | u \in U\right\} \cup \left\{\left\langle u, 1 - h_{\bar{B}}(u)\right\rangle | u \in U\right\}\\
&= \left\{\left\langle u, \left(1 - h_{\bar{A}}\right) \cup \left(1 - h_{\bar{B}}\right)\right\rangle | u \in U\right\}\\
&= \left\{\left\langle u, \left\{(1 - h) \in \left(1 - h_{\bar{A}}\right) \cup \left(1 - h_{\bar{B}}\right) | (1 - h) \geq \max\left(\left(1 - h_{\bar{A}}\right)^{-}, \left(1 - h_{\bar{B}}\right)^{-}\right)\right\}\right\rangle | u \in U\right\}
\end{aligned}$$

Hence, $\widetilde{J}^c(\alpha, \beta) = \widetilde{O}(\alpha, \beta)$ is proved.
(2) Similarly we can prove that $((\widetilde{F}, A) \vee (\widetilde{G}, B))^c = (\widetilde{F}, A)^c \wedge (\widetilde{G}, B)^c$. □

One may easily obtain the proofs of the following theorems related to hesitant fuzzy soft sets also.

Theorem 5.6 *Let* $(\widetilde{F}, A)$, $(\widetilde{G}, B)$, *and* $(\widetilde{J}, C)$ *be three hesitant fuzzy soft sets over* U*. Then the associative law of hesitant fuzzy soft sets holds as follows:*

$$\begin{aligned}
(\widetilde{F}, A) \wedge ((\widetilde{G}, B) \wedge (\widetilde{J}, C)) &= ((\widetilde{F}, A) \wedge (\widetilde{G}, B)) \wedge (\widetilde{J}, C)\\
(\widetilde{F}, A) \vee ((\widetilde{G}, B) \vee (\widetilde{J}, C)) &= ((\widetilde{F}, A) \vee (\widetilde{G}, B)) \vee (\widetilde{J}, C).
\end{aligned}$$

Theorem 5.7 *Let* $(\widetilde{F}, A)$ *and* $(\bar{G}, B)$ *be two hesitant fuzzy soft sets over U.Then,*
(1)$(\widetilde{F}, A)\widetilde{\cup}(\widetilde{F}, A) = (\widetilde{F}, A)$
(2) $(\widetilde{F}, A)\widetilde{\cap}(\widetilde{F}, A) = (\widetilde{F}, A)$
(3) $(\widetilde{F}, A)\widetilde{\cup}\widetilde{\Phi}_A = (\widetilde{F}, A)$
(4) $(\widetilde{F}, A)\widetilde{\cap}\widetilde{\Phi}_A = \widetilde{\Phi}_A$
(5) $(\widetilde{F}, A)\widetilde{\cup}\widetilde{U}_A = \widetilde{U}_A$
(6) $(\widetilde{F}, A)\widetilde{\cap}\widetilde{U}_A = (\widetilde{F}, A)$
(7) $(\widetilde{F}, A)\widetilde{\cup}(\widetilde{G}, B) = (\widetilde{G}, B)\widetilde{\cup}(\widetilde{F}, A)$
(8) $(\widetilde{F}, A)\widetilde{\cap}(\widetilde{G}, B) = (\widetilde{G}, B)\widetilde{\cap}(\widetilde{F}, A)$.

Theorem 5.8 *Let* $(\widetilde{F}, A)$ *and* $(\widetilde{G}, B)$ *be two hesitant fuzzy soft sets over U. Then,*
(1) $((\widetilde{F}, A)\widetilde{\cup}(\widetilde{G}, B))^c\widetilde{\subseteq}(\widetilde{F}, A)^c\widetilde{\cup}(\widetilde{G}, B)^c$
(2) $(\widetilde{F}, A)^c\widetilde{\cap}(\widetilde{G}, B)^c\widetilde{\subseteq}((\widetilde{F}, A)\widetilde{\cap}(\widetilde{G}, B))^c$.

Theorem 5.9 *Let* $(\widetilde{F}, A)$ *and* $(\widetilde{G}, B)$ *be two hesitant fuzzy soft sets over U. Then,*
(1) $(\widetilde{F}, A)^c\widetilde{\cap}(\widetilde{G}, B)^c\widetilde{\subseteq}((\widetilde{F}, A)\widetilde{\cup}(\widetilde{G}, B))^c$
(2) $((\widetilde{F}, A)\widetilde{\cap}(\widetilde{G}, B))^c\widetilde{\subseteq}(\widetilde{F}, A)^c\widetilde{\cup}(\widetilde{G}, B)^c$.

Theorem 5.10 *Let* $(\widetilde{F}, A)$ *and* $(\widetilde{G}, A)$ *be two hesitant fuzzy soft sets over* U; *then*
(1) $((\widetilde{F}, A)\widetilde{\cup}(\widetilde{G}, A))^c = (\widetilde{F}, A)^c\widetilde{\cap}(\widetilde{G}, A)^c$
(2) $((\widetilde{F}, A)\widetilde{\cap}(\widetilde{G}, A))^c = (\widetilde{F}, A)^c\widetilde{\cup}(\widetilde{G}, A)^c$.

Remark 5.3 The distribution law of hesitant fuzzy soft sets does not hold. That is,

$$\begin{aligned}(\widetilde{F}, A) \wedge ((\widetilde{G}, B) \vee (\widetilde{J}, C)) &\neq ((\widetilde{F}, A) \wedge (\widetilde{J}, C)) \vee ((\widetilde{G}, B) \wedge (\widetilde{J}, C)) \\ (\widetilde{F}, A) \vee ((\widetilde{G}, B) \wedge (\widetilde{J}, C)) &\neq ((\widetilde{F}, A) \vee (\widetilde{J}, C)) \wedge ((\widetilde{G}, B) \vee (\widetilde{J}, C)).\end{aligned}$$

For all $\alpha \in A, \beta \in B$, and $\delta \in C$, because $\tilde{F}(\alpha) \cap (\widetilde{G}(\beta) \cup \widetilde{J}(\delta)) \neq (\widetilde{F}(\alpha) \cap \widetilde{G}(\beta)) \cup (\widetilde{F}(\alpha) \cap \widetilde{J}(\delta))$ for the hesitant fuzzy sets, we can conclude that $\widetilde{F}, A) \wedge ((\widetilde{G}, B) \vee (\widetilde{J}, C)) \neq ((\widetilde{F}, A) \wedge (\widetilde{J}, C)) \vee ((\widetilde{G}, B) \wedge (\widetilde{J}, C)$. A similar argument is valid for the other version also.

5.4 Soft Rough Sets and Rough Soft Sets

Two hybrid models involving rough sets and soft sets will be discussed in this section. Fundamentals related to Pawlak Rough sets can be seen in [97–99, 106].

5.4.1 Soft Rough Sets

This subsection gives the notion of soft rough sets, which can be seen as a generalized rough set model based on soft sets. The standard soft set model is used to form the granulation structure of the universe, namely the soft approximation space. Based on this granulation structure, one can define soft rough approximations, soft rough sets and some related notions. As a hybrid model combining rough sets with soft sets, soft rough sets could be exploited to extend many practical applications based on rough sets or sets. The major results and examples in this subsection are taken from Feng et al. [41].

Definition 5.59 An information system (or a knowledge representation system) is a pair $\mathscr{I} = (U, A)$ of non-empty finite sets U and A, where U is a set of objects and A is a set of attributes; each attribute $a \in A$ is a function $a : U \to V_a$, where V_a is the set of values (called domain) of attribute a. Let U be a non-empty finite universe and R be an equivalence relation on U. The pair (U, R) is called a Pawlak approximation space. The equivalence relation R is often called an indiscernibility relation and related to an information system. Specifically, if $\mathscr{I} = (U, A)$ is an information system and $B \subseteq A$, then an indiscernibility relation $R = I(B)$ can be defined by

$$(x, y) \in I(B) \Longleftrightarrow a(x) = a(y), \quad \forall a \in B$$

where $x, y \in U$, and $a(x)$ denotes the value of attribute a for object x. Using the indiscernibility relation R, one can define the following two operations

$$R_*X = \{x \in U : [x]_R \subseteq X\}, \quad R^*X = \{x \in U : [x]_R \cap X \neq \emptyset\}$$

assigning to every subset $X \subseteq U$ two sets R_*X and R^*X called the R -lower and the R-upper approximation of X, respectively. Moreover, the sets $Pos_R\ X = R_*X$, $\text{Neg}_R\, X = U - R^*X$, $\text{Bnd}\ _R X = R^*X - R_*X$ are referred to as the R-positive, the R-negative and the R-boundary region of X, respectively. If the R-boundary region of X is empty, i.e., $R^*X = R_*X$, then X is crisp (or exact) with respect to R; in the opposite case, i.e., $Bnd_R X \neq \emptyset$, X is said to be rough (or inexact) with respect to R.

Note that sometimes the pair (R_*X, R^*X) is also referred to as the rough set of X with respect to R. If $X \subseteq U$ is defined by a predicate P and $x \in U$, we have the following interpretation:

$x \in \text{Pos}_R\, X = R_*X$ means that x certainly has property P
$x \in R^*X$ means that x possibly has property P
$x \in \text{Neg}_R\, X$ means that x definitely does not have property P.

Definition 5.60 Let $S = (F, A)$ be a soft set over U. Then the pair $P = (U, S)$ is called a soft approximation space. Based on the soft approximation space P, we define the following two operations:

$$\underline{apr}_P(X) = \{u \in U : \exists a \in A, [u \in F(a) \subseteq X]\}$$
$$\overline{apr}_P(X) = \{u \in U : \exists a \in A, [u \in F(a), F(a) \cap X \neq \emptyset]\}$$

assigning to every subset $X \subseteq U$ two sets $\underline{apr}_P(X)$ and $\overline{apr}_P(X)$, which are called the soft P-lower appoximation and the soft P upper approximation of X, respectively.

In general, we refer to $\underline{apr}_P(X)$ and $\overline{apr}_p(X)$, as soft rough approximations of X with respect to P. Moreover, the sets

$$\begin{aligned}\mathrm{Pos}_P(X) &= \underline{apr}_P(X)\\ \mathrm{Neg}_P(X) &= U - \overline{\mathrm{apr}}_P(X)\\ \mathrm{Bnd}_P(X) &= \overline{\mathrm{apr}}_P(X) - \underline{\mathrm{apr}}_P(X)\end{aligned}$$

are called the soft P-positive region, the soft P-negative region and the soft P-boundary region of X, respectively. If $\underline{apr}_P(X) = \overline{apr}_P(X)$, X is said to be soft P-definable; otherwise X is called a soft P-rough set.

Remark 5.4 We say that $X \subseteq U$ is a soft P-definable set if the soft P-boundary region Bndp (X) of X is empty. Also, it is clear that $\underline{apr}(X) \subseteq X$ and $\underline{apr}_P(X) \subseteq \overline{apr}_P(X)$ for all $X \subseteq U$. Nevertheless, it is worth noticing that $X \subseteq \overline{apr}_P(X)$ does not hold in general.

Example 71 Let $U = \{u_1, u_2, u_3, u_4, u_5, u_6\}$, $E = \{e_1, e_2, e_3, e_4, e_5, e_6\}$ and $A = \{e_1, e_2, e_3, e_4\} \subseteq E$. Let $S = (F, A)$ be a soft set over U given by table below and the soft approximation space $P = (U, S)$.

	u_1	u_2	u_3	u_4	u_5	u_6
e_1	1	0	0	0	0	1
e_2	0	0	1	0	0	0
e_3	0	0	0	0	0	0
e_4	1	1	0	0	1	0

For $X = \{u_3, u_4, u_5\} \subseteq U$, we have $\underline{apr}_P(X) = \{u_3\}$, and $\overline{apr}_P(X) = \{u_1, u_2, u_3, u_5\}$. Thus $\underline{\mathrm{apr}}_P(X) \neq \overline{apr}_P(X)$ and X is a soft P-rough set. Note that $X = \{u_3, u_4, u_5\} \subsetneqq \overline{apr}_P(X) = \{u_1, u_2, u_3, u_5\}$ in this case. Moreover, it is easy to see that $Pos_P(X) = \{u_3\}$ $\mathrm{Neg}_P(X) = \{u_4, u_6\}$ and $Bnd_P(X) = \{u_1, u_2, u_5\}$. On the other hand, one can consider $X_1 = \{u_3, u_4\} \subseteq U$. Since $\underline{apr}_P(X_1) = \{u_3\} = \overline{apr}_P(X_1)$, by definition, X_1 is a soft P-definable set.

The following results follows easily from the definition of soft rough approximations.

Proposition 5.8 *Let $S = (F, A)$ be a soft set over U and $P = (U, S)$ a soft approximation space. Then we have*

$$\underline{apr}_P(X) = \bigcup_{a\in A}\{F(a) : F(a) \subseteq X\}$$

and

$$\overline{apr}_p(X) = \bigcup_{a\in A}\{F(a) : F(a) \cap X \neq \emptyset\}$$

for all $X \subseteq U$.

Suppose that $S = (F, A)$ is a soft set over U and $P = (U, S)$ is the corresponding soft approximation space. One can verify that soft rough approximations satisfy the following properties:

$$\underline{apr}_P(\emptyset) = \overline{apr}_P(\emptyset) = \emptyset$$
$$\underline{apr}_P(U) = \overline{apr}_P(U) = \bigcup_{a\in A} F(a)$$
$$\underline{apr}_P(X \cap Y) \subseteq \underline{apr}_P(X) \cap \underline{apr}_P(Y)$$
$$\underline{apr}_P(X \cup Y) \supseteq \underline{apr}_P(X) \cup \underline{apr}_P(Y)$$
$$\overline{apr}_P(X \cup Y) = \overline{apr}_P(X) \cup \overline{apr}_P(Y)$$
$$\overline{apr}_P(X \cap Y) \subseteq \overline{apr}_P(X) \cap \overline{apr}_P(Y)$$
$$X \subseteq Y \Rightarrow \underline{apr}_P(X) \subset \underline{apr}_P(Y)$$
$$X \subseteq Y \Rightarrow \overline{apr}_P(X) \subseteq \overline{apr}_P(Y)$$

Proposition 5.9 *Let $S = (F, A)$ be a soft set over U and $P = (U, S)$ a soft approximation space. Then for any $X \subseteq U$, X is soft P definable if and only if $\overline{apr}_P(X) \subseteq X$.*

Proof Note first that if X is soft P-definable, then $apr_P(X) = \overline{apr}_P(X)$, and so $\overline{apr}_P(X) = \underline{apr}_P(X) \subseteq X$. Conversely, suppose that $\overline{apr}_P(X) \subseteq X$ for $X \subseteq U$. To show that X is soft P-definable, we only need to prove that $\overline{apr}_P(X) \subseteq \underline{apr}_P(X)$ since the reverse inequality is trivial. Let $u \in \overline{apr}_P(X)$. Then $u \in F(a)$ and $F(a) \cap X \neq \emptyset$ for some $a \in A$. It follows that $u \in F(a) \subseteq \overline{apr}_P(X) \subseteq X$. Hence $u \in \underline{apr}_P(X)$, and so $\overline{apr}_P(X) \subseteq \mathrm{apr}_P(X)$ as required. □

Theorem 5.11 *Let $S = (F, A)$ be a soft set over U and $P = (U, S)$ a soft approximation space. Then we have*
(1) $\underline{apr}_P\left(\overline{\mathrm{apr}}_P(X)\right) = \overline{\mathrm{apr}}_P(X)$
(2) $\overline{apr}_P\left(\underline{apr}_P(X)\right) \supseteq \underline{apr}_P(X)$
(3) $\underline{apr}_P\left(\underline{apr}_P(X)\right) = \underline{apr}_P(X)$
(4) $\overline{apr}_P\left(\overline{apr}_P(X)\right) \supseteq \overline{apr}_P(X)$
for all $X \subseteq U$.

Proof (1) Let $Y = \overline{apr}_P(X)$ and $u \in Y$. Then $u \in F(a)$ and $F(a) \cap X \neq \emptyset$ for some $a \in A$. By Proposition 5.8, $Y = \overline{apr}_P(X) = \bigcup_{a\in A}\{F(a) : F(a) \cap X \neq \emptyset\}$. There exists $a \in A$ such that $u \in F(a) \subseteq Y$. Hence $u \in \underline{apr}_P(Y)$, and so $Y \subseteq \underline{apr}_P(Y)$. On the other hand, we know that $\underline{apr}_P(Y) \subseteq Y$ holds for any $Y \subseteq U$. Thus it follows that $Y = \underline{apr}_P(Y)$ as required.
(2) Let $Y = \underline{apr}_P(X)$ and $u \in Y$. Then $u \in F(a) \subseteq X$ for some $a \in A$. Since $Y = \underline{apr}_P(X) = \cup_{a\in A}\{F(a) : F(a) \subseteq X\}$ by Proposition 5.8 we deduce that $u \in F(a)$ and $F(a) \cap Y = F(a) \neq \emptyset$. Hence $u \in \overline{apr}_P(Y)$, and so $Y \subseteq \overline{apr}_P(Y)$.

(3) Let $Y = \underline{apr}_p(X)$ and $u \in Y$. Then $u \in F(a) \subseteq X$ for some $a \in A$. But $Y = \underline{apr}_p(X) = \cup_{a\in A}\{F(a) : F(a) \subseteq X\}$, we deduce that $u \in F(a) \subseteq Y$ for $a \in A$. Thus $u \in \underline{apr}_p(Y)$, and so $Y \subseteq \underline{apr}_p(Y)$. Since $\underline{apr}_p(Y) \subseteq Y$ for any $Y \subseteq U$, we conclude that $Y = \underline{apr}_p(Y)$.
(4)Let $Y = \overline{apr}_p(X)$ and $u \in Y$. Then $u \in F(a)$ and $F(a) \cap X \neq \emptyset$ for some $a \in A$. But $Y = \overline{apr}_P(X) = \cup_{a\in A}\{F(a) : F(a) \cap X \neq \emptyset\}$, it follows that $u \in F(a)$ and $F(a) \cap Y = F(a) \neq \emptyset$. Hence, $u \in \overline{apr}_P(Y)$, and so $Y \subseteq \overline{apr}_P(Y)$. □

Definition 5.61 Let $S = (F, A)$ be a soft set over U. If $\cup_{acA} F(a) = U$, then S is said to be a full soft set.

Theorem 5.12 *Let $S = (F, A)$ be a soft set over U and $P = (U, S)$ a soft approximation space. Then the following conditions are equivalent:*

(1) S is a full soft set.
(2) $\underline{apr}_P(U) = U$.
(3) $\overline{apr}_p(U) = U$.
(4) $X \subseteq \overline{apr}_P(X)$ for all $X \subseteq U$.
(5) $\overline{apr_P}(\{u\}) \neq \emptyset$ for all $u \in U$.

Proof Note first that $\underline{apr}_p(U) = \cup_{a\in A}\{F(a) : F(a) \subseteq U\} = \bigcup_{a\in A} F(a)$.
Hence by definition, $S = (F, A)$ is a full soft set if and only if $\underline{apr}_p(U) = U$. That is, conditions (1) and (2) are equivalent. Similarly. we can show that (1) and (3) are equivalent conditions.

On the other hand, we can prove that conditions (4), (5) and (1) are also equivalent. Now assume that condition (4) holds. We prove that condition (5) is true. Given $u \in U$, by condition (4), we have $\{u\} \subseteq \overline{\mathrm{apr}}_P(\{u\})$. Thus $\overline{\mathrm{apr}}_P(\{u\}) \neq \emptyset$ since $u \in \overline{apr}(\{u\})$. This shows that the condition (4) implies the condition (5). Next, we prove that condition (5) implies condition (1) as well. So suppose that the condition (5) holds. For any $u \in U$, we have that $\overline{apr}_p(\{u\}) \neq \emptyset$. Let v be an element in $\overline{apr}_p(\{u\})$. Then there exists some $a \in A$ such that $v \in F(a)$ and $F(a) \cap \{u\} \neq 0$. It follows that $u = v \in F(a)$, and so we have that $u \in \cup_{a\in A}\{F(a)\}$. Hence $S = (F, A)$ is a full soft set.This shows that the condition (5) implies the condition (1). To complete the proof, it remains to show that (1) implies (4). Assume that $S = (F, A)$ is a full soft set and $X \subseteq U$. For any $x \in X$, since S is full. there exists some $a \in A$ such that $x \in F(a)$. It is clear that $X \cap F(a) \neq \emptyset$ since $x \in X \cap F(a)$. Hence we have that $x \in \overline{apr}_p(X)$ and so $X \subseteq \overline{apr}_p(X)$ as required. □

The following result is a direct consequence of Propositions 5.9 and 5.12.

Corollary 5.1 *Let $S = (F, A)$ be a full soft set over U and $P = (U, S)$ a soft approximation space. Then for any $X \subseteq U$, X is soft P definable if and only if $\overline{\mathrm{apr}}_P(X) = \bar{X}$.*

Further to this study, [41] explores the relationships between soft rough sets and Pawlak's rough sets and claims that soft rough set approximation is a worth considering alternative to the rough set approximation. Soft rough sets could provide a better approximation than rough sets do, depending on the structure of the equivalence classes and of the subsets.

5.4.2 Modified Soft Rough Sets

As given in section above, Feng et al. [41] introduced the notion of soft rough sets. In this model, some unusual situations have occurred, like upper approximation of a non-empty set may be empty. Upper approximation of a subset may not contain the set which does not occur in classical rough set theory. To overcome these difficulties, Shabir et al. [126] redefined a soft rough set model called MSR set. Roy and Bera [117] used this MSR approximation concept to define Rough soft sets. The following discussion in this subsection is adapted from [117].

Definition 5.62 Let (F, A) be a soft set over U, where F is a map $F : A \to P(U)$. Let $\varphi : U \to P(A)$ be another map defined as $\varphi(x) = \{a : x \in F(a)\}$. Then the pair (U, φ) is called MSR-approximation space and for any $X \subseteq U$, lower MSR-approximation is defined as:
$\underline{X}_\varphi = \{x \in X : \varphi(x) \neq \varphi(y) \text{ for all } y \in X^c\}$, where $X^c = U - X$ and its upper MSR-approximation is defined as:
$\bar{X}_\varphi = \{x \in U : \varphi(x) = \varphi(y) \text{ for some } y \in X\}$.
If $\underline{X}_\varphi \neq \bar{X}_\varphi$, then X is said to be MSR-set.

Example 72 Let $U = \{u_1, u_2, u_3, u_4, u_5\}$ be the set of schools considered as universal set and an attribute set $A = \{e_1, e_2, e_3, e_4\}$. Here e_1 denotes good location, e_2 denotes sufficient teachers, e_3 denotes good maintenance of discipline, e_4 denotes good relation in teacher-student. Let the soft set (F, A) over U is given by the following table:

	e_1	e_2	e_3	e_4
u_1	1	1	0	0
u_2	0	0	1	1
u_3	1	1	1	0
u_4	0	1	1	1
u_5	0	1	0	1

Here 1 and 0 denote 'yes' and 'no' respectively. Then from the definition of MSR set, $\psi : U \to P(A)$ is defined as follows:

$$\psi(u_1) = \{e_1, e_2\};\ \psi(u_2) = \{e_2\};\ \psi(u_3) = \{e_4\};\ \psi(u_4) = \{e_1, e_3\} = \psi(u_5)$$

Let $X = \{u_1, u_3, u_5\}$. Therefore for the MSR-approximation space (U, ψ), we can write

$$\underline{X}_\psi = \{u_1, u_3\} \text{ and } \bar{X}_\psi = \{u_1, u_3, u_4, u_5\}.$$

Clearly, $\underline{X}_\psi \neq \bar{X}_\psi$, so X is a modified soft rough set.

The following proposition establishes a link between lower soft rough approximation given by Feng et al. [41] and lower MSR approximation of any subset X of U.

Proposition 5.10 *Let $S = (F, A)$ be a soft set over U, such that $P = (U, S)$ be a soft approximation space and (U, φ) be an MSR-approximation space. Then for any subset X of U, $\underline{apr}_P(X) \subseteq \underline{X}_\varphi$.*

Proof Let $u \in \underline{apr}_P(X)$ then $u \in F(a) \subseteq X$ for some $a \in A$. That is $u \in X$ and $a \in \varphi(u)$. If possible let $u \notin \underline{X}_\varphi$. Then $\varphi(u) = \varphi(v)$ for some $v \in X^c$. As $a \in \varphi(u)$, $a \in \varphi(v)$ which implies that $v \in F(a)$. But $F(a) \subseteq X$ so $v \in X$ which is a contradiction. Hence $\underline{apr}_P(X) \subseteq \underline{X}_\varphi$. □

Remark 5.5 From Proposition 5.10, we have $\underline{apr}_P(X) \subseteq \underline{X}_\varphi$. This means that granules of information in MSR-sets are finer than soft rough sets. Thus MSR-sets are more robust than soft rough sets.

Theorem 5.13 *Let (F, A) be a soft set over some set U, then (U, φ) be an MSR-approximation space. For any $X, Y \subseteq U$, we have the following:*
(1)$\underline{X}_\varphi \subseteq X \subseteq \bar{X}_\varphi$
(2)$\underline{\emptyset}_\varphi = \emptyset = \bar{\emptyset}_\varphi$
(3)$\underline{U}_\varphi = U = \bar{U}_\varphi$
(4)$\underline{X \cap Y}_\varphi = \underline{X}_\varphi \cap \underline{Y}_\varphi$
(5)$\overline{X \cap Y}_\varphi \subseteq \bar{X}_\varphi \cap \bar{Y}_\varphi$
(6)$\underline{X \cup Y}_\varphi \supseteq \underline{X}_\varphi \cup \underline{Y}_\varphi$
(7)$\overline{X \cup Y}_\varphi = \bar{X}_\varphi \cup \bar{Y}_\varphi$
(8)$X \subseteq Y \Rightarrow \underline{X}_\varphi \subseteq \underline{Y}_\varphi, \bar{X}_\varphi \subseteq \bar{Y}_\varphi$
(9)$\bar{X}_\varphi = (\underline{(X^c)_\varphi})^c$
(10) $\left(\overline{(X^c)}_\varphi\right)^c = \underline{X}_\varphi$
(11)$\overline{\left(\underline{X}_\varphi\right)}_\varphi = \underline{X}_\varphi$
(12)$\underline{\left(\underline{X}_\varphi\right)_\varphi} = \underline{X}_\varphi$
(13)$\underline{\left(\bar{X}_\varphi\right)}_\varphi = \bar{X}_\varphi$
(14)$\overline{\left(\bar{X}_\varphi\right)}_\varphi = \bar{X}_\varphi$.

5.4.3 Rough Soft Sets

The notion of rough soft sets will be introduced in MSR-approximation space as given in [117].

Definition 5.63 Let (F, A) be a soft set over U and (U, ψ) be an MSR-approximation space with respect to A. Let (G, B) be another soft set over U. (G, B) is said to be rough soft set with respect to a parameter $e \in B$, if $\underline{G(e)}_\psi \neq$

$\overline{G(e)}_{\psi}$, (G, B) is said to be a full rough soft set or a simply rough one if $\underline{G(e)}_{\psi} \neq \overline{G(e)}_{\psi} \forall e \in B$ and we denote it by $RsG\,(e_B)$. We denote rough soft set with respect to e by $RsG(e) = \left(\underline{G(e)}_{\psi}, \overline{G(e)}_{\psi}\right)$.

Example 73 Considering a universal set of batsman $U = \{u_1, u_2, u_3, u_4, u_5, u_6, u_7, u_8\}$ and an attribute set $A = \{e_1, e_2, e_3\}$ where e_1 denotes Bold out, e_2 denotes Catch out, e_3 denotes LBW. Let (F, A) be a soft set representing the record of the player given by the following table:

	e_1	e_2	e_3
u_1	1	1	1
u_2	1	0	1
u_3	0	1	1
u_4	1	0	1
u_5	0	0	1
u_6	1	0	0
u_7	1	1	0
u_8	0	0	1

Here 1 and 0 denotes 'yes' and 'no' respectively. Then from the definition of MSR set, $\psi : U \to P(A)$ is defined as follows:

$$\psi\,(u_1) = \{e_1, e_2, e_3\}\,;\ \psi\,(u_2) = \{e_1, e_3\}\,;\ \psi\,(u_3) = \{e_2, e_3\}\,;\ \psi\,(u_4) = \{e_1, e_3\}$$
$$\psi\,(u_5) = \{e_3\}\,;\ \psi\,(u_6) = \{e_1\}\,;\ \psi\,(u_7) = \{e_1, e_2\}\,;\ \psi\,(u_8) = \{e_2\}.$$

Let (G, B) be another soft set defined as $G\,(e_1) = \{u_1, u_2, u_4, u_6, u_7\}\,;\ G\,(e_2) = \{u_1, u_3, u_4, u_6\}\,;\ G\,(e_3) = \{u_2, u_3, u_5, u_6, u_7, u_8\}\,;\ G\,(e_4) = \{u_1, u_2, u_3, u_5, u_6, u_7\}$, where e_1 denotes Bold out, e_2 denotes Catch out, e_3 denotes LBW and e_4 denotes Run out. Lower MSR-approximation set and upper MSR-approximation set of (G, B) are

$\underline{G\,(e_1)}_{\psi} = \{u_1, u_2, u_4, u_6, u_7\}\,;\ \overline{G\,(e_1)}_{\psi} = \{u_1, u_2, u_4, u_6, u_7\}$

$\underline{G\,(e_2)}_{\psi} = \{u_1, u_3, u_6\}\,;\quad \overline{G\,(e_2)}_{\psi} = \{u_1, u_2, u_3, u_4, u_6\}$

$\underline{G\,(e_3)}_{\psi} = \{u_3, u_5, u_6, u_7, u_8\}\,;\ \overline{G\,(e_3)}_{\psi} = \{u_2, u_3, u_4, u_5, u_6, u_7, u_8\}$

$\underline{G\,(e_4)}_{\psi} = \{u_1, u_3, u_5, u_6, u_7\}\,;\ \overline{G\,(e_4)}_{\psi} = \{u_1, u_2, u_3, u_4, u_5, u_6, u_7\}.$

Clearly, (G, B) is a rough soft set with respect to parameters e_2, e_3 and e_4.

Proposition 5.11 *Let (F, A) be a soft set over U and (U, ψ) be an MSR approximation space with respect to A. Let $(G_1, B_1)\,, (G_2, B_2)$ be two rough soft sets. Then*

$$H(e) = \begin{cases} G_1(e), & \text{if } e \in B_1 - B_2 \\ G_2(e), & \text{if } e \in B_2 - B_1 \\ G_1(e) \cup G_2(e), & \text{if } e \in B_1 \cap B_2 \end{cases}$$

is a rough soft set if $B_1 \cap B_2 = \phi$.

Proposition 5.12 *Let (F, A) be a soft set over U and (U, ψ) be an MSR approximation space with respect to A. Let $(G_1, B_1), (G_2, B_2)$ be two rough soft sets. Then $\forall e \in B_1 \cap B_2$, $H(e) = G_1(e) \cap G_2(e)$ is a rough soft set.*

Definition 5.64 Let (F, A) be a soft set over U and (U, ψ) be an MSR approximation space with respect to A. Let $(G_1, B_1), (G_2, B_2)$ be two rough soft sets. (G_1, B_1) is said to be rough soft subset of (G_2, B_2) if (i) $B_1 \subseteq B_2$, and (ii) $\forall e \in B_1$, $\underline{G_1(e)}_{\psi} = \underline{G_2(e)}_{\psi}$ and $\overline{G_1(e)}_{\psi} = \overline{G_2(e)}_{\psi}$ We write $(G_1, B_1) \sqsubseteq (G_2, B_2)$, where $\sqsubseteq$ denotes soft rough subset.

Definition 5.65 The union of rough soft sets $RsG(e_1)$ and $RsG(e_2)$ with respect to the parameters e_1 and e_2 respectively in MSR-approximation space (U, ψ) is denoted by RsG $(e_1) \sqcup$ Rs $G(e_2)$ and is defined as Rs $G(e_1) \sqcup (e_2) = \left(\underline{G(e_1)}_{\psi} \cup \underline{G(e_2)}_{\psi}, \overline{G(e_1)}_{\psi} \cup \overline{G(e_2)}_{\psi}\right)$.

The union of rough soft sets $RsG(e_A)$ and $RsG(e_B)$ is defined as $RsG(e_A) \sqcup RsG(e_B) = \left(\underline{G(e)}_{\psi} \cup \underline{G(f)}_{\psi}, \overline{G(e)}_{\psi} \cup \overline{G(f)}_{\psi}\right)$ for all $e \in A$ and $f \in B$.

Definition 5.66 The intersection of rough soft sets $RsG(e_1)$ and $RsG(e_2)$ with respect to parameters e_1 and e_2 respectively in MSR-approximation space (U, ψ) is denoted by Rs $G(e_1) \sqcap$ Rs $G(e_2)$ and is defined as Rs $G(e_1) \sqcap$ Rs $G(e_2) = \left(\underline{G(e_1)}_{\psi} \cap \underline{G(e_2)}_{\psi}, \overline{G(e_1)}_{\psi} \cap \overline{G(e_2)}_{\psi}\right)$.

The intersection of rough soft sets Rs $G(e_A)$ and Rs $G(e_B)$ is defined as $RsG(e_A) \sqcap RsG(e_B) = \left(\underline{G(e)}_{\psi} \cap \underline{G(f)}_{\psi}, \overline{G(e)}_{\psi} \cap \overline{G(f)}_{\psi}\right)$ for all $e \in A$ and $f \in B$.

Proposition 5.13 *Let (G_1, B_1) be a soft subset of (G_2, B_2). If (G_2, B_2) is a rough soft set, then (G_1, B_1) is rough soft subset of (G_2, B_2).*

Proof Since (G_1, B_1) is soft subset of (G_2, B_2), $G_1(e) = G_2(e)$ for all $e \in B_1$. Therefore $\underline{G_1(e)}_{\psi} = \underline{G_2(e)}_{\psi}$ and $\overline{G_1(e)}_{\psi} = \overline{G_2(e)}_{\psi}$ for all $e \in B_1$. Hence (G_1, B_1) is rough soft subset. □

Proposition 5.14 *Let (G, B) be a soft set over U and (U, ψ) be an MSR-approximation space. Then set $(RsG(e), \sqcup, \sqcap)$, $\forall e \in B$ together with (U, U) and (ϕ, ϕ) form a lattice where the order relation $\subseteq$ is defined as* RsG $(e_1) \subseteq$ RsG $(e_2) \Rightarrow$ $\underline{G(e_1)}_{\psi} \subseteq \underline{G(e_2)}_{\psi}$*, and* $\overline{G(e_1)}_{\psi} \subseteq \overline{G(e_2)}_{\psi}$.

Chapter 6
Applications and Future Directions of Research

Uncertainty is inherently present in problems of decision making and data analysis. And the role of Soft sets in modelling these is already well known and there are many such cases in literature. A review of this is given by Tripathy and Arun [145]. In this chapter we discuss various applications of soft sets in many real problems; like decision making, parameter reduction, game theory and data filling in incomplete information systems. We conclude the chapter by pointing out current status of work in the area and giving some fruitful areas of further study in both theoretical and practical aspects of soft sets.

6.1 Parameter Reduction and Decision Making Problems

Parameter reduction of soft sets was proposed by Maji et al. [79] in order to solve decision making problems. Later Chen et al. [32] pointed out some problems in the approach of Maji et al. [79] and presented another concept of parameter reduction of soft sets. To overcome the problem of suboptimal choice, in [32], Kong et al. [67] introduced the concept of normal parameter reduction of soft sets. However, the concept is too complex, the algorithm is hard to understand and involves a great amount of computation. Ma et al. [76] investigated the normal parameter reduction and an improved algorithm is given in [67].

In this section, we first see the method adopted in Maji et al. [79] together with the modification given by Chen et al. [32]. Further we discuss the parameter reduction of soft sets by means of attribute reductions in information systems introducd by Ning-Xin Xie [153]. Results and illustrative cases discussed here are taken mainly from [32, 67, 79, 153].

Definition 6.1 Assume that we have a binary operation, denoted by $*$, for subsets of the set U. Let (F, A) and (G, B) be soft sets over U. Then, the operation $*$ for soft sets is defined in the following way: $(F, A) * (G, B) = (H, A \times B)$, where

S. J. John, *Soft Sets*, Studies in Fuzziness and Soft Computing 400,
https://doi.org/10.1007/978-3-030-57654-7_6

$H(\alpha, \beta) = F(\alpha) * G(\beta), \alpha \in A, \beta \in B$, and $A \times B$ is the Cartesian product of the sets A and B.

Definition 6.2 A knowledge representation system can be formulated as a pair $S = (U, A)$, where U is a nonempty finite set called the universe, and A is a nonempty finite set of primitive attributes. Every primitive attribute $a \in A$ is a total function $a : U \to V_a$, where V_a is the set of values of a, called the domain of a.

Definition 6.3 With every subset of attributes $B \subseteq A$, we associate a binary relation $\mathrm{IND}(B)$, called an indiscernibility relation, defined by

$$\mathrm{IND}(B) = \{(x, y) \in U \times U : a(x) = a(y), \forall a \in B\}$$

Obviously, IND (B) is an equivalence relation and $\mathrm{IND}(B) = \bigcap_{a \in B} \mathrm{IND}(a)$. Suppose $V_a = \left\{\varepsilon_a^1, \varepsilon_a^2, \ldots, \varepsilon_a^{n(a)}\right\}$. Define $F_a : V_a \to P(U)$ as $F_a\left(\varepsilon_a^i\right) = \left\{x \in U : a(x) = \varepsilon_a^i\right\}$ then (F_a, V_a) is a soft set.

Suppose $A = \{a_1, a_2, \ldots, a_m\}$, then $S = (U, A)$ can be expressed as a soft set $\left(F, V_{a_1} \times V_{a_2} \times \cdots \times V_{a_m}\right) = \left(F_{a_1}, V_{a_1}\right) \cap \left(F_{a_2}, V_{a_2}\right) \cap \cdots \cap \left(F_{a_m}, V_{a_m}\right)$. For every $(p_1, p_2, \ldots, p_m) \in V_{a_1} \times V_{a_2} \times \cdots \times V_{a_m}$, $F(p_1, p_2, \ldots, p_m) = F_{a_1}(p_1) \cap F_{a_2}(p_2) \cap \cdots \cap F_{a_m}(p_m)$. All nonempty sets of $F(p_1, p_2, \ldots, p_m)$ form the collection of the equivalence classes of $\mathrm{IND}(A)$. Thus, the soft set can be applied to express a knowledge representation system.

Definition 6.4 Let R be a family of equivalence relations and let $A \in R$. We say that A is dispensable in R if $\mathrm{IND}(R) = \mathrm{IND}(R - \{A\})$; otherwise A is indispensable in R. The family R is independent if each $A \in R$ is indispensable in R; otherwise R is dependent. $Q \subset P$ is a reduction of P if Q is independent and $\mathrm{IND}(Q) = \mathrm{IND}(P)$, that is to say Q is the minimal subset of P that keeps the classification ability. The set of all indispensable relations in P will be called the core of P, and will be denoted as $\mathrm{CORE}(P)$. Clearly, $\mathrm{CORE}(P) = \cap \mathrm{RED}(P)$, where $\mathrm{RED}(P)$ is the family of all reductions of P.

6.1.1 A Decision Making Problem

Maji et al. [79] presented an application of soft set theory in a decision making problem with the help of rough approach. The problem is described as follows.

Let $U = \{h_1, h_2, h_3, h_4, h_5, h_6\}$ be a set of six houses, E = {expensive; beautiful; wooden; cheap; in green surroundings; modern; in good repair; in bad repair }, be a set of parameters. Consider the soft set (F, E) which describes the 'attractiveness of the house', given by

(F, E) = {expensive houses $= \phi$, beautiful houses $= \{h_1, h_2, h_3, h_4, h_5, h_6\}$

wooden houses $= \{h_1, h_2, h_6\}$, modern houses $= \{h_1, h_2, h_6\}$

Table 6.1 Tabular representation of the soft set (F, P)

U	e_1	e_2	e_3	e_4	e_5
h_1	1	1	1	1	1
h_2	1	1	1	1	0
h_3	1	0	1	1	1
h_4	1	0	1	1	0
h_5	1	0	1	0	0
h_6	1	1	1	1	1

houses in bad repair $= \{h_2, h_4, h_5\}$, cheap houses $= \{h_1, h_2, h_3, h_4, h_5, h_6\}$ houses in good repair $= \{h_1, h_3, h_6\}$, houses in green surroundings $= \{h_1, h_2, h_3, h_4, h_6\}\}$.

Suppose that, Mr. X is interested in buying a house on the basis of his choice parameters 'beautiful', 'wooden', 'cheap', 'in green surroundings', 'in good repair', etc., which constitute the subset
$P = \{$ beautiful, wooden, cheap, in green surroundings,in good repair$\}$ of the set E. That means, out of available houses in U, he is to select that house which qualifies with all (or with maximum number of) parameters of the soft set P.

To solve this problem, the soft set (F, P) is firstly expressed as a binary table.

If $h_i \in F\left(e_j\right)$ then $h_{ij} = 1$, otherwise $h_{ij} = 0$, where h_{ij} are the entries in Table 6.1.

Thus, a soft set can now be viewed as a knowledge representation system where the set of attributes is replaced by a set of parameters. Consider the tabular representation of the soft set (F, P). If Q is a reduction of P, then the soft set (F, Q) is called the reduct-soft-set of the soft set (F, P) The choice value of an object $h_i \in U$ is c_i, given by $c_i = \sum_j h_{ij}$, where h_{ij} are the entries in the table of the reduct-soft-set.

The algorithm [79] for Mr. X to select the house he wishes is listed as follows.

1. Input the soft set (F, E)
2. Input the set P of choice parameters of Mr. X which is a subset of E
3. Find all reduct-soft-sets of (F, P)
4. Choose one reduct-soft-set say (F, Q) of (F, P)
5. Find k, for which $c_k = \max c_i$

Then h_k is the optimal choice object. If k has more than one value, then any one of them could be chosen by Mr. X using his option.

It is claimed in [79] that $\{e_1, e_2, e_4, e_5\}$ and $\{e_2, e_3, e_4, e_5\}$ are two reductions of $P = \{e_1, e_2, e_3, e_4$ es $\}$. But $\{e_1, e_2, e_4, e_5\}$ and $\{e_2, e_3, e_4, e_5\}$ are not really the reductions of $P = \{e_1, e_2, e_3, e_4, e_5\}$ The following computing results given in [32] will illustrate this.

Suppose R_P is the indiscernibility relation induced by $P = \{e_1, e_2, e_3, e_4, e_5\}$, then the partition defined by R_P is $\{\{h_1, h_6\}, \{h_2\}, \{h_3\}, \{h_4\}, \{h_5\}\}$. If we delete $\{e_1, e_3\}$ from P, then the indiscernibility relation and the partition are invariant, so both of e_1 and e_3 are dispensable in P. If we delete one of $\{e_2, e_4, e_5\}$ from P, then the indiscernibility relation and the partition would be changed, thus all of these three parameters are indispensable. For example, suppose we delete $\{e_2\}$ from P, then the partition is changed to $\{\{h_1, h_3, h_6\}, \{h_2, h_4\}, \{h_5\}\}$. So $\{e_2, e_4, e_5\}$ is in fact the reduction of $P = \{e_1, e_2, e_3, e_4, e_5\}$. From Table 6.1 we can also conclude that e_1 and e_3 are not relevant and will not affect the choices of the house since they take the same values for every house.

On the other hand, this algorithm computes the reduction of the soft set in Step 3 before computing the choice value in Step 5, which would lead to two problems. First, after reduction, the objects that take max choice value may be changed, so it is possible that the decision after reduction is not the best one. Second, since the reductions of soft set are not unique, it is possible that there would be a difference between the objects that take max choice value obtained using different reductions. In these two cases, the choice object may not be optimal or may be quite difficult to select. Furthermore, even if these two problems do not appear in the example presented, it is highly possible that they appear in other situations. The following example illustrates this.

Example 74 Suppose we have a soft set (F, E) with the tabular representation as (Table 6.2)

Clearly, $c_2 = 5$ is the biggest choice value, thus h_2 takes the max choice value and will be the optimal choice object. Suppose R_E is the indiscernibility relation induced by E, then the partition induced by R_E is $\{\{h_1\}, \{h_2\}, \{h_3\}, \{h_4\}, \{h_5\}, \{h_6\}\}$. The partition obtained from $\{e_1, e_4, e_5\}$ is invariant. If we delete one of $\{e_1, e_4, e_5\}$ then the partition is changed, so $\{e_1, e_4, e_5\}$ is a reduction of (F, E). For example, if we delete e_1 from $\{e_1, e_4, e_5\}$, then the partition will be changed to $\{\{h_1, h_6\}, \{h_2, h_4\}, \{h_3\}, \{h_5\}\}$. Similarly we can examine that $\{e_2, e_4, e_5\}$ is the reduction of (F, E). For $\{e_1, e_4, e_5\}$, h_1 takes the max choice value and will be the optimal choice object with respect to $\{e_1, e_4, e_5\}$, while h_6 takes the max choice value for $\{e_2, e_4, e_5\}$ and will be the optimal choice object with respect to $\{e_2, e_4, e_5\}$. This means that the optimal object is changed after reduction (the optimal choice object is not h_2) and that different reductions decide different optimal objects. If we select the choice objects according to $\{e_1, e_4, e_5\}$ and $\{e_2, e_4, e_5\}$, we will miss the real optimal one. In other words, both of the predicted difficulties do in fact appear.

Table 6.2 Soft set (F, E)

U	e_1	e_2	e_3	e_4	e_5	e_6	e_7
h_1	1	0	1	1	1	0	0
h_2	0	1	1	1	0	1	1
h_3	0	0	1	0	1	0	1
h_4	1	0	1	1	0	0	0
h_5	1	0	1	0	0	1	0
h_6	0	1	1	1	1	0	0

As Example 74 shows that the algorithm presented in [79] which first computes the reduct soft-set then computes the choice value, is not error-free. For the application found in [79] the choice values of objects are obtained by the number of parameters the object belongs to, thus there is a straight forward relationship between the choice values and the conditional parameters. But for the rough set theory there is no such kind of straightforward relationship between the decision attributes and the conditional attributes, i.e., the decision attributes values are not briefly computed by the conditional attributes values. This statement is the key difference between soft sets and rough sets. In [79] they make the choice values as the decision parameter and try to find minimal subset of conditional parameter set by using reduction in rough set theory to keep the optimal choice object. However, the attributes reduction in rough set theory is designed to find a minimal attributes set that retains the classification ability of the indiscernibility relation. since choice values in soft set is not decided by the classification ability of the indiscernibility relation, the attributes reduction can not be applied to reduce the number of parameters to keep the optimal choice objects in soft set. Otherwise it is possible that the optimal choice object may be changed after reduction as indicated by Example 74. If the parameters set E is divided into two parts, i.e., $E = E_1 \cup E_2$, where E_1 is the conditional parameters set and E_2 is the decision parameters set, here E_2 is not computed by E_1, that is to say there is no straight forward relationship between E_1 and E_2, and either E_1 and E_2 induce indiscernibility relation or partition on the universe, then E_1 and E_2 can be viewed as conditional and decision attributes in rough set theory respectively. We can find the minimal subset of E_1 to keep the classification ability of E_1 relative to E_2 invariant. This is just the concept of relative reduction in rough set theory and is totally different from the decision-making problem in [79]. There the authors did not distinguish between these two cases. One should notice in the soft set as shown in Table 6.1 that no matter how the parameters are reduced, h_1 and h_6 could be selected as optimal objects. So this application problem is a very special case and the method of introducing reduct soft set in [79] is meaningless to deal with this application problem, which could only result in possibly misleading/wrong final decision.

Further, a weighted table of a soft set is presented in [79] by having $d_{ij} = w_j \times h_{ij}$ instead of 0 and 1 only, where h_{ij} are the entries in the table of the soft set and w_j are weights of e_j. The weighted choice value of an object $h_i \in U$ is c_i, given by $c_i = \sum_j d_{ij}$. By imposing weights on his choice parameters, Mr. X could now use the following revised algorithm for arriving at his final decisions.

1. Input the soft set (F, E)
2. Input the set P of choice parameters of Mr. X which is a subset of E
3. Find all reduct-soft-sets of (F, P)
4. Choose one reduct-soft-set say (F, Q) of (F, P)
5. Find weighted table of the soft set (F, Q), according to the weights decided by Mr. X
6. Find k, for which $c_k = \max c_i$

Then h_k is the optimal choice object. If k has more than one value, then any one of them could be chosen by Mr. X using his option.

Clearly this revised algorithm still suffers from the same two problems discussed earlier. This analysis seems to show that soft set theory is quite different from rough set theory and that attributes reduction in rough set theory usually cannot be applied to the decision problems as mentioned in [79]. Now we give the parameterization reduction of soft sets to deal with the decision problems as introducd by Chen et al. [32].

6.1.2 Parameterization Reduction of Soft Sets

Suppose $U = \{h_1, h_2, \ldots, h_n\}$, $E = \{e_1, e_2, \ldots, e_m\}$, (F, E) is a soft set with tabular representation. Define $f_E(h_i) = \sum_j h_{ij}$ where h_{ij} are the entries in the table of (F, E). Denote M_E as the collection of objects in U which takes the max value of f_E. For every $A \subset E$, if $M_{E-A} = M_E$, then A is called a dispensable set in E, otherwise A is called an indispensable set in E. Roughly speaking, $A \subset E$ is dispensable means that the difference among all objects according to the parameters in A does not influence the final decision. The parameter set E is called independent if every $A \subset E$ is indispensable in E, otherwise E is dependent. $B \subseteq E$ is called a reduction of E if B is independent and $M_B = M_E$, i.e., B is the minimal subset of E that keeps the optimal choice objects invariant. Clearly, after the reduction of the parameter set E, we have less parameters and the optimal choice objects have not been changed.

The reduction of parameter sets in soft set theory and attributes reduction in rough set theory are in some ways similar to the approach of finding minimal parameters sets or attributes sets in decision-making but they use different methods. In rough set theory, they define single dispensable attribute while in soft set theory we cannot define a single dispensable parameter as the dispensable attribute. This is because in soft set the decision value $f_E(h_i)$ is computed by the number of parameters that h_i takes the value of 1, the optimal choice objects is obtained by the order of $f_E(h_i)$. Although a single parameter may influence the order of $f_E(h_i)$, it is possible there is another parameter, such that these two parameters do not influence the order of $f_E(h_i)$. That is for a parameter $e \in E$ satisfying $M_{E-\{e\}} \neq M_E$, it is possible that there exists an $e' \in E$, such that $M_{E-\{e,e'\}} = M_E$. For instance, in Example 74, $M_E = \{h_1, h_6\}$ $M_{E-\{e_2\}} = \{h_1, h_3, h_6\}$ and $M_{E-\{e_2\}} \neq M_E$, but $M_{E-\{e_1,e_2\}} = \{h_1, h_6\} = M_E$. This case is not shared by rough set theory, i.e., in rough set theory if an attribute is indispensable, any set of attributes containing this attribute will also be indispensable. This means that without this set the ability of the knowledge representation system for solving classification problems will be changed.

However, in rough set theory the attributes reduction is designed to keep the classification ability of conditional attributes relative to the decision attributes. There is not straightforward connection between the conditional attributes and the decision attributes. But for the soft set, the connection between the decision values and the conditional parameters are straightforward, i.e., the decision values are computed by the conditional parameters, and the reduction of parameters is designed to offer minimal subset of the conditional parameters set to keep the optimal choice objects. Now we know that the problems tackled by attributes reduction in rough set theory and parameters reduction in soft set theory are different and their methods are also different, which has been analyzed in previous paragraph. The reduction of parameter sets in soft set theory and the reduction of attributes in rough set theory are different tools for different purposes. In general, one cannot be applied in the place of the other.

For the soft set in Table 6.1, if we delete e_1, e_3 and e_4 from P, then the optimal choice objects are unchanged. If we delete any subset of P which include at least one of e_2 and e_5, then the optimal choice objects will be changed. For example, if we delete $\{e_2, e_4\}$ from P, then the optimal choice objects will be $\{h_1, h_3, h_6\}$. For $\{e_2, e_5\}$ the optimal choice objects are not changed. Thus the soft set in Table 6.1 has a parameter reduction $\{e_2, e_5\}$. This means $\{e_2, e_5\}$ are the key parameters in Mr. X's selection of a house. However, $\{e_2, e_5\}$ is not the attributes reduction of P and the attributes reduction $\{e_2, e_4, e_5\}$ is not the parameter reduction of P since it is not the minimal parameter set to maintain the optimal choice objects.

In what follows, we employ our Example 74 to illustrate our idea of parameterization reduction. As we mentioned before, for the soft soft set of Example 74, the optimal choice object is h_2. By our definition of parameterization reduction we can examine that $\{e_2, e_6\}$ and $\{e_6, e_7\}$ are two parameterization reductions (not all) of E since they agree to the optimal choice object as h_2 but they are really not the attributes of E since they induce different partitions. As analyzed in previous section, we know $\{e_2, e_4, e_5\}$ is an attributes reduction of E, but it is really not the parameterization reduction of E since it presents another optimal choice object h_6. Thus the analysis of the soft soft set in Table 6.1 and Example 74 confirm the difference between the parameterization reduction of soft sets and attributes reduction of rough sets.

For the weighted soft sets, we just need to change h_{ij} to $w_j \times h_{ij}$ in the soft set table, then it is possible to propose a similar idea for presenting the reduction of parameter sets for weighted soft sets and this can be applied to improve the decision problem with the weighted soft set in [79].

In a fixed-decision problem where the final decision is unknown, the parameter reduction has only one application, i.e., to present the key parameters. However, if we want to discover knowledge from a data set using a soft set with tabular representation where the decision attribute is given, the parameter reduction can offer optimal parameter sets for newly input/testing objects. This is due to the fact that the complexity of computing the decisions can be reduced by the action of attributes reduction in rough set theory.

6.1.3 Parameter Reduction of Soft Sets by Means of Attribute Reductions in Information Systems

As a continuation of our discussion in Sect. 4.6, we see further applications of parameter reduction of soft sets in information systems. Results and algorithms given are taken from [153]. We recall that

Definition 6.5 Suppose that U is a finite set of objects and A a finite set of attributes. The pair (U, A, V, g) is called an information system, if g is an information function from $U \times A$ to $V = \bigcup_{a \in A} V_a$ where every $V_a = \{g(x, a) | a \in A \text{ and } x \in U\}$ is the values of the attribute a. An information system (U, A, V, g) is called 2-value if $V = \{0, 1\}$.

Definition 6.6 Suppose that (f, A) is a soft set over U. Then $I_S = (U, A, V, g_s)$ is called the 2-value information system induced by S, where $g_s : U \times A \rightarrow V$ For any $x \in U$ and $a \in A$

$$g_s(x, a) = \begin{cases} 1, & x \in f(a) \\ 0, & x \notin f(a) \end{cases}$$

Definition 6.7 Let $I = (U, A, V, g)$ be a 2-value information system. Then $S_I = (f_I, A)$ is called the soft set over U induced by I, where $f_I : A \rightarrow 2^U$ and for any $x \in U$ and $a \in A$, $f_I(a) = \{x \in U | g(x, a) = 1\}$.

Lemma 6.1 *Suppose that* $S = (F, A)$ *is a soft set over* U, $I_S = (U, A, V, g_s)$ *the 2-value information system induced by S over U and* $S_{I_S} = \left(f_{I_S}, A\right)$ *the soft set over* U *induced by Is. Then* $S = S_{I_s}$

Proof By Definition, for any $a \in A$, $f_{Is}(a) = \{x \in U | g_s(x, a) = 1\}$, and for any $x \in U$ and $a \in A$

$$g_s(x, a) = \begin{cases} 1, & x \in f(a) \\ 0, & x \notin f(a) \end{cases}$$

This implies that $g_s(x, a) = 1 \Leftrightarrow x \in f(a)$. So, for any $x \in U, a \in A$, $f(a) = f_{Is}(a)$ Hence $f_A = (f_{Is}, A)$. This implies $S = S_{Is}$. □

Lemma 6.2 *Let* $I = (U, A, V, g)$ *be a 2-value information system,* $S_I = \left(f_{I,A}\right)$ *the soft set over* U *induced by I and* $I_{S_i} = \left(U, A, V, g_{S_i}\right)$ *the 2-value information system induced by* S_I. *Then* $I = I_{S_I}$.

Proof For any $x \in U$ and $a \in A$

$$g_{S_I}(x, a) = \begin{cases} 1, & x \in f_I(a) \\ 0, & x \notin f_I(a) \end{cases}$$

For any $x \in U$ and $a \in A$, by Definition, $f_I(a) = \{x \in U | g(x, a) = 1\}$. Since $I = (U, A, V, g)$ is a 2-value information system, $g(x, a) = 0$ for $x \notin f_I(a)$, This implies that

$$g(x,a)=\begin{cases}1, & x\in f_I(a)\\ 0, & x\notin f_I(a)\end{cases}$$

So for any $x \in U$ and $a \in A$, $g_{S_I}(x,a) = g(x,a)$. Hence $g_{S_I} = g$. This implies $I = I_{S_I}$. □

From Theorem 4.35, next theorem follows clearly

Theorem 6.1 *Suppose that*

$$\Sigma = \{S \mid S = f_A \text{ is a soft set over } U\}$$

and

$$\Gamma = \{I \mid I = (U, A, V, g) \text{ is a } 2-\text{value information system}\}$$

Then there exists a one-to-one correspondence between Σ *and* Γ

Definition 6.8 Let (f, A) be a soft set over U and let (U, A, V, g) be the 2-value information system induced by (f, A) over U. For any $B \subseteq A$, R_B is defined as follows:

$$R_B = \{(x,y) \in U \times U \mid g(x,a) = g(y,a)(\forall a \in B)\}$$

We denote $R_{[a]} = R_a$. Obviously, $R_B = \bigcap_{a\in B} R_a$.

Proposition 6.1 *Suppose that* (f, A) *is a soft set over* U *and* $B \subseteq A$. *Then the following properties hold.*

(1) $R_B = \{(x,y) \in U \times U \mid \{x,y\} \subseteq f(a) \text{ or } \{x,y\} \cap f(a) = \emptyset(\forall a \in B)\} = \bigcap_{a\in B} R_a$.
(2) R_B *is an equivalence relation.*
(3) If $B_1 \subseteq B_2 \subseteq A$, *then* $R_{B_1} \supseteq R_{B_2} \supseteq R_A$.

Proof (1) Obviously, $R_B = \bigcap_{a\in B} R_a$.
By Definition,

$$g(x,a)=\begin{cases}1, & x\in f(a)\\ 0, & x\notin f(a)\end{cases}$$

$$g(y,a)=\begin{cases}1, & y\in f(a)\\ 0, & y\notin f(a)\end{cases}$$

Suppose $(x,y) \in R_B$. Then $\forall a \in B$, $g(x,a) = g(y,a)$. So $\forall a \in B$, $g(x,a) = g(y,a) = 1$ or $g(x,a) = g(y,a) = 0$. This implies that $\forall a \in B$,

$$x \in f(a), y \in f(a) \text{ or } x \notin f(a), y \notin f(a)$$

Thus

$$(x, y) \in \{(x, y) \in U \times U | \{x, y\} \subseteq f(a) \text{ or } \{x, y\} \cap f(a) = \emptyset \quad (\forall a \in B)\}$$

On the other hand, suppose
$(x, y) \in \{(x, y) \in U \times U | \{x, y\} \subseteq f(a) \text{ or } \{x, y\} \cap f(a) = \emptyset \quad (\forall a \in B)\}$, then $\forall a \in B$,

$$\{x, y\} \subseteq f(a) \text{ or } \{x, y\} \cap f(a) = \emptyset$$

If $\{x, y\} \subseteq f(a)$, then $x \in f(a), y \in f(a)$. So $g(x, a) = 1 = g(y, a)$. If $\{x, y\} \cap f(a) = \emptyset$, then $x \notin f(a), y \notin f(a)$. So $g(x, a) = 0 = g(y, a)$. Thus $\forall a \in B$, $g(x, a) = g(y, a)$. This implies that $(x, y) \in R_B$. Hence

$$R_B = \{(x, y) \in U \times U | \{x, y\} \subseteq f(a) \text{ or } \{x, y\} \cap f(a) = \emptyset (\forall a \in B)\}$$

(2)and (3) are obvious. □

Definition 6.9 Let (f, A) be a soft set over U. Then for any $B \subseteq A$, the pair (U, R_B) is an approximation space. Based on (U, R_B), we define a pair of operations $\underline{R_B}, \overline{R_B}: 2^U \longrightarrow 2^U$ as follows:

$$\underline{R_B}(X) = \{x \in U | R_B(x) \subseteq X\}, \overline{R_B}(X) = \{x \in U | R_B(x) \cap X \neq \emptyset\},$$

where $X \in 2^U$ and $R_B(x) = \{y \in U | x R_B y\}$. $\underline{R_B}(X)$ and $\overline{R_B}(X)$ are called lower approximation and upper approximation of X with respect to B, respectively. X is called a definable set with respect to the parameter set B if $\underline{R_B}(X) = \overline{R_B}(X)$. X is called a rough set with respect to B if $\underline{R_B}(X) \neq \overline{R_B}(X)$. Moreover, the sets

$$\begin{aligned} \text{Pos}_P(X) &= \underline{R_B}(X) \\ \text{Neg}_P(X) &= U - \overline{R_B}(X) \\ \text{Bnd}_P(X) &= \overline{R_B}(X) - \underline{R_B}(X) \end{aligned}$$

are called the positive region, the negative region and the boundary region of X with respect to B, respectively.

Proposition 6.2 *Suppose that (f, A) is a soft set over U. Then for any $B \subseteq A$ and $X, Y \in 2^U$*

(1) $\underline{R_B}(\emptyset) = \overline{R_B}(\emptyset) = \emptyset$; $\underline{R_B}(U) = \overline{R_B}(U) = U$
(2) $\underline{R_B}(X) \subseteq X \subseteq \overline{R_B}(X)$
(3) $X \subseteq Y \Rightarrow \underline{R_B}(X) \subseteq \underline{R_B}(Y)$; $X \subseteq Y \Rightarrow \overline{R_B}(X) \subseteq \overline{R_B}(Y)$
(4) $\underline{R_B}(X \cap Y) = \underline{R_B}(X) \cap \underline{R_B}(Y)$; $\overline{R_B}(X \cup Y) = \overline{R_B}(X) \cup \overline{R_B}(Y)$
(5) $\underline{R_B}(X \cup Y) \supseteq \underline{R_B}(X) \cup \underline{R_B}(Y)$; $\overline{R_B}(X \cap Y) \subseteq \overline{R_B}(X) \cap \overline{R_B}(Y)$
(6) $\underline{R_B}(U - X) = U - \overline{R_B}(X)$; $\overline{R_B}(U - X) = U - \underline{R_B}(X)$
(7) $\overline{R_B}(\underline{R_B}(X)) \subseteq X \subseteq \underline{R_B}(\overline{R_B}(X))$

Parameter reduction of soft sets means deleting parameters of soft sets which are no or less influence for obtaining an optimal decision, and reducing number of parameters in decision making. Specific approach is first classifying parameters of soft sets according to their importance and then finding the minimum or minor set of parameters which is no or less influence for getting the optimal decision. Since there exists a one-to-one correspondence between "the set of all soft sets" and "the set of all 2-value information systems", we can do the parameter reduction of soft sets with the help of attribute reductions in information systems. The parameter reduction of soft sets plays an important role in decision-making problems, which can save expensive tests and time.

Definition 6.10 Suppose that (f, A) is a soft set over U.

(1) $B \subseteq A$ is called a parameter reduction of (f, A) if $R_A = R_B$ and $R_A \neq R_{B-[a]}$ for any $a \in B$.
(2) The intersection set of all the parameter reduction of f_A is called the soft core. We denote it by core (f_A).

we denote the set of all the parameter reduction of (f, A) by $pr\ (f_A)$.

Proposition 6.3 *Let (f, A) be a soft set over U. Then $pr\ (f_A) \neq \emptyset$.*

Proof (1) If $R_A \neq R_{A-|a|}$ for any $a \in A$, then A is a parameter reduction of f_A.
(2) If $R_A = R_{A-[a]}$ for some $a \in A$, then we consider $B_1 = A - \{a\}$. If $R_A \neq R_{B_1-[b_1]}$ for any $b_1 \in B_1$, then B_1 is a parameter reduction of f_A. Otherwise, we consider $B_1 - \{b_1\}$ again and repeat the above-mentioned process. Since A is a finite set, we can find at least a parameter reduction of f_A. Thus, $pr\ (f_A) \neq \emptyset$. □

Definition 6.11 Suppose that (f, A) is a soft set over U and pr $(f_A) = \{A_k | 1 \leq k \leq n\}$ Then

(1) $a \in A$ is called core if $a \in \bigcap_{i=1}^{k} A_k =$ core (f_A)
(2) $a \in A$ is called relative indispensable if $a \in \bigcup_{i=1}^{n} A_i -$core$(f_A)$
(3) $a \in A$ is called absolutely dispensable if $a \in A - \bigcup_{i=1}^{n} A_i$
(4) $a \in A$ is called dispensable if $a \in A-$core(f_A)

Obviously, $a \in A$ is dispensable if and only if a is relative indispensable or absolutely dispensable.

We have the following proposition.

Proposition 6.4 *Let (f, A) be a soft set over U. Then*

(1) $|pr\ (f_A)| = 1$ *if and only if* core $(f_A) \in$ pr (f_A)
(2) $a \in$ core (f_A) *if and only if* $R_A \neq R_{A-[a]}$
(3) $a \in A$ *is dispensable if and only if* $R_A = R_{A-|a|}$

Definition 6.12 Let (f, A) be a soft set over U with $|U| = n$ and let (U, A, V, g) be the 2 -value information system induced by f_A over U. For $x, y \in U$, we define $d(x, y)$ as follows:

$$d(x, y) = \{a \in A | g(x, a) \neq g(y, a)\}$$

(1) $d(x, y)$ is called the set of parameters which can discern the objects x and y.
(2) $\mathfrak{D}(f_A) = (d_{ij})_{n \times n}$ is called the discernibility matrix of (f, A), where $U = \{x_1, x_2, \ldots x_n\}$ and $d_{ij} = d(x_i, x_j)$.

Remark 6.1 Suppose that (f, A) is a soft set over U

(1) $d(x, y) = \{a \in A | x \in f(a), y \notin f(a) \text{ or } x \notin f(a), y \in f(a)\}$ for any $x, y \in U$
(2) $d(x, y) = d(y, x)$ for any $x, y \in U$
(3) $d(x, x) = \emptyset$ for any $x \in U$

Proposition 6.5 *Let (f, A) be a soft set over U, let (U, A, V, g) be the 2-value information system induced by f_A over U and let $B \subseteq A$. Then*

$$R_A = R_B \text{ if and only if } B \cap d(x, y) \neq \emptyset \text{ for any } x, y \in U$$

Proof Sufficiency. Suppose $R_A \neq R_B$. By Proposition 6.1, $R_A \subseteq R_B$. So $R_A \not\supseteq R_B$ This implies $R_B - R_A \neq \emptyset$. Pick $(x_0, y_0) \in R_B - R_A$. Then $(x_0, y_0) \in R_B$ and $(x_0, y_0) \notin R_A$. This implies $A - B \supseteq d(x_0, y_0) \neq \emptyset$. So $B \cap d(x_0, y_0) = \emptyset$. This is a contradiction.

Necessity. Suppose $B \cap d(x_0, y_0) = \emptyset$ for some $d(x_0, y_0) \neq \emptyset$. This implies $g(x_0, a) = g(y_0, a)$ for any $a \in B$. So $(x_0, y_0) \in R_B$. Since $d(x_0, y_0) \neq \emptyset$, $g(x_0, a_0) \neq g(y_0, a_0)$ for some $a_0 \in A$. So $(x_0, y_0) \notin R_A$. Thus $R_A \neq R_B$. This is a contradiction. □

Clearly we have

Theorem 6.2 *Suppose that (f, A) is a soft set over U. Then $B \subseteq A$ is a parameter reduction of (f, A) if and only if*

(1) $B \cap d(x, y) \neq \emptyset$ for any $x, y \in U$
(2) for any $a \in B$, there are $x_a, y_a \in U$ such that $(B - \{a\}) \cap d(x_a, y_a) = \emptyset$

Theorem 6.3 *Let (f, A) be a soft set over U. Then*

$$\text{core}(f_A) = \{a \in A | d(x, y) = \{a\} \text{ for any } x, y \in U\}$$

Proof Denote

$$D = \{a \in A | d(x, y) = \{a\} \text{ for any } x, y \in U\}$$

and

$$\text{pr}(f_A) = \{B_k | k \leq q\}$$

(1) Let $a \in D$. Then there exist $x_0, y_0 \in U$ such that $d(x_0, y_0) = \{a\}$. Since B_k is a parameter reduction of f_A, $B_k \cap d(x_0, y_0) \neq \emptyset$. This implies $a \in B_k (k \leq q)$. So $a \in \bigcap_{i=1}^{n} B_k = \text{core}(f_A)$. Thus $\text{core}(f_A) \supseteq D$
(2) Let $a \in \text{core}(f_A)$. Suppose there are not x and $y \in U$ such that $d(x, y) = \{a\}$. Then we have the following three cases:

Case 1: $d(x, y) = \emptyset$ for any $x, y \in U$. $a \in \text{core}(f_A)$ implies $a \in B_k$ for any $k \leq q$. By Proposition 6.5, $B_k \cap d(x, y) \neq \emptyset$. This is a contradiction.

Case 2: There exist x_0 and $y_0 \in U$ such that $|d(x_0, y_0)| = 1$, but $d(x_0, y_0) \neq \{a\}$. We suppose $d(x_0, y_0) = \{b\}(b \neq a)$.

Similar to the proof of (1), $b \in \text{core}(f_A)$. Thus b is equal to some $a \in \text{core}(f_A)$ This is a contradiction.

Case 3: For any $a \in d(x, y)$, $|d(x, y)| \geq 2$. Put $B = \cup(d(x, y) - \{a\})$. Then $B \cap d(x, y) \neq \emptyset$. By Proposition 6.5, $R_A = R_B$. Thus there exists $C \subseteq B$ such that C is a parameter reduction of (f, A). But $a \notin C$. This is a contradiction.

This show that $\text{core}(f_A) \subseteq D$. Hence, $\text{core}(f_A) = D$. □

It is more convenient to calculate the parameter reductions and the core of soft sets by using discernibility functions when there are many parameters in soft sets.

Here, we use the following propositional connectives in mathematical logic. They are disjunction, conjunction, implication and biimplication.

For any $a \in A$, we specify a Boolean variable "a". If $d(x, y) = \{a_1, a_2, \ldots, a_k\}$ with $x, y \in U$, then we specify a Boolean function $a_1 \vee a_2 \vee \cdots \vee a_k$. Denote

$$\bigvee\{a_1, a_2, \ldots, a_k\} = \bigvee_{i=1}^{k} a_i = a_1 \vee a_2 \vee \cdots \vee a_k$$
$$\bigwedge\{a_1, a_2, \ldots, a_k\} = \bigwedge_{i=1}^{k} a_i = a_1 \wedge a_2 \wedge \cdots \wedge a_k$$

We stipulate that $\vee\emptyset = 1$ and $\wedge\emptyset = 0$ where 0 and 1 are two Boolean constants.

Definition 6.13 Let $U = \{x_1, x_2, \ldots, x_n\}$, let (f, A) be a soft set over U and let $\mathfrak{D}(f_A) = (d_{ij})_{n\times n}$ be the discernibility matrix of (f, A). We define the discernibility function $\Delta(f_A)$ of (f, A) as follows:

$$\Delta(f_A) = \bigwedge\left(\bigvee d_{ij}\right)$$

Let $U = \{x_1, x_2, \ldots, x_n\}$ and let (f, A) be a soft set over U. Denote

$$Q(f_A) = \left\{\bigvee d_{ij} | 1 \leq i, j \leq n\right\}$$

Define a binary relation "$\leq$" on Q(f_A) as follows: $\bigvee d_{ij} \leq \bigvee d_{kl}$ if and only if $d_{ij} \subseteq d_{kl}$ for any $\bigvee d_{ij}, \bigvee d_{kl} \in Q(f_A)$. For any $\vee d_{ij}, \vee d_{kl} \in Q(f_A)$, we denote
$\left(\bigvee d_{ij}\right) \bigsqcup \left(\bigvee d_{kl}\right) = \bigvee\left(d_{ij} \cup d_{kl}\right)$
$\left(\bigvee d_{ij}\right) \prod \left(\bigvee d_{kl}\right) = \bigvee\left(d_{ij} \cap d_{kl}\right)$

Proposition 6.6 *$(Q(f_A), \leq)$ is a poset.*

Proof (1) $\vee d_{ij} \leq \vee d_{ij}$ for any $\vee d_{ij} \in Q(f_A)$.
(2) Let $\vee d_{ij}, \vee d_{kl} \in Q(f_A)$. Suppose $\vee d_{ij} \leq \vee d_{kl}$ and $\vee d_{kl} \leq \vee d_{ij}$. Then $d_{ij} \subseteq d_{kl}$ and $d_{kl} \subseteq d_{ij}$. This implies $d_{ij} = d_{kl}$. So $\vee d_{ij} = \vee d_{kl}$.
(3) Let $\vee d_{ij}, \vee d_{kl}, \vee d_{hv} \in Q(f_A)$. Suppose $\vee d_{ij} \leq \vee d_{kl}$ and $\vee d_{kl} \leq \vee d_{hv}$ Then $d_{ij} \subseteq d_{kl}$ and $d_{kl} \subseteq d_{hv}$. This implies $d_{ij} \subseteq d_{hv}$. So $\vee d_{ij} \leq \vee d_{hv}$ Thus $(Q(f_A), \leq)$ is a poset. □

Proposition 6.7 *Let $U = \{x_1, x_2, \ldots, x_n\}$ and let (f, A) be a soft set over U. If $\{d_{ij} | 1 \leq i, j \leq n\}$ is a topology on A, then $\left(Q(f_A), \leq, L, \bigsqcup, \prod\right)$ is a lattice with top element and bottom element.*

Proof Denote $\tau = \{d_{ij} | 1 \leq i, j \leq n\}$ By Proposition 6.6, $(Q(f_A), \leq)$ is a poset. Let $\vee d_{ij}, \vee d_{kl} \in Q(f_A)$. since τ is a topology, $d_{ij} \cup d_{kl} \in \tau$. This implies

$$\left(\bigvee d_{ij}\right) \bigcup \left(\bigvee d_{kl}\right) = \bigvee \left(d_{ij} \cup d_{kl}\right) \in Q(f_A)$$

Similarly, $\left(\bigvee d_{ij}\right) \prod \left(\bigvee d_{kl}\right) = \bigvee \left(d_{ij} \cap d_{kl}\right) \in Q(f_A)$. Obviously, $1_{Q(f_A)} = \vee A$, $0_{Q(f_A)} = \vee \emptyset$. Thus $(Q(f_A), \leq, L, \Pi)$ is a lattice with top element and bottom element. □

Definition 6.14 Let (f, A) be a soft set over U and let $\Delta(f_A)$ be the discernibility function of f_A. If $\Delta(f_A) = \bigvee_{k=1}^{q} \left(\bigwedge_{l=1}^{p_k} a_{kl}\right)$, where every $B_k = \{a_{kl} | l \leq p_k\} \subseteq A$ has not repetitive elements, then $\bigvee_{k=1}^{q} \left(\bigwedge_{l=1}^{p_k} a_{kl}\right)$ is called the standard minimum formula of $\Delta(f_A)$. We denote it by $\Delta^*(f_A)$, that is,

$$\Delta^*(f_A) = \bigvee_{k=1}^{q} \left(\bigwedge_{l=1}^{p_k} a_{kl}\right)$$

Theorem 6.4 *Suppose that (f, A) is a soft set over U, $\Delta(f_A)$ discernibility function of (f, A) and $\Delta^*(f_A) = \bigvee_{k=1}^{q} \left(\bigwedge_{l=1}^{p_k} a_{kl}\right)$ the standard minimum formula of $\Delta(f_A)$. Then $B_k = \{a_{kl} | l \leq p_k\}$ $(k \leq q)$ are all parameter reductions of f_A.*

Proof (1) Let $B_{k_0} \in \{B_k | k \leq q\}$
(i) Obviously, $\Delta^*(f_A) = \bigvee_{k=1}^{q} \left(\bigwedge_{l=1}^{p_k} a_{kl}\right) = \bigvee_{k=1}^{q} (\wedge B_k)$. So, $\wedge B_{k_0} \longrightarrow \Delta^*(f_A)$. Since $\Delta^*(f_A) = \Delta(f_A) = \wedge \left(\vee d_{ij}\right)$, we have $\Delta^*(f_A) \Longleftrightarrow \vee d_{ij}$ for any $1 \leq i, j \leq n$. Thus for any $x, y \in U$, $\wedge B_{k_0} \longrightarrow \bigvee d(x, y)$. Now, $\wedge B_{k_0} \Longleftrightarrow a_{k_0 l}$ for any $l \leq p_{k_0}$ and $\bigvee d(x, y) \Longleftrightarrow a$ for some $a \in d(x, y)$. Then for any $x, y \in U$, a_{k_0}, for any $l \leq p_{k_0} \longrightarrow a$ for some $a \in d(x, y)$. So for any $x, y \in U$, there exists $l_0 \leq p_{k_0}$ such that $a = a_{k_0 l_0}$, that is $a \in B_{k_0} \cap d(x, y)$. Thus for any $x, y \in U$, $B_{k_0} \cap d(x, y) \neq \emptyset$.

(ii) To prove that B_{k_0} is a parameter reduction of (f, A), we need to show that for any $a \in B_{k_0}$, there are $x_a, y_a \in U$ such that

$$\left(B_{k_0} - \{a\}\right) \cap d(x_a, y_a) = \emptyset$$

Suppose that there exists $a_0 \in B_{k_0}$ such that $\left(B_{k_0} - \{a_0\}\right) \cap d(x, y) \neq \emptyset$ for any $x, y \in U$. Pick $b_{xy} \in \left(B_{k_0} - \{a_0\}\right) \cap d(x, y)$. Then $\wedge \left(B_{k_0} - \{a_0\}\right) \longrightarrow b_{xy}$ and $b_{xy} \to \vee d(x, y)$ Thus for any $x, y \in U$

$$\bigwedge \left(B_{k_0} - \{a_0\}\right) \longrightarrow \bigvee d(x, y)$$

since $\Delta^* (f_A)$ contains all true explanations of $\Delta (f_A)$, we have $B_{k_0} - \{a_0\} \in \{B_k | k \leq q\}$. Then
$\left(\wedge B_{k_0}\right) \vee \left(\wedge \left(B_{k_0} - \{a_0\}\right)\right) = \left(\left(\wedge \left(B_{k_0} - \{a_0\}\right)\right) \wedge \{a_0\}\right) \vee \left(\left(\wedge \left(B_{k_0} - \{a_0\}\right)\right) \wedge 1\right) = \left(\wedge \left(B_{k_0} - \{a_0\}\right)\right) \wedge (\{a_0\} \vee 1) = \left(\wedge \left(B_{k_0} - \{a_0\}\right)\right) \wedge 1 = \wedge \left(B_{k_0} - \{a_0\}\right)$. This implies $B_{k_0} \notin \{B_k | k \leq q\}$, a contradiction. Thus B_{k_0} is a parameter reduction of (f, A).

(2) Suppose that B is a parameter reduction of (f, A). We need to prove that there exists $B_{k_1} \in \{B_k | k \leq q\}$ such that $B = B_{k_1}$. Since B is a parameter reduction of (f, A), we have $R_A = R_B$. Then $B \cap d(x, y) \neq \emptyset$ for any $x, y \in U$. Similar to the proof of (1)(ii), we have

$$B \in \{B_k | k \leq q\}$$

Thus, there exists $B_{k_1} \in \{B_k | k \leq q\}$ such that $B = B_{k_1}$. Hence $B_k = \{a_{kl} | l \leq p_k\}$ $(k \leq q)$ are all the parameter reduction of (f, A). □

An algorithm for parameter reduction
Input: A soft set (f, A)
Output: All the parameter reduction of (f, A) and the core.
Step 1: Input a soft set (f, A)
Step 2: Calculate the discernibility matrix $\mathfrak{D} (f_A)$ of (f, A)
Step 3: Give discernibility function $\Delta (f_A)$ of (f, A)
Step 4: Calculate standard minimum formula $\Delta^* (f_A)$ of $\Delta (f_A)$
Step 5: Output all the parameter reduction of (f, A) and the core.

The parameter reduction of soft sets may play an important role in knowledge discovery.

6.2 Medical and Financial Diagnosis Problems

The technique of similarity measure of two soft sets can be applied to detect whether an ill person is suffering from a certain disease or not. Here we provide a simple example taken from Majumdar and Samantha [81] to show the possibility of using this method for diagnosis of diseases which could be improved by incorporating clinical results and other competing diagnosis.

Here we will try to estimate the possibility that an ill person having certain visible symptoms is suffering from pneumonia. For this, we first construct a model soft set for pneumonia and the soft set for the ill person. Next we find the similarity measure of these two sets. If they are significantly similar, then we conclude that the person is possibly suffering from pneumonia.

Table 6.3 Model Soft set (F, E) representing pneumonia

(F, E)	e_1	e_2	e_3	e_4	e_5	e_6	e_7	e_8
y	1	0	1	1	1	0	0	1
n	0	1	0	0	0	1	1	0

Table 6.4 Soft set (G, E) representing ill person

(G, E)	e_1	e_2	e_3	e_4	e_5	e_6	e_7	e_8
y	0	1	0	1	0	1	1	0
n	1	0	1	0	1	0	0	1

Table 6.5 Soft set (H, E) representing ill person

(H, E)	e_1	e_2	e_3	e_4	e_5	e_6	e_7	e_8
y	1	0	1	1	1	1	1	1
n	0	1	0	0	0	0	0	0

Let our universal set contain only two elements yes and no, i.e., $U = \{y, n\}$ Here the set of parameters E is the set of certain visible symptoms. Let $E = \{e_1, e_2, e_3, e_4, e_5, e_6, e_7, e_8\}$, where e_1 = high body temperature, e_2 = low body temperature, e_3 = cough with chest congestion, e_4 = cough with no chest congestion, e_5 = body ache, e_6 = headache, e_7 = loose motion, and e_8 = breathing trouble.

Our model soft set for pneumonia (F, E) is given and this can be prepared with the help of a medical person (Table 6.3):

Now the ill person is having fever, cough and headache. After talking to him, we can construct his soft set (G, E) as (Table 6.4):

Then we find the similarity measure of these two sets as:

$$S(F, G) = \frac{\sum_i \mathbf{F}(e_i) \bullet \mathbf{G}(e_i)}{\sum_i \left[\mathbf{F}(e_i)^2 \vee \mathbf{G}(e_i)^2\right]} = \frac{1}{8} < \frac{1}{2}$$

Hence the two soft soft sets, i.e., two symptoms (F, E) and (G, E) are not significantly similar. Therefore, we conclude that the person is not possibly suffering from pneumonia. A person suffering from the following symptoms whose corresponding soft set (H, E) is given in (Table 6.5):

Then $S(F, H) = \frac{6}{8} > \frac{1}{2}$. Here the two soft sets, i.e., two symptoms (F, E) and (H, E) are significantly similar. Therefore, we conclude that the person is possibly suffering from pneumonia.

The assumption of Majumdar and Samantha [81] that matrix representation is suitable for mathematical manipulation of soft sets has encountered some problems as shown in Athar Kharal [64] and to rectify this, set operations based measure was introduced in [64].

Definition 6.15 For two soft sets (F, A) and (G, B) in a soft space $S(X)_E$, where A and B are not identically void, we define Euclidean distance as:

$$e((F, A), (G, B)) = \|A \Delta B\| + \sqrt{\sum_{\varepsilon \in A \cap B} \|F(\varepsilon) \Delta G(\varepsilon)\|^2}$$

Normalized Euclidean distance as:

$$q((F, A), (G, B)) = \frac{\|A \Delta B\|}{\sqrt{\|A \cup B\|}} + \sqrt{\sum_{\varepsilon \in A \cap B} \chi(\varepsilon)}$$

$$\text{where } \chi(\varepsilon) = \begin{cases} \frac{\|F(\varepsilon) \Delta G(\varepsilon)\|^2}{\|F(\varepsilon) \cup G(\varepsilon)\|}, & \text{if } F(\varepsilon) \cup G(\varepsilon) \neq \phi \\ 0, & \text{otherwise} \end{cases}$$

where all the radicals yield non-negative values only.

Remark 6.2 The mappings e, q: $S(X)_E \times S(X)_E \to \mathbb{R}^+$, as defined above, are metrics.

Lemma 6.3 *For the soft sets* $\widetilde{\Phi}_E$, $\widetilde{X}_E$ *and an arbitrary soft set* (F, A) *in a soft space* $S(X)_E$, *we have:*
(1) $e((F, A), (F, A)^r) = 2\|A\|$
(2) $q((F, A), (F, A)^r) = \sqrt{2\|A\|}$
(3) $e(\widetilde{\Phi}_E, \widetilde{X}_E) = \sqrt{\|E\|\|X\|}$
(4) $q(\widetilde{\Phi}_E, \widetilde{X}_E) = \sqrt{\|E\|}$

Definition 6.16 Based on distances e and q defined, two similarity measures may be defined as:

$$S_K^e((F, A), (G, B)) = \frac{1}{1+e((F,A),(G,B))}$$
$$S_K^q((F, A), (G, B)) = \frac{1}{1+q((F,A),(G,B))}$$

Further, we may define another pair of similarity measures as:

$$S_W^e((F, A), (G, B)) = e^{-\alpha \cdot e((F,A),(G,B))}$$
$$S_W^q((F, A), (G, B)) = e^{-\alpha \cdot q((F,A),(G,B))}$$

where α is a positive real number (parameter) called the steepness measure.

Definition 6.17 Two soft sets (F, A) and (G, B) in a soft space (X, E) are said to be α -similar, denoted as $(F, A) \approx_\alpha (G, B)$, if

$$S((F, A), (G, B)) \geq \alpha \quad \text{for } \alpha \in (0, 1)$$

where S is a similarity measure.

Proposition 6.8 $\approx_\alpha$ *is reflexive and symmetric.*

We now present a financial diagnosis problem from [64] where this modified similarity measures can be applied.

The notion of similarity measure of two soft sets can be applied to detect whether a firm is suffering from a certain economic syndrome or not. In the following example, we estimate if two firms with observed profiles of financial indicators are suffering from serious liquidity problem. Example 75 Suppose the firm profiles are given as:

Profile 1 The firm ABC maintains a beerish future outlook as well as the same behaviour in trading of its share prices. During the last fiscal year, the profit-earning ratio continued to rise. Inflation is increasing continuously. ABC has a low amount of paid-up capital and a similar situation is seen in foreign direct investment flowing into ABC.

Profile 2 The firm XYZ showed a fluctuating share price and hence a varying future outlook. Like ABC, profit-earning ratio remained beerish. As both firms are in the same economy, inflation is also rising for XYZ and may be considered even high in view of XYZ. Competition in the business area of XYZ is increasing. Debit level went high but the paid-up capital lowered.

For this, we first construct a model soft set for liquidity-problem and the soft sets for the firm profiles. Next, we find similarity measure of these soft sets. If they are significantly similar, then we conclude that the firm is possibly suffering from liquidity problem.

Let $X =$ {inflation, profit-earning ratio, share price, paid-up capital, competitiveness, business diversification, future outlook, debt level, foreign direct investment, fixed income } be the collection of financial indicators which are given in both profiles. Further let $E =$ { fluctuating, medium, rising, high, beerish } be the universe of parameters, which are basically linguistic labels commonly used to describe the state of financial indicators.

The profile of a firm by observing its financial indicators may easily be coded into a soft set using appropriate linguistic labels. Let (F, A) and (G, B) be soft sets coding profiles of firms ABC and XYZ, respectively, and are given as:

$$(F, A) = \left\{ \begin{array}{c} \text{bearish } = \{ \text{ future outlook, share price } \} \\ \text{rising } = \{ \text{ profit earning ratio, inflation } \}, \\ \text{low } =\{ \text{ paid-up capital, foreign direct investment } \} \end{array} \right\}$$

$$(G, B) = \left\{ \begin{array}{rl} \text{fluctuating} & = \{ \text{ share price, future outlook } \} \\ \text{beerish} & = \{ \text{ profit earning ratio } \} \\ \text{rising} & = \{ \text{ inflation, compitition } \} \\ \text{high} & = \{ \text{ inflation, debit level } \} \\ \text{low} & = \{ \text{ paid-up capital } \} \end{array} \right\}$$

The model soft set for a firm suffering from liquidity problem can easily be prepared in a similar manner by help of a financial expert. In our case, we take it to be as follows:

$$(H, C) = \left\{ \begin{array}{c} \text{fluctuating } = \{ \text{ share price, future outlook} \} \\ \text{low } = \{ \text{ fixed income, paid-up capital} \} \\ \text{beerish } = \{ \text{profit earning ratio, foreign direct investment} \} \\ \text{high } = \{ \text{ inflation, debt level} \} \end{array} \right\}$$

For the sake of ease in mathematical manipulation, we denote the indicators and labels by symbols as follows:

$p =$ profit-earning ratio
$s =$ share price
$i =$ inflation
$c =$ paid-up capital
$m =$ competition
$d =$ business diversification
$o =$ future work
$l =$ debt level
$f =$ foreign direct investment
$x =$ fixed income

$$\text{and} \quad \begin{aligned} e_1 &= \text{fluctuating} \\ e_2 &= \text{low} \\ e_3 &= \text{rising} \\ e_4 &= \text{high} \\ e_5 &= \text{beerish} \end{aligned}$$

Thus we have $X = \{i, p, s, c, m, d, o, l, f, x\}, E = \{e_1, e_2, e_3, e_4, e_5\}$ and the soft sets of firm profiles become:

$$\begin{aligned} (F, A) &= \{e_5 = \{o, s\}, e_3 = \{p, i\}, e_2 = \{s, f\}\} \\ (G, B) &= \{e_1 = \{s, o\}, e_2 = \{c\}, e_3 = \{i, m\}, e_4 = \{i, l\}, e_5 = \{p, f\}\} \\ (H, C) &= \{e_1 = \{o, s\}, e_2 = \{c\}, e_4 = \{i, l\}, e_5 = \{p, f\}\} \end{aligned}$$

As the calculations give:

$$S_K^e((F, A), (H, C)) = \frac{1}{4 + \sqrt{7}} = 0.15$$

$$S_K^e((G, B), (H, C)) = \frac{1}{2} = 0.5$$

Hence, we conclude that the firm with profile (G, B) i.e., XYZ is suffering from a liquidity problem as its soft set profile is significantly similar to the standard liquidity problem profile, whereas the firm ABC is very less likely to be suffering from the same problem.

6.3 Soft Sets in Game Theory

In classical as well as fuzzy games, modelling of payoff functions are done using real valued functions and arithmetic operations. Moreover fuzzy games use membership functions, whose setting up is often difficult, for modelling uncertainty involved. Soft Games which is being introduced is free some this difficulty and is based on soft set theory. Payoff functions of the soft game are set valued function and solution of the soft games obtained by using the operations of sets that make this game very convenient and easily applicable in practice.

The work given in this section is taken from Deli and Cagman [36].

6.3.1 Two Person Soft Games

In this section, we construct two person soft games with soft payoffs. Four solution methods for the games will be given. For basic definitions and preliminaries of the game theory we refer [8, 43, 91, 92, 148].

Definition 6.18 Let X, Y are a sets of strategies. A choice of behavior is called an action. The elements of $X \times Y$ are called action pairs. That is, $X \times Y$ is the set of available actions.

Definition 6.19 Let U be a set of alternatives, $P(U)$ be the power set of U, X, Y are sets of strategies. Then, a set valued function

$$f_S : X \times Y \to P(U)$$

is called a soft payoff function. For each $(x, y) \in X \times Y$, the value $f_S(x, y)$ is called a soft payoff.

Definition 6.20 Let X and Y be a set of strategies of Player 1 and 2, respectively, U be a set of alternatives and $f_{S_k} : X \times Y \to P(U)$ be a soft payoff function for player k, $(k = 1, 2)$. Then, for each Player k, a two person soft game (tps-game) is defined by a soft set over U as

$$S_k = \left\{\left((x, y), f_{S_k}(x, y)\right) : (x, y) \in X \times Y\right\}$$

The tps-game is played as follows: at a certain time Player 1 chooses a strategy $x_i \in X$, simultaneously Player 2 chooses a strategy $y_j \in Y$ and once this is done each player **k** (k $= 1, 2$) receives the soft payoff $f_{S_k}\left(x_i, y_j\right)$. If $X = \{x_1, x_2, \ldots, x_m\}$ and $Y = \{y_1, y_2, \ldots, y_n\}$, then the soft payoffs of S_k can be arranged in the form of the $m \times n$ matrix shown in Table below:

Example 75 Let $U = \{u_1, u_2, u_3, u_4, u_5, u_6, u_7, u_8, u_9, u_{10}\}$ be a set of alternatives, $P(U)$ be the power set of U, $X = \{x_1, x_3, x_5\}$ and $Y = \{x_1, x_2, x_4\}$ be a set of the strategies Player 1 and 2, respectively. If Player 1 constructs a tps-games as follows,

$$S_1 = \{((x_1, x_1), \{u_1, u_2, u_5, u_8\}), ((x_1, x_2), \{u_1, u_2, u_3, u_4, u_5, u_8\}), ((x_1, x_4), \{u_3, u_8\}), ((x_3, x_1), \{u_1, u_3, u_7\}), ((x_3, x_2), \{u_1, u_2, u_3, u_5, u_6, u_7\}), ((x_3, x_4), \{u_1, u_2, u_3\}), ((x_5, x_1), \{u_3, u_4, u_5, u_8\}), ((x_5, x_2), \{u_1, u_2, u_3 u_4, u_5, u_6, u_8\}), ((x_5, x_4), \{u_1, u_2, u_3, u_8\})\}$$

then the soft payoffs of the game can be arranged as:

S_1	x_1	x_2	x_4
x_1	$\{u_1, u_2, u_5, u_8\}$	$\{u_1, u_2, u_3, u_4, u_5, u_8\}$	$\{u_3, u_8\}$
x_3	$\{u_1, u_3, u_7\}$	$\{u_1, u_2, u_3, u_5, u_6, u_7\}$	$\{u_1, u_2, u_3\}$
x_5	$\{u_3, u_4, u_5, u_8\}$	$\{u_1, u_2, u_3, u_4, u_5, u_6, u_8\}$	$\{u_1, u_2, u_3, u_8\}$

Let us explain some elements of this game; if Player 1 select x_3 and Player 2 select x_2, then the value of game will be a set $\{u_1, u_2, u_3, u_5, u_6, u_7\}$, that is,

$$f_{S_1}(x_3, x_2) = \{u_1, u_2, u_3, u_5, u_6, u_7\}$$

In this case, Player 1 wins the set of alternatives $\{u_1, u_2, u_3, u_5, u_6, u_7\}$ and Player 2 lost the same set of alternatives.

Similarly, if Player 2 constructs a tps-game as follows (Table 6.6),

$$S_2 = \{((x_1, x_1), \{u_3, u_4, u_6, u_7\}), ((x_1, x_2), \{u_6, u_7\}), (x_1, x_4), \{u_1, u_2, u_4 u_5, u_6, u_7\}), ((x_3, x_1), \{u_2, u_4, u_5, u_6, u_8\}), ((x_3, x_2), \{u_4, u_8\}), ((x_3, x_4) \{u_4, u_5, u_6, u_7, u_8\}), ((x_5, x_1), \{u_1, u_2, u_6, u_7\}), (x_5, x_2), \{u_7\}) (x_5, x_4), \{u_4, u_5, u_6, u_7\})\}$$

then the soft payoffs of the game can be arranged as:

S_2	x_1	x_2	x_4
x_1	$\{u_3, u_4, u_6, u_7\}$	$\{u_6, u_7\}$	$\{u_1, u_2, u_4, u_5, u_6, u_7\}$
x_3	$\{u_2, u_4, u_5, u_6, u_8\}$	$\{u_4, u_8\}$	$\{u_4, u_5, u_6, u_7, u_8\}$
x_5	$\{u_1, u_2, u_6, u_7\}$	$\{u_7\}$	$\{u_4, u_5, u_6, u_7\}$

Table 6.6 Soft payoff matrix

S_k	y_1	y_2	...	y_n
x_1	$f_{S_k}(x_1, y_1)$	$f_{S_k}(x_1, y_2)$	...	$f_{S_k}(x_1, y_n)$
x_2	$f_{S_k}(x_2, y_1)$	$f_{S_k}(x_2, y_2)$	...	$f_{S_k}(x_2, y_n)$
$\vdots$		$\vdots$	$\ddots$	$\vdots$
x_m	$f_{S_k}(x_m, y_1)$	$f_{S_k}(x_m, y_2)$	...	$f_{S_k}(x_m, y_n)$

Let us explain some element of this tps-game; if Player 1 select x_3 and Player 2 select x_2, then the value of game will be a set $\{u_4, u_8\}$, that is,

$$f_{S_2}(x_3, x_2) = \{u_4, u_8\}$$

In this case, Player 1 wins the set of alternatives $\{u_4, u_8\}$ and Player 2 lost $\{u_4, u_8\}$.

Definition 6.21 Let $S_k = \{((x, y), f_{S_k}(x, y)) : (x, y) \in X \times Y\}$ be a two person soft game and $(x_i, y_j), (x_r, y_s) \in X \times Y$. Then, Player k is called rational, if the player's soft payoff satisfies the following conditions:

(1) Either $f_{S_k}(x_i, y_j) \supseteq f^k_{X\times Y}(x_r, y_s)$ or $f_{S_k}(x_r, y_s) \supseteq f^k_{X\times Y}(x_i, y_j)$
(2) When $f_{S_k}(x_i, y_j) \supseteq f^k_{X\times Y}(x_r, y_s)$ and $f_{S_k}(x_r, y_s) \supseteq f^k_{X\times Y}(x_i, y_j)$, then $f_{S_k}(x_i, y_j) = f^k_{X\times Y}(x_r, y_s)$.

Definition 6.22 Let $S_k = \{((x, y), f_{S_k}(x, y)) : (x, y) \in X \times Y\}$ be a two person soft game. Then, an action $(x^*, y^*) \in X \times Y$ is called an optimal action if

$$f_{S_k}(x^*, y^*) \supseteq f_{S_k}(x, y) \text{ for all } (x, y) \in X \times Y$$

Definition 6.23 Let $S_k = \{((x, y), f_{S_k}(x, y)) : (x, y) \in X \times Y\}$ be a two person soft game. Then,

(1) if $f_{S_k}(x_i, y_j) \supset f_{S_k}(x_r, y_s)$, we say that a player strictly prefers action pair (x_i, y_j) over action (x_r, y_s).
(2) if $f_{S_k}(x_i, y_j) = f_{S_k}(x_r, y_s)$, we say that a player is indifferent between the two actions.
(3) if $f_{S_k}(x_i, y_j) \supseteq f_{S_k}(x_r, y_s)$, we say that a player either prefers (x_i, y_j) to (x_r, y_s) or is indifferent between the two actions.

Definition 6.24 Let $S_k = \{((x, y), f_{S_k}(x, y)) : (x, y) \in X \times Y\}$ be a two person soft game for $k = 1, 2$. Then,

(1) If $f_{S_k}(x, y) = \emptyset$ for all $(x, y) \in X \times Y$, then S_k is called a empty soft game, denoted by $\tilde{S}_\Phi$
(2) If $f_{S_k}(x, y) = U$ for all $(x, y) \in X \times Y$, then S_k is called a full soft game, denoted by $\tilde{\mathcal{S}}_E$

Definition 6.25 A tps-game is called a two person disjoint soft game if intersection of the soft payoff of players is empty set for each action pair.

Definition 6.26 A tps-game is called a two person universal soft game if union of the soft payoff of players is universal set for each action pair.

Proofs of the following Propositions are straight forward.

Proposition 6.9 *Let $S_k = \{((x, y), f_{S_k}(x, y)) : (x, y) \in X \times Y\}$ be a two person disjoint soft game for $k = 1, 2$. Then,*
(1) $(S_1^c)^c = S_1$
(2) $(S_2^c)^c = S_2$
(3) $S_1 \backslash S_2 = S_1$
(4) $S_2 \backslash S_1 = S_2$
(5) $S_1 \cap S_2 = \check{S}_\phi$
where S^C is the complement of S whose approximate function f_{S^C} is given by $f_{S^C}(x) = U \backslash f_S(x)$ $\forall x \in E$ and approximate function of $S_1 \cap S_2$ is given by $f_{S_1 \cap S_2}(x) = f_{S_1}(x) \cap f_{S_2}(x) \forall x \in E$.

Proposition 6.10 *Let $S_k = \{((x, y), f_{S_k}(x, y)) : (x, y) \in X \times Y\}$ be a two person universal soft game for $k = 1, 2$. Then,*
(1) $(S_k^c)^c = S_k, k = 1, 2$
(2) $S_1 \cup S_2 = \tilde{S}_E$
(3) $(S_1^c)^c = S_1$
(4) $(S_2^c)^c = S_2$
(5) $S_1^c = S_2$
(6) $S_2^c = S_1$
where approximate function of $S_1 \cup S_2$ is given by $f_{S_1 \cup S_2}(x) = f_{S_1}(x) \cup f_{S_2}(x) \forall x \in E$.

Proposition 6.11 *Let $S_k = \{((x, y), f_{S_k}(x, y)) : (x, y) \in X \times Y\}$ be a two person both universal and disjoint soft game for $k = 1, 2$. Then,*
(1) $S_1 \backslash S_2 = S_1$
(2) $S_2 \backslash S_1 = S_2$
(3) $S_1 \cap S_2 = \check{S}_\phi$
(4) $S_1 \cup S_2 = \check{S}_E$

Definition 6.27 Let f_{S_k} be a soft payoff function of a tps-game S_k. If the following properties hold
(1) $\bigcup_{i=1}^m f_{S_k}(x_i, y_j) = f_{S_k}(x, y)$
(2) $\cap_{j=1}^n f_{S_k}(x_i, y_j) = f_{S_k}(x, y)$
then $f_{S_k}(x, y)$ is called a a soft saddle point value and (x, y) is called a a soft saddle point of Player k's in the tps-game.

Note that if $f_{S_1}(x, y)$ is a soft saddle point of a tps -game S_1, then Player 1 can then win at least by choosing the strategy $x \in X$ and Player 2 can keep her/his loss to at most $f_{S_1}(x, y)$ by choosing the strategy $y \in Y$. Hence the soft saddle point is a value of the tps-game.

Example 76 Let $U = \{u_1, u_2, u_3, u_4, u_5, u_6, u_7, u_8, u_9, u_{10}\}$ be a set of alternatives, $X = \{x_1, x_2, x_3, x_4\}$ and $Y = \{y_1, y_2, y_3\}$ be the strategies for Player 1 and 2 respectively. Then, tps-game of Player 1 is given as:

S_1	y_1	y_2	y_3
x_1	$\{u_2, u_4, u_7\}$	$\{u_4\}$	$\{u_4\}$
x_2	$\{u_5\}$	$\{u_7\}$	$\{u_4, u_7\}$
x_3	$\{u_2, u_4, u_5, u_7, u_8, u_{10}\}$	$\{u_4, u_8\}$	$\{u_7, u_8\}$
x_4	$\{u_2, u_4, u_5, u_7, u_8\}$	$\{u_1, u_4, u_7, u_8\}$	$\{u_4, u_7, u_8\}$

Clearly
$$\bigcup_{i=1}^{4} f_{S_1}(x_i, y_1) = \{u_2, u_4, u_5, u_7, u_8, u_{10}\}$$
$$\bigcup_{i=1}^{4} f_{S_1}(x_i, y_2) = \{u_1, u_4, u_7, u_8\}$$
$$\bigcup_{i=1}^{4} f_{S_1}(x_i, y_3) = \{u_4, u_7, u_8\}$$
$$\bigcap_{j=1}^{3} f_{S_1}(x_1, y_j) = \{u_4\}$$
$$\bigcap_{j=1}^{3} f_{S_1}(x_2, y_j) = \phi$$
and
$$\bigcap_{j=1}^{3} f_{S_1}(x_3, y_j) = \{u_8\}$$
$$\bigcap_{j=1}^{3} f_{S_1}(x_4, y_j) = \{u_4, u_7, u_8\}$$

Therefore, $\{u_4, u_7, u_8\}$ is a soft saddle point of the *tps* -game, since the intersection of the forth row is equal to the union of the third column. So, the value of the tps-game is $\{u_4, u_7, u_8\}$
Note that every tps-game has not a soft saddle point. For instance, in the above example, if $\{u_4, u_7, u_8\}$ is replaced with $\{u_4, u_7, u_8, u_9\}$ in soft payoff $f_{S_1}(x_4, y_3)$ then a soft saddle point of the game can not be found. For such tps-games, soft upper and soft lower values of the tps-game may be used, which is given in the following definition.

Definition 6.28 Let f_{S_k} be a soft payoff function of a tps-game S_k. Then,

(1) Soft upper value of the tps-game, denoted $\bar{v}$, is defined by
$$\bar{v} = \cap_{y\in Y}\left(\cup_{x\in X}\left(f_{S_k}(x, y)\right)\right)$$

(2) Soft lower value of the *tps*− game, denoted $\underline{v}$, is defined by
$$\underline{v} = \cup_{x\in X}\left(\cap_{y\in Y}\left(f_{S_k}(x, y)\right)\right)$$

(3) If soft upper and soft lower value of a tps-game are equal, they are called value of the tps-game, denoted by v. That is $v = \underline{v} = \bar{v}$.

Example 77 In Example 76, it is clear that soft upper value $\bar{v} = \{u_4, u_7, u_8\}$ and soft lower value $\underline{v} = \{u_4, u_7, u_8\}$, hence $\underline{v} = \bar{v}$. It means that value of the tps-game is $\{u_4, u_7, u_8\}$.

Theorem 6.5 *If $\underline{v}$ and $\bar{v}$ be a soft lower and soft upper value of a tps-game, respectively. Then, the soft lower value is subset or equal to the soft upper value, that is* $\underline{v} \subseteq \bar{v}$.

Proof Assume that $\underline{v}$ be a soft lower value, $\bar{v}$ be a soft upper value of a tps-game and $X = \{x_1, x_2, \ldots, x_m\}$ and $Y = \{y_1, y_2, \ldots, y_n\}$ are sets of the strategies for Player 1 and 2, respectively. We choose $x_i^* \in X$ and $y_j^* \in Y$. Then,

$$\begin{aligned}
\underline{v} &= \cup_{x \in X} \left(\bigcap_{y \in Y} (f_{X \times Y}(x, y)) \right) \\
&\subseteq \cap_{y \in Y} \left(f_{X \times Y}(x^*, y) \right) \\
&\subseteq f_{X \times Y} \left(x^*, y^* \right) \\
&\subseteq \cup_{x \in X} \left(f_{X \times Y}(x, y^*) \right) \\
&\subseteq \cap_{y \in Y} \left(\cup_{x \in X} (f_{X \times Y}(x, y)) \right)
\end{aligned}$$

i.e., $\underline{v} = \cup_{x \in X} \left(\bigcap_{y \in Y} (f_{X \times Y}(x, y)) \right) \subseteq \bar{v} = \cap_{y \in Y} (\cup_{x \in X} (f_{X \times Y}(x, y)))$. □

Example 78 Let us consider soft upper value $\bar{v}$ and soft lower value $\underline{v}$ in Example 76. It is clear that $\bar{v} = \{u_4, u_7, u_8\} \subseteq \underline{v} = \{u_4, u_7, u_8\}$, hence $\underline{v} \subseteq \bar{v}$.

Theorem 6.6 *Let $f_{S_k}(x, y)$ be a soft saddle point, $\underline{v}$ be a soft lower value and $\bar{v}$ be a soft upper value of a tps-game. Then,*

$$\underline{v} \subseteq f_{S_k} \left(x^*, y^* \right) \subseteq \bar{v}.$$

Proof Assume that $f_{S_k}(x^*, y^*)$ be a soft saddle point, $\underline{v}$ be a soft lower value, $\bar{v}$ be a soft upper value of a tps-game and $X = \{x_1, x_2, \ldots, x_m\}$ and $Y = \{y_1, y_2, \ldots, y_n\}$ are sets of the strategies for Player 1 and 2, respectively. We choose $x_i^* \in X$ and $y_j^* \in Y$. Then, since $f_{S_k}(x^*, y^*)$ is a soft saddle point, we have
$\bigcup_{i=1}^{m} f_{S_k}(x_i, y_j) = \bigcap_{j=1}^{n} f_{S_k}(x_i, y_j) = f_{S_k}(x^*, y^*)$
Clearly, $\underline{v} = \cup_{x \in X} \left(\cap_{y \in Y} (f_{X \times Y}(x, y)) \right) \subseteq \bigcup_{i=1}^{m} f_{S_k}(x_i, y_j) = f_{S_k}(x^*, y^*)$
and, $f_{S_k}(x^*, y^*) = \bigcap_{j=1}^{n} f_{S_k}(x_i, y_j) \subseteq \bar{v} = \cap_{y \in Y} (\cup_{x \in X} (f_{X \times Y}(x, y))$
Thus $\underline{v} \subseteq f_{S_k}(x^*, y^*) \subseteq \bar{v}$. □

Corollary 6.1 *Let $f_{S_k}(x, y)$ be a soft saddle point, $\underline{v}$ be a soft lower value and $\bar{v}$ be a soft upper value of a tps-game. If $v = \underline{v} = \bar{v}$, then $f_{S_k}(x, y)$ is exactly v.*

Note that in every tps-game, the soft lower value $\underline{v}$ can not be equals to the soft upper value $\bar{v}$. If in a tps-game, $\underline{v} \neq \bar{v}$, then to get the solution of the game, soft dominated strategy may be used. We define soft dominated strategy for tps-game as follows.

Definition 6.29 Let S_1 be a tps-game with its soft payoff function f_{S_1}. Then,

(1) a strategy $x_i \in X$ is called a soft dominated to another strategy $x_r \in X$, if $f_{S_1}(x_i, y) \supseteq f_{S_1}(x_r, y)$ for all $y \in Y$.
(2) a strategy $y_j \in Y$ is called a soft dominated to another strategy $y_s \in Y$, if $f_{S_1}(x, y_j) \subseteq f_{S_1}(x, y_s)$ for all $x \in X$.

By using soft dominated strategy, tps-games may be reduced by deleting rows and columns that are obviously bad for the player who uses them. This process of eliminating soft dominated strategies sometimes leads us to a solution of a tps-game. Such a method of solving tps-game is called a soft elimination method.

Note that the soft elimination method cannot be used for some tps-games which do not have soft dominated strategies. In this case, we can use soft Nash equilibrium that is defined as follows.

Definition 6.30 Let S_k be a tps-game with its soft payoff function f_{S_k} for $k = 1, 2$ If the following properties hold

(1) $f_{S_1}(x^*, y^*) \supseteq f_{S_1}(x, y^*)$ for each $x \in X$
(2) $f_{S_2}(x^*, y^*) \supseteq f_{S_2}(x^*, y)$ for each $y \in Y$ then, $(x^*, y^*) \in X \times Y$ is called a soft Nash equilibrium of a tps-game.

Note that if $(x^*, y^*) \in X \times Y$ is a soft Nash equilibrium of a tps-game, then Player 1 can then win at least $f_{S_1}(x^*, y^*)$ by choosing strategy $x^* \in X$ and Player 2 can win at least $f_{S_2}(x^*, y^*)$ by choosing strategy $y^* \in Y$. Hence the soft Nash equilibrium is an optimal action for $tps-$ game, therefore, $f_{S_k}(x^*, y^*)$ is the solution of the $tps-$ game for Player k, $k = 1, 2$.

6.3.2 An Application

In this section, we give a financial problem that is solved by using both soft dominated strategy and soft saddle point methods as in [36].

There are two companies, say Player 1 and Player 2, who competitively want to increase sale of produces in the country. Therefore, they give advertisements. Assume that two companies have a set of different products $U = \{u_1, u_2, u_3, u_4, u_5, u_6, u_7, u_8\}$ where for $i = 1, 2, \ldots, 8$, the product u_i stand for "oil", "salt", "honey", "jam", "cheese", "sugar", "cooker" and "jar", respectively. The products can be characterized by a set of strategy $X = Y = \{x_i : i = 1, 2, 3\}$ which contains styles of advertisement where for $j = 1, 2, 3$, the strategies x_j stand for "TV^n, "radio" and "newspaper", respectively.

Suppose that $X = \{x_1, x_2, x_3\}$ and $Y = \{y_1 = x_1, y_2 = x_2, y_3 = x_3\}$ are strategies of Player 1 and 2, respectively. Then, a tps-game of Player 1 is given as:

S_1	y_1	y_2	y_3
x_1	$\{u_1, u_2, u_3, u_5, u_8\}$	$\{u_1, u_2, u_3, u_4, u_5, u_8\}$	$\{u_3\}$
x_2	$\{u_1, u_3, u_7\}$	$\{u_1, u_2, u_3, u_5, u_6, u_7\}$	$\{u_2, u_3\}$
x_3	$\{u_1, u_2, u_3, u_4, u_5\}$	$\{u_1, u_2, u_3, u_4, u_5, u_6, u_8\}$	$\{u_1, u_2, u_3\}$

In the Table, let us explain action pair (x_1, y_1); if Player 1 select $x_1 =$ "TV'' and Player 2 select $y_1 =$ "TV'', then the soft payoff of Playoff of Player 1 is a set $\{u_1, u_2, u_3, u_5, u_8\}$ that is,

$$f_{S_1}(x_1, y_1) = \{u_1, u_2, u_3, u_5, u_8\}$$

In this case, Player 1 increase sale of $\{u_1, u_2, u_3, u_5, u_8\}$ and Player 2 decrease sale of $\{u_1, u_2, u_3, u_5, u_8\}$. We can now solve the game. It is seen in Table above that,

$$\{u_1, u_2, u_3, u_5, u_8\} \subseteq \{u_1, u_2, u_3, u_4, u_5, u_8\}$$
$$\{u_1, u_3, u_7\} \subseteq \{u_1, u_2, u_3, u_5, u_6, u_7\}$$
$$\{u_1, u_2, u_3, u_4, u_5\} \subseteq \{u_1, u_2, u_3, u_4, u_5, u_6, u_8\}$$

the middle column is dominated by the right column. Deleting the middle column we obtain the following table.

S_1	y_1	y_3
x_1	$\{u_1, u_2, u_3, u_5, u_8\}$	$\{u_3\}$
x_2	$\{u_1, u_3, u_7\}$	$\{u_2, u_3\}$
x_3	$\{u_1, u_2, u_3, u_4, u_5\}$	$\{u_1, u_2, u_3\}$

In the Table there is no another soft dominated strategy, we can use soft saddle point method.

$$\bigcup_{i=1}^{3} f_{S_1}(x_i, y_1) = \{u_1, u_2, u_3, u_4, u_5, u_7, u_8\}$$
$$\bigcup_{i=1}^{3} f_{S_1}(x_i, y_3) = \{u_1, u_2, u_3\}$$
$$\bigcap_{j=1,3} f_{S_1}(x_1, y_j) = \{u_3\}$$
$$\bigcap_{j=1,3} f_{S_1}(x_2, y_j) = \{u_3\}$$
$$\bigcap_{j=1,3} f_{S_1}(x_3, y_j) = \{u_1, u_2, u_3\}$$

Here, optimal strategy of the game is (x_3, y_3) since

$$\bigcup_{i=1}^{3} f_{S_1}(x_i, y_3) = \bigcap_{j=1,3} f_{S_1}(x_3, y_j)$$

Therefore, value of the tps -game is $\{u_1, u_2, u_3\}$.

6.4 Soft Matrix Theory and Decision Making

Soft matrices which will turn out to be a more suitable tool for theoretical studies in soft set theory and their operations are discussed in this section. Products of soft matrices and their properties are also studied. Finally a soft max- min decision making method is constructed and successfully applied to the problems that contain uncertainties. The content in this section is taken from Cagman and Enginoglu [27]. In this section, in order to specify the parameter set E, the soft set (f, A) will be equivalently denoted by (f_A, E), for $A \subset E$.

6.4.1 Soft Matrices

Definition 6.31 Let (f_A, E) be a soft set over U. Then a subset of $U \times E$ is uniquely defined by

$$R_A = \{(u, e) : e \in A, u \in f_A(e)\}$$

which is called a relation form of (f_A, E). The characteristic function of R_A is written by

$$\chi_{R_A} : U \times E \to \{0, 1\}, \quad \chi_{R_A(u,e)} = \begin{cases} 1, & (u, e) \in R_A \\ 0, & (u, e) \notin R_A \end{cases}$$

If $U = \{u_1, u_2, \ldots, u_m\}$, $E = \{e_1, e_2, \ldots, e_n\}$ and $A \subseteq E$, then R_A can be presented by a table as in the following form

R_A	e_1	e_2	$\cdots$	e_n
u_1	$\chi_{R_A}(u_1, e_1)$	$\chi_{R_A}(u_1, e_2)$	$\cdots$	$\chi_{R_A}(u_1, e_n)$
u_2	$\chi_{R_A}(u_2, e_1)$	$\chi_{R_A}(u_2, e_2)$	$\cdots$	$\chi_{R_A}(u_2, e_n)$
$\vdots$	$\vdots$	$\vdots$	$\ddots$	$\vdots$
u_m	$\chi_{R_A}(u_m, e_1)$	$\chi_{R_A}(u_m, e_2)$	$\cdots$	$\chi_{R_A}(u_m, e_n)$

If $a_{ij} = \chi_{R_A}(u_i, e_j)$, we can define a matrix

$$[a_{ij}]_{m \times n} = \begin{bmatrix} a_{11} & a_{12} & \cdots & a_{1n} \\ a_{21} & a_{22} & \cdots & a_{2n} \\ \vdots & \vdots & \ddots & \vdots \\ a_{m1} & a_{m2} & \cdots & a_{mn} \end{bmatrix}$$

which is called an $m \times n$ soft matrix of the soft set (f_A, E) over U.

According to this definition, a soft set (f_A, E) is uniquely characterized by the matrix $[a_{ij}]_{m \times n}$. It means that a soft set (f_A, E) is formally equal to its soft matrix $[a_{ij}]_{\text{mxn}}$. Therefore, we shall identify any soft set with its soft matrix and use these two concepts as interchangeable.

The set of all $m \times n$ soft matrices over U will be denoted by $\text{SM}_{m \times n}$. From now on we shall delete the subscripts $m \times n$ of $[a_{ij}]_{m \times n}$, we use $[a_{ij}]$ instead of $[a_{ij}]_{m \times n}$, since $[a_{ij}] \in SM_{m \times n}$ means that $[a_{ij}]$ is an $m \times n$ soft matrix for $i = 1, 2, \ldots, n$ and $j = 1, 2, \ldots, m$.

Example 79 Assume that $U = \{u_1, u_2, u_3, u_4, u_5\}$ is a universal set and $E = \{e_1, e_2, e_3, e_4\}$ is a set of all parameters. If $A = \{e_2, e_3, e_4\}$ and $f_A(e_2) = \{u_2, u_4\}$, $f_A(e_3) = \emptyset$, $f_A(e_4) = U$, then we write a soft set $(f_A, E) = \{(e_2, \{u_2, u_4\}), (e_4, U)\}$ and then the relation form of (f_A, E) is written by

$$R_A = \{(u_2, e_2), (u_4, e_2), (u_1, e_4), (u_2, e_4), (u_3, e_4), (u_4, e_4), (u_5, e_4)\}$$

Hence, the soft matrix $[a_{ij}]$ is written by

$$[a_{ij}] = \begin{bmatrix} 0\,0\,0\,1 \\ 0\,1\,0\,1 \\ 0\,0\,0\,1 \\ 0\,1\,0\,1 \\ 0\,0\,0\,1 \end{bmatrix}$$

Definition 6.32 Let $[a_{ij}] \in SM_{m\times n}$. Then $[a_{ij}]$ is called

(a) a zero soft matrix, denoted by [0], if $a_{ij} = 0$ for all i and j.
(b) an A -universal soft matrix, denoted by $[\tilde{a}_{ij}]$, if $a_{ij} = 1$ for all $j \in I_A = \{j : e_j \in A\}$ and i.
(c) universal soft matrix, denoted by [1], if $a_{ij} = 1$ for all i and j.

Definition 6.33 Let $[a_{ij}], [b_{ij}] \in SM_{m\times n}$. Then

(a) $[a_{ij}]$ is a soft submatrix of $[b_{ij}]$, denoted by $[a_{ij}] \tilde{\subseteq} [b_{ij}]$, if $a_{ij} \leqslant b_{ij}$ for all i and j.
(b) $[a_{ij}]$ is a proper soft sub matrix of $[b_{ij}]$, denoted by $[a_{ij}] \tilde{\subset} [b_{ij}]$, if $a_{ij} \leqslant b_{ij}$ for all i and j and for at least one term $a_{ij} < b_{ij}$ for some i and j.
(c) $[a_{ij}]$ and $[b_{ij}]$ are soft equal matrices, denoted by $[a_{ij}] = [b_{ij}]$, if $a_{ij} = b_{ij}$ for all i and j.

Definition 6.34 Let $[a_{ij}], [b_{ij}] \in SM_{m\times n}$. Then the soft matrix $[c_{ij}]$ is called

(a) union of $[a_{ij}]$ and $[b_{ij}]$, denoted $[a_{ij}] \tilde{\cup} [b_{ij}]$, if $c_{ij} = \max\{a_{ij}, b_{ij}\}$ for all i and j.
(b) intersection of $[a_{ij}]$ and $[b_{ij}]$, denoted $[a_{ij}] \tilde{\cap} [b_{ij}]$, if $c_{ij} = \min\{a_{ij}, b_{ij}\}$ for all i and j.
(c) complement of $[a_{ij}]$, denoted by $[a_{ij}]^\circ$, if $c_{ij} = 1 - a_{ij}$ for all i and j.

Definition 6.35 Let $[a_{ij}], [b_{ij}] \in SM_{m\times n}$. Then $[a_{ij}]$ and $[b_{ij}]$ are disjoint, if $[a_{ij}] \tilde{\cap} [b_{ij}] = [0]$ for all i and j.

The following propositions follows directly.

Proposition 6.12 *Let* $[a_{ij}] \in SM_{m\times n}$. *Then*
i. $[[a_{ij}]^\circ]^\circ = [a_{ij}]$
ii. $[0]^\circ = [1]$

Proposition 6.13 *Let* $[a_{ij}], [b_{ij}] \in SM_{m\times n}$. *Then*
i. $[a_{ij}] \tilde{\subseteq} [1]$
ii. $[0] \tilde{\subseteq} [a_{ij}]$
iii. $[a_{ij}] \tilde{\subseteq} [a_{ij}]$
iv. $[a_{ij}] \tilde{\subseteq} [b_{ij}]$ *and* $[b_{ij}] \tilde{\subseteq} [c_{ij}] \Rightarrow [a_{ij}] \tilde{\subseteq} [c_{ij}]$

Proposition 6.14 *Let* $[a_{ij}], [b_{ij}], [c_{ij}] \in SM_{m\times n}$. *Then*
i. $[a_{ij}] = [b_{ij}]$ *and* $[b_{ij}] = [c_{ij}] \Leftrightarrow [a_{ij}] = [c_{ij}]$
ii. $[a_{ij}] \tilde{\subseteq} [b_{ij}]$ *and* $[b_{ij}] \tilde{\subseteq} [a_{ij}] \Leftrightarrow [a_{ij}] = [b_{ij}]$

Proposition 6.15 *Let* $[a_{ij}], [b_{ij}], [c_{ij}] \in SM_{m\times n}$. *Then*
i. $[a_{ij}] \tilde{\cup} [a_{ij}] = [a_{ij}]$
ii. $[a_{ij}] \tilde{\cup} [0] = [a_{ij}]$
iii. $[a_{ij}] \tilde{\cup} [1] = [1]$
iv. $[a_{ij}] \tilde{\cup} [a_{ij}]^{\circ} = [1]$
v. $[a_{ij}] \tilde{\cup} [b_{ij}] = [b_{ij}] \tilde{\cup} [a_{ij}]$
vi. $([a_{ij}] \tilde{\cup} [b_{ij}]) \tilde{\cup} [c_{ij}] = [a_{ij}] \tilde{\cup} ([b_{ij}] \tilde{\cup} [c_{ij}])$

Proposition 6.16 *Let* $[a_{ij}], [b_{ij}], [c_{ij}] \in SM_{m\times n}$. *Then*
i. $[a_{ij}] \tilde{\cap} [a_{ij}] = [a_{ij}]$
ii. $[a_{ij}] \tilde{\cap} [0] = [0]$
iii. $[a_{ij}] \tilde{\cap} [1] = [a_{ij}]$
iv. $[a_{ij}] \tilde{\cap} [a_{ij}]^{\circ} = [0]$
v. $[a_{ij}] \tilde{\cap} [b_{ij}] = [b_{ij}] \tilde{\cap} [a_{ij}]$
vi. $([a_{ij}] \tilde{\cap} [b_{ij}]) \tilde{\cap} [c_{ij}] = [a_{ij}] \tilde{\cap} ([b_{ij}] \tilde{\cap} [c_{ij}])$

Proposition 6.17 *Let* $[a_{ij}], [b_{ij}] \in SM_{m\times n}$. *Then De Morgan's laws are valid*
i. $([a_{ij}] \tilde{\cup} [b_{ij}])^{\circ} = [a_{ij}]^{\circ} \tilde{\cap} [b_{ij}]^{\circ}$
ii. $([a_{ij}] \tilde{\cap} [b_{ij}])^{\circ} = [a_{ij}]^{\circ} \tilde{\cup} [b_{ij}]^{\circ}$

Proof

$$\begin{aligned} ([a_{ij}] \tilde{\cup} [b_{ij}])^{\circ} &= [\max\{a_{ij}, b_{ij}\}]^{\circ} \\ &= [1 - \max\{a_{ij}, b_{ij}\}] \\ &= [\min\{1 - a_{ij}, 1 - b_{ij}\}] \\ &= [a_{ij}]^{\circ} \tilde{\cap} [b_{ij}]^{\circ} \end{aligned}$$

Other one can be proved similarly. □

Proposition 6.18 *Let* $[a_{ij}], [b_{ij}], [c_{ij}] \in SM_{m\times n}$. *Then*
i. $[a_{ij}] \tilde{\cup} ([b_{ij}] \tilde{\cap} [c_{ij}]) = ([a_{ij}] \tilde{\cup} [b_{ij}]) \tilde{\cap} ([a_{ij}] \tilde{\cup} [c_{ij}])$
ii $[a_{ij}] \tilde{\cap} ([b_{ij}] \tilde{\cup} [c_{ij}]) = ([a_{ij}] \tilde{\cap} [b_{ij}]) \tilde{\cup} ([a_{ij}] \tilde{\cap} [c_{ij}])$

6.4.2 Products of Soft Matrices

Four special products of soft matrices will be defined to construct soft decision making methods.

Definition 6.36 Let $[a_{ij}], [b_{ik}] \in SM_{m\times n}$. Then And-product of $[a_{ij}]$ and $[b_{ik}]$ is defined by

$$\wedge : SM_{m\times n} \times SM_{m\times n} \to SM_{m\times n^2}, \quad [a_{ij}] \wedge [b_{ik}] = [c_{ip}]$$

where $c_{ip} = \min\{a_{ij}, b_{ik}\}$ such that $p = n(j-1)+k$

Definition 6.37 Let $[a_{ij}], [b_{ik}] \in SM_{m\times n}$. Then Or-product of $[a_{ij}]$ and $[b_{ik}]$ is defined by

$$\vee : SM_{m\times n} \times SM_{m\times n} \to SM_{m\times n^2}, \quad [a_{ij}] \vee [b_{ik}] = [c_{ip}]$$

where $c_{ip} = \max\{a_{ij}, b_{ik}\}$ such that $p = n(j-1)+k$

Definition 6.38 Let $[a_{ij}], [b_{ik}] \in SM_{m\times n}$. Then *And*− Not-product of $[a_{ij}]$ and $[b_{ik}]$ is defined by

$$\bar{\wedge} : SM_{m\times n} \times SM_{m\times n} \to SM_{m\times n^2}, \quad [a_{ij}] \bar{\wedge} [b_{ik}] = [c_{ip}]$$

where $c_{ip} = \min\{a_{ij}, 1-b_{ik}\}$ such that $p = n(j-1)+k$

Definition 6.39 Let $[a_{ij}], [b_{ik}] \in SM_{m\times n}$. Then Or-Not-product of $[a_{ij}]$ and $[b_{ik}]$ is defined by

$$\underline{\vee} : SM_{m\times n} \times SM_{m\times n} \to SM_{m\times n^2}, \quad [a_{ij}] \underline{\vee} [b_{ik}] = [c_{ip}]$$

where $c_{ip} = \max\{a_{ij}, 1-b_{ik}\}$ such that $p = n(j-1)+k$

Note Commutativity will not hold for the products of soft matrices.

Proposition 6.19 *Let* $[a_{ij}], [b_{ik}] \in SM_{m\times n}$. *Then the following De Morgan's types of results are true.*
i. $([a_{ij}] \vee [b_{ik}])^\circ = [a_{ij}]^\circ \wedge [b_{ik}]^\circ$
ii. $([a_{ij}] \wedge [b_{ik}])^\circ = [a_{ij}]^\circ \vee [b_{ik}]^\circ$
iii. $([a_{ij}] \underline{\vee} [b_{ik}])^\circ = [a_{ij}]^0 \bar{\wedge} [b_{ik}]^0$
iv. $([a_{ij}] \bar{\wedge} [b_{ik}])^\circ = [a_{ij}]^\circ \underline{\vee} [b_{ik}]^\circ$

6.4.3 Soft Max−Min Decision Making

In this section, we define soft max-min decision function and develop a soft max-min decision making (SMmDM) method using this. This method will select optimum alternatives from the set of all alternatives.

Definition 6.40 Let $\left[c_{ip}\right] \in SM_{m\times n^2}$, $I_k = \left\{p : \exists i, c_{ip} \neq 0, (k-1)n < p \leqslant kn\right\}$ for all $k \in I = \{1, 2, \ldots, n\}$. Then soft max-min decision function, denoted Mm, is defined as follows

$$Mm : SM_{m\times n^2} \to SM_{m\times 1}, \quad Mm\left[c_{ip}\right] = \left[\max_{k\in I}\{t_k\}\right]$$

where

$$t_k = \begin{cases} \min_{p\in I_k}, \{C_{ip}\} & \text{if } I_k \neq \emptyset \\ 0, & \text{if } I_k = \emptyset \end{cases}$$

The one column soft matrix $Mm\left[c_{ip}\right]$ is called max-min decision soft matrix.

Definition 6.41 Let $U = \{u_1, u_2, \ldots, u_m\}$ be an initial universe and $Mm\left[c_{ip}\right] = [d_{i1}]$. Then a subset of U can be obtained by using $\left[d_{i_1}\right]$ as in the following way

$$\text{opt}_{[d_{i1}]}(U) = \{u_i : u_i \in U, d_{i1} = 1\}$$

which is called an optimum set of U.

Now, by using the definitions, we can construct a SMmDM method by the following algorithm.
Step 1: Choose feasible subsets of the set of parameters,
Step 2: construct the soft matrix for each set of parameters.
Step 3: find a convenient product of the soft matrices.
Step 4: find a max-min decision soft matrix,
Step 5: find an optimum set of U.

Note that by the similar way, we can define soft min-max, soft min-min and soft max-max decision making methods which may be denoted by (SmMDM), (SmmDM), (SMMDM), respectively. One of them may be more useful than others according to the type of the problems.

Application
Assume that a real estate agent has a set of different types of houses $U = \{u_1, u_2, u_3, u_4, u_5\}$ which may be characterized by a set of parameters $E = \{e_1, e_2, e_3, e_4\}$. For $j = 1, 2, 3, 4$ the parameters e_j stand for "in good location", "cheap", "modern". "large", respectively. Then we can give the following examples.

Suppose that a married couple, Mr. X and Mrs. X, come to the real estate agent to buy a house. If each partner has to consider their own set of parameters, then we select a house on the basis of the sets of partners' parameters by using the SMmDM as follows.

Assume that $U = \{u_1, u_2, u_3, u_4, u_5\}$ is a universal set and $E = \{e_1, e_2, e_3, e_4\}$ is a set of all parameters.
Step 1: First, Mr. X and Mrs. X have to choose the sets of their parameters,

$A = \{e_2, e_3, e_4\}$ and $B = \{e_1, e_3, e_4\}$ respectively.
Step 2: Then we can write the following soft matrices which are constructed according to their parameters.

$$[a_{ij}] = \begin{bmatrix} 0&0&1&1 \\ 0&1&1&1 \\ 0&1&1&0 \\ 0&1&0&0 \\ 0&1&0&1 \end{bmatrix}$$

$$[b_{ik}] = \begin{bmatrix} 1&0&1&1 \\ 1&0&0&1 \\ 0&0&1&1 \\ 0&0&1&1 \\ 0&0&0&1 \end{bmatrix}$$

Step 3: Now, we can find a product of the soft matrices $[a_{ij}]$ and $[b_{ik}]$ by using And-product as follows

$$[a_{ij}] \wedge [b_{ik}] = \begin{bmatrix} 0&0&0&0&0&0&0&0&1&0&1&1&1&0&1&1 \\ 0&0&0&0&1&0&0&1&1&0&0&1&1&0&0&1 \\ 0&0&0&0&0&0&1&1&0&0&1&1&0&0&0&0 \\ 0&0&0&0&0&0&1&1&0&0&0&0&0&0&0&0 \\ 0&0&0&0&0&0&0&1&0&0&0&0&0&0&0&1 \end{bmatrix}$$

Here, we use And-product since both Mr. X and Mrs. X's choices have to be considered.
Step 4: We can find a max-min decision soft matrix as

$$Mm\left([a_{ij}] \wedge [b_{ik}]\right) = \begin{bmatrix} 1 \\ 0 \\ 0 \\ 0 \\ 0 \end{bmatrix}$$

Step 5: Finally, we can find an optimum set of U according to $Mm\left([a_{ij}] \wedge [b_{ik}]\right)$

$$\text{opt}_{Mm([a_{ij}] \wedge [b_{ik}])}(U) = \{u_1\}$$

where u_1 is an optimum house to buy for Mr. X and Mrs. X.

Note that the optimal set of U may contain more than one element. Similarly, we can also use the other products $[a_{ij}] \vee [b_{ik}]$, $[a_{lj}] \bar{\wedge} [b_{ik}]$ and $[a_{ij}] \underline{\vee} [b_{ik}]$ for the other problems conveniently.

6.5 Soft Sets in Incomplete Information Systems

A major problem that restricts the usage of Soft sets is partial or incomplete information. All the studies considered in earlier sections, either theoretical or practical, are based on complete information only. However, incomplete information naturally occurs in practical problems. This may happen due to the fact that attendees omit some questions or may not understand the meaning of question in questionnaire. Additionally, mistake in collection and processing of data, some official restrictions and technical non feasibility in data collection, can also lead to such situations. Hence, soft sets under incomplete information become incomplete soft sets.

Basically, there are two methods one which deletes objects with incomplete data and another one which fills up the missing data using some scientific techniques [170]. However, probably the first technique makes valuable information missing. Data filling converts an incomplete soft set into a complete soft set, which makes the soft set more useful. So far, few researches focus on data filling approaches for incomplete soft sets. Discussions, results and examples provided in this section are taken from [108].

Definition 6.42 A pair (F, E) is called an incomplete soft set over U, if there exists $x_i \in U \quad (i = 1, 2 \ldots, n)$ and $e_j \in E \quad (j = 1, 2 \ldots, m)$, making $x_i \in F\left(e_j\right)$ unknown, that is, $F\left(e_j\right)(x_i) = \text{null}$, where $F\left(e_j\right)(x_i)$ represents the $(i, j)^t h$ entry in the tabular representation of soft set (F, E).

In tabular representation, null is represented by "*".

In a soft set, associations between parameters will be very useful for filling incomplete data. If we have already found that parameter e_i is associated with parameter e_j and there are missing data in $F(e_i)$, we can fill the missing data according to the corresponding data in $F\left(e_j\right)$ based on the association between e_i and e_j. To measure these associations, we define the notion of association degree and some relative notions.

Let U be a universe set and E be a set of parameters. U_{ij} denotes the set of objects that have specified values 0 or 1 both on parameter e_i and parameter e_j such that

$$U_{ij} = \left\{x | F(e_i)(x) \neq \text{"*" and } F\left(e_j\right)(x) \neq \text{"*"}, x \in U\right\}$$

In other words, U_{ij} stands for the set of objects that have known data both on e_i and e_j. Based on U_{ij}, we have the following definitions.

Definition 6.43 Let E be a set of parameters and $e_i, e_j \in E, (i, j = 1, 2, \ldots m)$. Consistent Association Number between parameter e_i and parameter e_j is denoted by CN_{ij} and defined as

$$CN_{ij} = \left|\left\{x | F(e_i)(x) = F\left(e_j\right)(x), x \in U_{ij}\right\}\right|$$

where m denotes the number of parameters, $||$ denotes the cardinality of set.

Definition 6.44 Let E be a set of parameters and $e_i, e_j \in E, (i, j = 1, 2, \ldots, m)$. Consistent Association Degree between parameter e_i and parameter e_j is denoted by CD_{ij} and defined as

$$CD_{ij} = \frac{CN_{ij}}{|U_{ij}|}$$

Obviously, the value of CD_{ij} is in [0, 1]. Consistent Association Degree measures the extent to which the value of parameter e_i keeps consistent with that of parameter e_j over U_{ij}. Similarly, we can define Inconsistent Association Number and Inconsistent Association Degree as follows.

Definition 6.45 Let E be a set of parameters and $e_i, e_j \in E, (i, j = 1, 2, \ldots, m)$. Inconsistent Association Number between parameter e_i and parameter e_j is denoted by IN_{ij} and defined as

$$IN_{ij} = \left|\left\{x | F(e_i)(x) \neq F(e_j)(x), x \in U_{ij}\right\}\right|$$

Definition 6.46 Let E be a set of parameters and $e_i, e_j \in E, (i, j = 1, 2, \ldots m)$. Inconsistent Association Degree between parameter e_i and parameter e_j is denoted by ID_{ij} and defined as

$$ID_{ij} = \frac{IN_{ij}}{|U_{ij}|}$$

Obviously, the value of ID_{ij} is also in [0, 1]. Inconsistent Association Degree measures the extent to which parameters e_i and e_j is inconsistent.

Definition 6.47 Let E be a set of parameters and $e_i, e_j \in E, (i, j = 1, 2, \ldots, m)$. Association Degree between parameter e_i and parameter e_j is denoted by D_{ij} and defined as

$$D_{ij} = \max\left\{CD_{ij}, ID_{ij}\right\}$$

If $CD_{ij} > ID_{ij}$, then $D_{ij} = CD_{ij}$, it means most of objects over U_{ij} have consistent values on parameters e_i and e_j. If $CD_{ij} < ID_{ij}$, then $D_{ij} = ID_{ij}$, it means most of objects over U_{ij} have inconsistent values on parameters e_i and e_j. If $CD_{ij} = ID_{ij}$, it means that there is the lowest association degree between parameters e_i and e_j.

Theorem 6.7 *For any parameters e_i and e_j, $D_{ij} \geq 0.5$. $(i, j = 1, 2, \ldots m)$*

Proof For any parameters e_i and e_j, from the definitions of CD_{ij} and ID_{ij}, we have

$$CD_{ij} + ID_{ij} = 1$$

Therefore, at least one of CD_{ij} and ID_{ij} is more than 0.5, namely, $D_{ij} = \max\left\{CD_{ij}, ID_{ij}\right\} \geq 0.5$. □

Definition 6.48 Let E be a set of parameters and $e_i \in E(i = 1, 2, \ldots m)$. Maximal Association Degree of parameter e_i is denoted by D_i and defined as

$$D_i = \max D_{ij}, \quad \mathrm{j} = 1, 2, \ldots \mathrm{m}$$

where m is the number of parameters.

6.5.1 Algorithm for Data Filling

Now we can propose the data filling method based on the association degree between the parameters. Suppose the mapping set $F(e_i)$ of parameter e_i includes missing data. At first, calculate association degrees between parameter e_i and each of other parameters respectively over existing complete information, and then find the parameter e_j which has the maximal association degree with parameter e_i. Finally the missing data in $F(e_i)$ will be filled according to the corresponding data in mapping set $F(e_j)$. However, sometimes a parameter perhaps has a lower maximal association degree, that is, the parameter has weaker association with other parameters. In this case, the association is not reliable any more and we have to find other methods. Inspired by the data analysis approach in [170], we can use the probability of objects appearing in the $F(e_i)$ to fill the missing data. In this method we give priority to the association between the parameters instead of the probability of objects appearing in the $F(e_i)$ to fill the missing data due to the fact that the relation between the parameters are more reliable than that between the objects in soft set. Therefore, we can set a threshold, if the maximal association degree equals or exceeds the predefined threshold, the missing data in $F(e_i)$ will be filled according to the corresponding data in $F(e_j)$, or else the missing data will be filled in terms of the probability of objects appearing in the $F(e_i)$. Below given is the the details of the algorithm.

Algorithm

1. Input the incomplete soft set (F, E).
2. Find e_i, which includes missing data $F(e_i)(x)$.
3. Compute D_{ij}, $j = 1, 2, \ldots, m$, where m is the number of parameters in E.
4. Compute maximal association degree D_i.
5. If $D_i \geq \lambda$ (a fixed threshold value), find the parameter e_j which has the maximal association degree D_i with parameter e_i.
6. If there is consistent association between e_i and e_j, $F(e_i)(x) = F(e_j)(x)$. If there is inconsistent association between e_i and e_j, $F(e_i)(x) = 1 - F(e_j)(x)$
7. If $D_i < \lambda$, compute the probabilities P_1 and P_0 that stand for object x belongs to and does not belong to $F(e_i)$, respectively.

$$P_1 = \frac{n_1}{n_1 + n_0}, \quad P_0 = \frac{n_0}{n_1 + n_0}$$

where n_1 and n_0 stand for the number of objects that belong to and does not belong to $F(e_i)$ respectively.
8. If $P_1 > P_0$, $F(e_i)(x) = 1$. If $P_0 > P_1$, $F(e_i)(x) = 0$. If $P_1 = P_0$, 0 or 1 may be assigned to $F(e_i)(x)$.
9. If all of the missing data is filled, algorithm end, or else go to step 2.

In order to make the computation of association degree easier, we construct an association degree table in which rows are labeled by the parameters including missing data and columns are labeled by all of the parameters in parameter set, and the entries are association degree D_{ij}. To distinguish the inconsistent association degree from consistent degree, we add a minus sign before the inconsistent association degree.

Example 80 Consider the incomplete soft set (F, E) whose tabular representation is given in Table 6.7 below. There are missing data in $F(e_2)$, $F(e_3)$, $F(e_5)$ and $F(e_6)$. We will fill the missing data in (F, E) by using Algorithm. Firstly, we construct an association degree table as Table 6.8.

For parameter e_2, we can see from the table, the association degree $D_{21} = 0.86$, $D_{23} = 0.83$, $D_{24} = 0.71$, $D_{25} = 0.67$, $D_{26} = 0.67$, where D_{21}, D_{23} and D_{24} are from inconsistent association degree, D_{25} and D_{26} are from consistent association degree. The maximal association degree $D_2 = 0.86$. We set the threshold $\lambda = 0.8$. Therefore, in terms of the Algorithm, we can fill $F(e_2)(c_7)$ according to $F(e_1)(c_7)$. Because $F(e_1)(c_7) = 1$ and there is inconsistent association between parameters e_2 and e_1, so we fill 0 into $F(e_2)(c_7)$. Similarly, we can fill 0, 1 into $F(e_3)(c_4)$ and $F(e_5)(c_6)$ respectively.

For parameter e_6, we have the maximal association degree $D_6 = 0.67 < \lambda$. That means there is not reliable association between parameter e_6 and other parameters. So we can not fill the data $F(e_6)(c_4)$ according to other parameters. In terms of the steps 8 and 9 in Algorithm, we have $P_0 = 1$, $P_1 = 0$. Therefore, we fill 0 into $F(e_6)(c_4)$. Table 6.9 shows the tabular representation of the filled soft set (F, E).

Table 6.7 Incomplete soft set (F,E)

U	e_14	e_2	e_3	e_4	e_5	e_6
c_1	1	0	1	0	1	0
c_2	1	0	0	1	0	0
c_3	0	1	0	0	1	0
c_4	0	1	*	1	0	*
c_5	1	0	1	1	0	0
c_6	0	1	0	0	*	0
c_7	1	*	1	0	1	0
c_8	0	0	1	1	0	0

Table 6.8 Association degree table for incomplete soft set (F, E)

	e_1	e_2	e_3	e_4	e_5	e_6
e_2	−0.86	–	−0.83	−0.71	0.67	0.67
e_3	0.71	−0.83	–	0.57	0.5	−0.57
e_5	0.57	0.67	0.5	−1	–	0.5
e_6	−0.57	0.67	0.57	0.57	0.5	–

Table 6.9 Tabular representation of the filled up soft set (F, E))

U	e_1	e_2	e_3	e_4	e_5	e_6
c_1	1	0	1	0	1	0
c_2	1	0	0	1	0	0
c_3	0	1	0	0	1	0
c_4	0	1	0	1	0	0
c_5	1	0	1	1	0	0
c_6	0	1	0	0	1	0
c_7	1	0	1	0	1	0
c_8	0	0	1	1	0	0

6.6 Future Directions of Research and Developments in Soft Set Theory

Data mining and knowledge discovery is a promising new field of active research of current day significance, while Molodtsov soft sets is a novel mathematical technique for handling uncertainties associated with knowledge acquisition and decision making problems. The current literature and carried out research promptly justifies the complementary relationship between soft set theory and other well established technologies of soft computing. The aim of this section in this monograph is to proceed further exploring this connection and find out the scope of integrating the advantages of soft set theory with other theories related to soft technologies. Obviously before going out this direction, the theoretical tools needed for this investigation should also be considered in detail. The significance of this discussion is that we can broaden the scope of soft sets theory along with applications and provide a wide range of scientific and standard methods for decision making under uncertain conditions.

6.6.1 Background of Soft Computing Technology

Soft computing refers to a collection of computational techniques which study, model, and analyze very complex systems for which the conventional methods have not yielded the best solutions. The rationale behind introducing soft computing techniques is to help deal with real life situations. With the advent of world wide web, advanced database technologies and cloud computing like methodologies, the volume of data that is needed to be handled in various scientific, technological, industrial and social fields witnessed tremendous increment. The traditional statistical techniques of data management and analysis are of little or not effective in extracting useful information from these huge chunks of data, which are often regarded as rich in data but poor in knowledge. As a consequence, starting from the last decade of previous century there evolved a new generation of technologies to deal with this large volume of data. These techniques aim at finding hidden, previously unknown, and potentially useful knowledge from the data. These tools generally falls under the umbrella of data mining and knowledge discovery.

Most of the management activities need rapid decisions in a series of processes which often contains situations involving randomness, fuzziness, preferences, roughness, greyness etc. Further, uncertainty in decision making is a characteristic of many other areas like information science, systems science, computer science etc. also. Due to the complicated nature of real world uncertainties, traditional or hard computing technologies are not always effective in dealing with them. In such situations, an effective means to deal with is to emulate the reasoning in human thinking and human intuition which is based on human's extraordinary capacity of learning. Many of the computing technologies like rough systems, grey systems, fuzzy systems and neural networks take the full advantage of this and are effective in addressing real life problems in its domain by exploiting the tolerance for imprecision, uncertainty and partial truth to achieve tractability, robustness and better rapport with reality.

6.6.2 Current Status of Soft Set Theory Research

Soft set theory was initiated by Russian researcher Molodtsov [88] in 1999. Molodtsov proposed soft set as a completely generic mathematical tool for modelling uncertainties. The guiding force behind the consolidation of ideas led to the innovation of this theory was the concept of parameterization with the aim of making a certain discretization of those fundamental mathematical concepts which are effectively continuous in nature. Certainly it is a deviation from the traditional out look and gives a totally different flavour to this theory. A soft set can be considered as an approximate description of an object precisely consists of two parts, namely predicate and approximate value set. There is no limited condition to the description of objects; so researchers can choose the form of parameters they needed, which greatly simplifies the decision-making process and make the process more efficient in the presence of partial information.

Mainly there are three theories: theory of probability, theory of interval mathematics and theory of fuzzy sets which can be considered as the mathematical tools to handle impreciseness and vagueness. But all these theories have their own difficulties. Theory of probability can handle only stochastically stable problem. For such a system one should check the existence of the limit of sample mean and have to perform a large number of trials. This can be done in engineering but not in social or economical situations. Interval mathematics is an efficient method to explain the errors of calculation by constructing an interval estimate for the exact solution of the problem. But this method is not sufficiently adaptable for problems with different uncertainties. They cannot exactly describe a smooth change of information or unreliable and defective information. Fuzzy theory which is the most appropriate theory to deal with uncertainty developed by Zadeh [160] is always facing the problem of setting up membership function. All these techniques lack in parameterization of the tools and hence they could not be applied successfully in tackling in real life problems. Soft set theory is standing in a unique way in the sense that it is relatively free from the above difficulties and has a wider scope for applications.

Apart from its huge potential for applications in many directions, theoretical studies involving soft sets also attained momentum in structural, order, graphical, categorical and many other directions. Maji et al. [79, 80] presented basic definitions on soft sets such as subset, complement of soft set etc. The same authors also extended soft sets to fuzzy soft sets in [78].The algebraic nature of generalized set theories dealing with uncertainties has been studied by many authors. Aktas and Cagman [5] defined the notion of soft groups and derived some properties. Feng [40] dealt with the algebraic structure of semi rings by applying soft set theory. Qiu Mei Sun [133] introduced a basic version of soft module theory, which extends the notion of a module by including some algebraic structures in soft set. Jayanta Ghosh et al. [45] introduced the notion of fuzzy soft rings and fuzzy soft ideals and studied some of their algebraic properties. Osman and Aslihan [10] introduced and studied soft subrings and soft ideals of a ring. Moreover, they introduced soft subfields of a field and soft submodule of a left R-module. Algebraic structures combining two or more structures where one among this is a soft set can be seen in [58, 93, 166] and many more.

Further to the theoretical study, topological structures on soft sets were introduced by Shabir and Naz [125] and a totally different approach for the same is given by Cagman et al. [28]. Topological structures involving hybridization of soft sets with other structures such as fuzzy, intuitionistic fuzzy, hesitant fuzzy sets are also introduced [18, 25, 73, 131, 132, 134, 146]. Those with respect to Pythagorean fuzzy [12, 13, 101], picture fuzzy [63], spherical fuzzy [103, 104] and Fermatean fuzzy [122] are also promising ones but not yet introduced. Combinations of more than two structures where one is a soft set and their topological extensions are also developed by many authors. For lattice theoretic works associated with soft sets see [57, 68, 72]. Adding to the scenario, soft analogous of generalized topologies, in the sense of Csaszar [34], are studied by Thomas and Sunil in a series of papers [137–141].

Apart from the hybrid structures discussed above, there are many other structures available also. Combination of graph theory, rough set theory, Category theory etc. with soft sets [4, 112, 121, 142, 167] yielded much useful structures with lots of realistic significance.

In the meantime soft set theory has been applied practically in many domains. Maji et al. [79] applied soft set based methods in decision making problems. Chen et al. [32] improved soft set based decision making by introducing parametrization reduction. Also the difference between parametrization reduction of soft sets and attributes reduction in rough sets is pointed out [32]. Mushrif et al. [90] presented a novel method for classification of natural textures using the notion of soft set theory. The concept of bijective soft set and some of its operations are proposed by Ke Gong et al. [47]. In [152], Zhi Xiao et al. defined exclusive disjunctive soft sets and it is applied to attribute reduction of incomplete information systems. Using reduct fuzzy soft sets and level soft sets, flexible schemes for decision making based on (weighted) interval valued fuzzy soft sets are proposed by Feng et al. [42]. Majumdar and Samantha [81] and Kharal [64] applied similarity measures of soft sets to medical diagnosis problem and financial diagnosis problem respectively. Deli and Cagman [36] applied Game theoretic methods to soft sets and soft matrices were introduced by Cagman and Enginoglu [27]. An application of data filling in incomplete information system using soft sets can be seen in Qin et al. [108].

6.6.3 Some Future Directions of Research in Soft Set Theory Applications

Comparing soft set theory with other theories dealing with uncertainty, from the discussions and deliberations provided in the earlier chapters, it is clear that soft set theory has some irreplaceable advantages, but still has some one-sidedness and short comings. Each soft computing technique will be having a better scope in some particular application and there is no method that can solve all the problems. In practical point of view, we often combine several soft computing techniques to get a 'hybrid' system, so that it could complement each other to overcome the limitations of individual techniques by avoiding the existing method's disadvantages or weaknesses when used separately. Such a hybrid system is superior to the use of a single method, and the word 'hybrid' in this context means combining with some advantages of the existing methods. These types of hybrid approach are much promising in knowledge discovery systems. Hybridization of various existing soft computing techniques with soft set theory can be applied to the fields of data mining and knowledge discovery. The following hybrid structures will be of quite interesting to be investigated.

(i) Hybrid of Soft sets and Probability Statistics: In the context of classical soft set theory, the authors Pie and Miao [102] investigated the relationships between soft sets and Information systems and observed the equivalence between soft sets and classical information systems. Therefore, if classification of data is done by classical soft set models, the information required regarding the data must be completely correct or unambiguous as in the case of classical information systems. In practical applications, the data of the knowledge base is often acquired by means of random or statistical methods, and it is most likely to be noisy, ambiguous, and incomplete, so that the patterns of classes often overlap, which is not sufficient to produce deterministic rules. The probability is an objective reflection of uncertain random events. The hybrid of soft sets and probability statistics can expand the functions of soft set method, and also can be used to acquire the probability decision-making rules from the noisy data. An initiation towards this direction can be seen in Ping Zhu and Wen [169] and Fatimah et al. [39]. Here the set-valued mapping in Molodtsov soft set is replaced by a mapping having probability distributions as its co domain.

Classical soft sets deals only those concepts that do not overlap and it does not depend on event occurrence. While probability depends on event occurrence and hence the hybrid of soft set models and probability can be used to identify strong non deterministic rules applicable for the estimation of decision probabilities from the noisy data.

(ii) Hybrid of Soft sets and Fuzzy Sets: In describing a 'humanistic' problem mathematically, bivalent set theory seems to have limitations. Fuzzy logic is a super set of conventional (Boolean) logic that has been extended to handle the concept of partial truth. It is the logic underlying modes of reasoning which are approximate rather than exact. The importance of fuzzy logic derives from the fact that most modes of human reasoning and especially common sense reasoning are approximate in nature and it can be used to obtain and simulate and even reason the fuzziness in practical information by mainly studying the fuzzy information granularity. Fuzzy sets can also be used as the membership function in an intelligent system. Besides having no mechanism for data learning, it is quite often difficult to set the membership values in fuzzy set theory. The knowledge should be given clearly by the designers in the applied fields, which would certainly bring some difficulties for the application of fuzzy set theory. A standard soft set may be redefined as the classification of objects in two distinct classes, thus confirming that soft sets can deal with a Boolean-valued information system. Molodtsov pointed out that one of the main advantages of soft set theory is that it is free from the inadequacy of parameterization tools, unlike the theories of fuzzy set, probability and interval mathematics. A hybrid of soft sets and fuzzy sets will give both adequacy of parameterization and adaptability for non-Boolean (infinite valued of fuzzy) logic. As already mentioned, there already exists tremendous amount of efforts in this direction, but it is still an untiring task as the complexity involved in real world problems is increasing exponentially and for tackling them, more and more efficient tools are essential.

(iii) Hybrid of Soft sets and Rough Sets: Rough set theory is another mathematical tool motivated by practical needs to deal with vagueness and uncertainty. The main idea of rough set theory is based on the indiscernibility relation, which is mathematically an equivalence relation. Every object is associated with a certain amount of information and objects with the same or similar information are indiscernible with respect to the available information. The indiscernible blocks or partitions formed by indiscernible objects of the universe are called the elementary cells or the knowledge granularity. Based on the knowledge used in approximation, the universe is divided into elementary cells of the indiscernible objects that are based on conditional attribute sets. Rough set is defined in terms of topological operations, interior and closure, called approximations. The lower approximation of the set X is the union of all elementary cells which are fully contained in X. The upper approximation of the set X is the union of all elementary cells which have non empty intersection with X. The difference of these will form the boundary region. If the boundary is non empty, set is rough otherwise it is exact.

As a tool to address uncertainty, in soft sets also there is an associated approximation space for each parameter because every parameter in a soft set induces an equivalence relation. Intersection of all induced equivalence relations give rise to indiscernibility associated with a soft set. Thus clearly these two concepts are complementary and the joint applications will definitely yield fruitful results.

(iv) Hybrid of Soft sets and Grey System theory: Grey system theory was introduced by Julong Deng, in 1982 as a mathematical tool to handle small sample and poor information. Grey system theory and Soft set theory are two different tools that are used to deal with uncertain or incomplete information, and yet they are relevant and complementary to a certain degree. They both improve the generality of data presentation by costing reduction in its accuracy. Advantage is that neither of them needs a priori knowledge, such as probability distributions or membership. Soft set theory researches into parametrization of objects while in grey system valuable information is extracted by mining the partially known information, so that the behavior and the law of system can be properly described and effectively controlled. The appropriate hybrid of the two theories can overcome the shortages of their definitions and applications and thus may result in more powerful functions.

(v) Hybrid of Soft sets and Neural Networks: Neural networks simulate knowledge acquisition and organizational skills of human brain with a relatively high adaptability, capability for errors, and universalizing ability. Neural network methods can make up the shortcomings in soft set theory like the sensitivity towards noise data and the weakness at universalizing ability. On the other hand, soft set methods can help make up the weak points that exist in neural network methods, such as difficulties in ascertaining relatively important attribute combinations, lacking general ways to build a network structure, opaque reasoning process, and lack of explanation ability. The hybrid of soft sets and neural networks, which is a combination of two thinking manners, logical thinking and visualized thinking, may make it possible to discover knowledge and make predictions from a fuzzy, incomplete, and noisy database to include the structural knowledge in the system.

By broadening the the soft structure through above discussed five hybrid structures, one can apply them to to all five kinds of analytical method of data mining: automatic prediction of trends and behavior, association analysis, cluster analysis, concept description, and deviation detection.

References

1. Acar, U., Koyuncu, F., Tanay, B.: Soft sets and soft rings. Comput. Math. Appl. **59**, 3458–3463 (2010)
2. Adamek, J., Herrlich, H., Strecker, G.: Abstract and Concrete Categories: The Joy of Cats. Dover Publications, Downers Grove (2009)
3. Ahmad, B., Hussain, S.: On some structures of soft topology. Math. Sci. **6**, 1–7 (2012)
4. Akram, M., Nawaz, S.: Operations on soft graphs. Fuzzy Inf. Eng. **7**(4), 423–449 (2015)
5. Aktas, H., Cagman, N.: Soft sets and soft groups. Inf. Sci. **177**, 2726–2735 (2007)
6. Alhazaymeh, K., Halim, S.A., Salleh, A.R., Hassan, N.: Soft intuitionistic fuzzy sets. Appl. Math. Sci. **6**(54), 2669–2680 (2012)
7. Ali, M.I., Feng, F., Liu, X., Min, W.K., Shabir, M.: On some new operations in soft set theory. Comput. Math. Appl. **57**, 1547–1553 (2009)
8. Aliprantis, C.D., Chakrabarti, S.K.: Games and Decision making. Oxford University Press, Oxford (2000)
9. Anderson, F.W., Fuller, K.R.: Rings and Categories of Modules. Springer, Heidelberg (1992)
10. Atagun, A.O., Sezgin, A.: Soft substructures of rings, fields and modules. Comput. Math. Appl. **61**, 592–601 (2011)
11. Atanassov, K.T.: Intuitionistic fuzzy sets. Fuzzy Sets Syst. **20**(1), 87–96 (1986)
12. Athira, T.M., Sunil, J.J., Garg, H.: Entropy and distance measures of Pythagorean fuzzy soft sets and their applications. J. Intell. Fuzzy Syst. **37**(3), 4071–4084 (2019)
13. Athira, T.M., Sunil, J.J., Garg, H.: A novel entropy measure of Pythagorean fuzzy soft sets. AIMS Math. **5**(2), 1050–1061 (2019)
14. Aygunoglu, A., Aygun, H.: Some notes on soft topological spaces. Neural Comput. Applic. **21**(Suppl 1), S113–S19 (2012)
15. Babitha, K.V., Sunil, J.J.: Soft set relations and functions. Comput. Math. Appl. **60**(7), 1840–1849 (2010)
16. Babitha, K.V.: Soft Sets: Relations, Topology, Hybrid Structures, Ph.D. Thesis, National Institute of Technology Calicut (2010)
17. Babitha, K.V., Sunil, J.J.: Transitive closures and orderings on soft sets. Comput. Math. Appl. **62**(5), 2235–2239 (2011)
18. Babitha, K.V., Sunil, J.J.: Hesitant fuzzy soft sets. J. New Results Sci. **3**, 98–107 (2013)
19. Babitha, K.V., Sunil, J.J.: Studies on soft topological spaces. J. Intell. Fuzzy Syst. **28**, 1713–1722 (2015)
20. Babitha, K.V., Sunil, J.J., Soft topologies generated by soft set relations. In: Handbook of Research on Generalized and Hybrid Set Structures and Applications for Soft Computing. IGI Global Pub., pp. 118–126 (2015)
21. Birkhoff, G.: Lattice Theory, vol. XXV. AMS Colloquim Pub., New York City (1948)

S. J. John, *Soft Sets*, Studies in Fuzziness and Soft Computing 400,
https://doi.org/10.1007/978-3-030-57654-7

22. Biswas, R., Nanda, S.: Rough groups and rough subgroups. Bull. Polish Acad. Math. **42**, 251–254 (1994)
23. Bonikowaski, Z.: Algebraic structures of rough sets. In: Ziarko, W.P. (ed.) Rough Sets, pp. 242–247. Fuzzy Sets and Knowledge Discovery, Springer, Berlin (1995)
24. Borzooei, R.A., Mobini, M., Ebrahimi, M.M.: The category of soft sets. J. Intell. Fuzzy Syst. **28**(1), 157–167 (2015)
25. Bayramov, S., Gunduz, C.: On intuitionistic fuzzy soft topological spaces. TWMS J. Pure Appl. Math. **5**(1), 66–79 (2014)
26. Cagman, N., Enginoglu, S., Cıtak, F.: Fuzzy soft set theory and its applications. Iran. J. Fuzzy Syst. **8**(3), 137–147 (2011)
27. Cagman, N., Enginoglu, S.: Soft matrix theory and its decision making. Comput. Math. Appl. **59**, 3308–3314 (2010)
28. Cagman, N., Karatas, S., Enginoglu, S.: Soft topology. Comput. Math. Appl. **62**, 351–358 (2011)
29. Chang, C.L.: Fuzzy topological spaces. J. Math. Anal. Appl. **24**, 182–190 (1968)
30. Chen, S.M.: Measures of similarity between vague sets. Fuzzy Sets Syst. **74**, 217–223 (1995)
31. Chen, S.M., Yeh, M., Hsiao, P.: A comparison of similarity measures of fuzzy values. Fuzzy Sets Syst. **72**(1), 79–89 (1995)
32. Chen, D., Tsang, E.C.C., Yeung, D.S., Wang, X.: The parameterization reduction of soft sets and its applications. Comput. Math. Appl. **49**, 757–763 (2005)
33. Cormen, T.H., Leiserson, C. E., Rivest, R. L.: Introduction to Algorithms (1st edn.). MIT Press and McGraw-Hill, Section 26.2, The Floyd Warshall algorithm, pp. 558–565 (1990)
34. Csaszar, A.: Generalized topology, generalized continuity. Acta Math. Hung. **96**(4), 351–357 (2002)
35. Cuong, B.C., Kreinovich, V.: Picture fuzzy sets—A new concept for computational intelligence problems. In: Third World Congress on Information and Communication Technologies (WICT 2013), Hanoi, pp. 1–6 (2013)
36. Deli, I., Cagman, N.: of soft sets in decision making based on game theory. Ann. Fuzzy Math. Inform. **11**(3), 425–438 (2016)
37. Demirci, M., Waterman, M.S.: Genuine sets. Fuzzy Sets Syst. **105**, 377–384 (1999)
38. Dugundji, J.: Topology. Universal Book Stall, New Delhi (1995)
39. Fatimah, F., Rosadi, D., Hakim, R.F.: Probabilistic soft sets and dual probabilistic soft sets in decision-making. Neural Comput. Appl. **31**, 397–407 (2019)
40. Feng, F., Jun, Y.B., Zhao, X.: Soft semirings. Comput. Math. Appl. **56**, 2621–2628 (2008)
41. Feng, F., Liu, X., Fotea, V.L., Jun, Y.B.: Soft sets and soft rough sets. Inf. Sci. **181**, 1125–1137 (2011)
42. Feng, F., Li, Y., Fotea, V.L.: Application of level soft sets in decision making based on interval-valued fuzzy soft sets. Comput. Math. Appl. **60**(6), 1756–1767 (2010)
43. Ferguson, T.S.: Game Theory. UCLA, Los Angeles (2008)
44. Gau, W.L., Buehrer, D.J.: Vague sets. IEEE Trans. Syst. Man Cybern. **23**(2), 610–614 (1993)
45. Ghosh, J., Dinda, B., Samanta, T.K.: Fuzzy soft rings and fuzzy soft ideals. Int. J. Pure Appl. Sci. Technol. **2**(2), 66–74 (2011)
46. Gouen, J.A.: L-fuzzy sets. J. Math. Anal. Appl. **18**(1), 145–174 (1967)
47. Gong, K., Xiao, Z., Zhang, X.: The bijective soft set with its operations. Comput. Math. Appl. **60**, 2270–2278 (2010)
48. Harary, F.: Graph Theory. Addison-Wesley, Boston (1969)
49. Hazra, H., Majumdar, P., Samanta, S.K.: Soft topology. Fuzzy Inf. Eng. **4**, 105–115 (2012)
50. Hilton, P.J., Stammbach, U.: A Course in Homological Algebra, vol. 4. Springer Science & Business Media, Berlin (2012)
51. Hrbacek, K., Jech, T.: Introduction to Set Theory. Marcel Dekker Inc., New York (1999)
52. Huang, Y.: BCI-algebra. Science Press, Beijing (2006)
53. Imai, Y., Iseki, K.: On axiom systems of propositional calculi, XIV. Proc. Japan Acad. Ser. A, Math. Sci., **42**, 19–22 (1966)

54. Iseki, K., Tanaka, S.: An introduction to the theory of BCK-algebras. Math. Jpn. **23**, 1–26 (1978)
55. Iseki, K.: An algebra related with a propositional calculus. Proc. Jpn. Acad. Ser. A, Math. Sci.**42**, 26–29 (1966)
56. Iwinski, T.: Algebraic approach to rough sets. Bull. Polish Acad. Sci. Math. **35**, 673–683 (1987)
57. Jobish, V.D., Babitha, K.V., Sunil, J.J.: On soft lattice operations. J. Adv. Res. Pure Math. **5**(2), 71–86 (2013)
58. Jun, Y.K.: Soft BCK/BCI-algebras. Comput. Math. Appl. **56**, 1408–1413 (2008)
59. Kashiwara, M., Schapira, P.: Categories and sheaves, vol. 332. Springer Science & Business Media, Berlin (2005)
60. Kassel, C.: Quantum Groups. Springer, Berlin (1991)
61. Kearfott, R.B., Kreinovich, E. (eds.): Applications of Interval Computations. Kluwer Academic Publishers, Dordrecht (1996)
62. Kelly, J.L.: General topology. Springer, New York (1955)
63. Khan, M.J., Kumam, P., Ashraf, S., Kumam, W.: Generalized picture fuzzy soft sets and their application in decision support systems. Symmetry **11**, 415 (2019)
64. Kharal, A.: Distance and similarity measure for soft sets. New Math. Nat. Comput. **6**(3), 321–334 (2010)
65. Kharal, A., Ahmad, B.: Mappings on soft classes. New Math. Nat. Comput. **7**(3), 471–481 (2011)
66. Klir, G., Yuan, B.: Fuzzy Sets and Fuzzy Logic: Theory and Applications. Prentice Hall, Upper Saddle River, NJ (1995)
67. Kong, Z., Gao, L., Wang, L., Li, S.: The normal parameter reduction of soft sets and its algorithm. Comput. Math. Appl. **56**, 3029–3037 (2008)
68. Kuppusamy, K., Nagarajan, R., Meenambigai, G.: An application of soft sets to lattices. Kragujevac J. Math. **3**(1), 75–87 (2011)
69. Kuratowski, K.: Topology (vol I). Academic Press, Cambridge (1966)
70. Kutlu Gundogdu, F., Kahraman, C.: Spherical fuzzy sets and spherical fuzzy TOPSIS method. J. Intell. Fuzzy Syst. **36**(1), 337–352 (2019)
71. Li, D., Cheng, C.: New similarity measures of intuitionistic fuzzy sets and application to pattern recognition. Patt. Recogn. Lett. **23**(1–3), 221–225 (2002)
72. Li, F.: Soft lattices. Glob. J. Sci. Front. Res. **10**(4), 56–58 (2010)
73. Li, Z., Cui, R.: On the topological structure of intuitionistic fuzzy soft sets. Ann. Fuzzy Math. Inform. **5**(1), 229–239 (2013)
74. Lowen, R.: Fuzzy topological spaces and fuzzy compactness. J. Math. Anal. Appl. **56**, 621–633 (1976)
75. Luca, D., Termini, S.: A definition of a nonprobability entropy in the setting of fuzzy sets theory. Inf. Control **20**, 301–312 (1972)
76. Ma, X., Sulaiman, N., Qin, H., Herawan, T., Zain, J.M.: A new efficient normal parameter reduction algorithm of soft sets. Comput. Math. Appl. **62**, 588–598 (2011)
77. Mac Lane, S.: Categories for the working mathematician, vol. 5. Springer Science & Business Media, Berlin (2013)
78. Maji, P.K., Biswas, R., Roy, A.R.: Fuzzy soft sets. J. Fuzzy Math. **9**(3), 589–602 (2001)
79. Maji, P.K., Roy, A.R., Biswas, R.: An application of soft sets in a decision making. Comput. Math. Appl. **44**, 1077–1083 (2002)
80. Maji, P.K., Biswas, R., Roy, A.R.: Soft set theory. Comput. Math. Appl. **45**(4–5), 555–562 (2003)
81. Majumdar, P., Samanta, S.K.: Similarity measure of soft sets. New Math. Nat. Comput. **4**(1), 1–12 (2008)
82. Majumdar, P., Samanta, S.K.: On similarity measures of fuzzy soft sets. Int. J. Adv. Soft Comput. Appl. **3**(2), 1–8 (2011)
83. Majumdar, P., Samanta, S.K.: Softness of a soft set: Soft set entropy. Ann. Fuzzy Math. Inform. **6**(1), 59–68 (2013)

84. Mathew, B., John, S.J., Garg, H.: Vertex rough graphs. Complex Intell. Syst. **6**, 347–353 (2020)
85. Ming, L.Y., Kang, L.M.: Fuzzy Topology, Advances in Fuzzy Systems- Application and Theory, vol. 9. World Scienctific Publishing Co., Singapore (1997)
86. Min, W.K.: A note on soft topological spaces. Comput. Math. Appl. **62**, 3524–3528 (2011)
87. Min, W.K.: Similarity in soft set theory. Appl. Math. Lett. **25**, 310–314 (2012)
88. Molodtsov, D.: Soft set theory - first results. Comput. Math. Appl. **37**(4–5), 19–31 (1999)
89. Munkres, J.R.: Topology—A first course. Prentice Hall of India (2000)
90. Mushrif, M.M., Sengupta, S., Ray, A.K.: Texture classification using a novel, soft-set theory based classification algorithm. In: Narayanan P.J., Nayar S.K., Shum HY. (eds.) Computer Vision—ACCV 2006. ACCV 2006. Lecture Notes in Computer Science, vol. 3851. Springer, Berlin, Heidelberg (2006)
91. Neumann, J.V., Morgenstern, O.: The Theory of Games and Economic Behavior. Princeton University Press, Princeton (1944)
92. Owen, G.: Game theory. Academic Press, San Diego (1995)
93. Pan, W., Zhan, J.: Rough fuzzy groups and rough soft groups. Ital. J. Pure Appl. Math. **36**, 617–628 (2016)
94. Patrick Suppes.: Axiomatic Set Theory, Dover Publications Inc., New York (1972)
95. Paul, R.: Halmos. Naive Set Theory. Springer, Berlin (1974)
96. Dwyer, P.S.: Linear Computations. Chapman & Hall, New York (1951)
97. Pawlak, Z.: Rough sets. Int. J. Comput. Inform. Sci. **11**, 341–356 (1982)
98. Pawlak, Z.: Rough Sets: Theoretical Aspects of Reasoning about Data. Kluwer Academic Publishers, Dordrecht (1991)
99. Pawlak, Z., Skowron, A.: Rudiments of rough sets. Inform. Sci. **177**, 3–27 (2007)
100. Pawlak, Z.: Hard and Soft Sets. In: Ziarko, W.P. (ed.) Rough Sets. Fuzzy Sets and Knowledge Discovery. Workshops in Computing. Springer, London (1994)
101. Peng, X., Yang, Y., Song, J., Jiang, Y.: Pythagorean fuzzy soft set and its application. Comp. Eng. **41**(7), 224–229 (2015)
102. Pei, D., Miao, D.: From soft sets to information systems. In: Hu, X., Liu, Q., Skowron, A., Lin, T.Y., Yager, R.R., Zhang, B. (eds.) Proceedings of Granular Computing, IEEE, vol. 2, pp. 617–621 (2005)
103. Perveen, F.A., Sunil, J.J., Babitha, K.V., Garg, H.: Spherical fuzzy soft sets and its applications in decision-making problems. J. Intell. Fuzzy Syst. **37**(6), 8237–8250 (2019)
104. Perveen, F.A., Sunil, J.J., Babitha, K.V.: Spherical Fuzzy Soft Sets. In: Kahraman, C., Kutlu Gundogdu, F. (eds.) Decision Making with Spherical Fuzzy Sets. Studies in Fuzziness and Soft Computing, vol 392. Springer, Cham (2021)
105. Peyghana, E., Samadia, B., Tayebib, A.: About soft topological spaces. J. New Results Sci. **2**, 60–75 (2013)
106. Polkowski.: Rough Sets: Mathematical Foundations. Physica-Verlag, Heidelberg (2002)
107. Preuss, G.: Foundations of topology: an approach to convenient topology. Springer Science & Business Media, Berlin (2011)
108. Qin, H., Ma, X., Herawan, T., Zain, J.M.: Data filling approach of soft sets under incomplete information. In: Nguyen, N.T., Kim, C.-G., Janiak, A. (eds.) ACIIDS 2011, LNAI 6592, pp. 302–311. Springer, Berlin, Heidelberg (2011)
109. Qin, K., Q. Liu, Q., Xu, Y.: Redefined soft relations and soft functions. Int. J. Comput. Intell. Syst.**8**(5), 819–828 (2015)
110. Qin, K.Y., Yang, J.L., Entropy of soft sets. In: Proceedings of International Conference on Computer Information Systems and Industrial Applications, Atlantis Press, pp. 788–792 (2015)
111. Ratheesh, K.P.: A Categorical Study of Soft sets and Soft graphs. Ph.D. Thesis, National Institute of Technology Calicut (2019)
112. Ratheesh, K.P., Sunil, J.J.: On the objects and morphisms of category of soft sets. Int. J. Pure Appl. Math. **117**(11), 447–453 (2017)
113. Roman, S.: An Introduction to the Language of Category Theory. Springer, Berlin (2017)

114. Rosenfeld, A.: Fuzzy groups. J. Math. Anal. Appl. **35**, 512–517 (1971)
115. Rosenfeld, A.: Fuzzy graphs In: Fuzzy sets and their applications to cognitive and decision processes. Elsevier, Amsterdam, pp. 77–95 (1975)
116. Roy, A.R., Maji, P.K.: A fuzzy soft set theoretic approach to decision making problems. J. Comput. Appl. Math. **203**, 412–418 (2007)
117. Roy, S.K., Bera, S.: Approximation of Rough Soft Set and Its Application to Lattice. Fuzzy Inf. Eng. **7**, 379–387 (2015)
118. Sabir, H., Ahmad, B.: Some properties of soft topological spaces. Comput. Math. Appl. **62**, 4058–4067 (2011)
119. Sabir, H.: A note on soft connectedness. J. Egypt. Math. Soc. **23**, 6–11 (2015)
120. Smarandache, F.: Plithogenic set, an extension of crisp, fuzzy, intuitionistic fuzzy, and neutrosophic sets - revisited. Neutrosophic Sets Syst. **21**, 153–166 (2018)
121. Sardar, S.K., Gupta, S.: Soft category theory-an introduction. J. Hyperstruct. **2**(2), 118–135 (2013)
122. Senapati, T., Yager, R.R.: Fermatean fuzzy sets. J. Ambient Intell. Human Comput. **11**, 663–674 (2020)
123. Sezgin, A., Atagun, A.O.: On operations on soft sets. Comput. Math. Appl. **61**, 1457–1467 (2011)
124. Sezgin, A., Atagun, A.O.: Soft groups and normalistic soft groups. Comput. Math. Appl. **62**, 685–698 (2011)
125. Shabir, M., Naz, M.: On Soft topological spaces. Comp. Math. Appl. **61**, 1786–1799 (2011)
126. Shabir, M., Ali, M.I., Shaheen, T.: Another approach to soft rough sets. Knowl. -Based Syst. **40**, 72–80 (2013)
127. Shijina, V., Sunil, J.J., Thomas, A.S.: Multiple sets: A unified approach towards modelling vagueness and multiplicity. J. New Theory **11**, 29–53 (2016)
128. Shijina, V., Sunil, J.J.: Multiple relations and its application in medical diagnosis. Int. J. Fuzzy Syst. Appl. (IJFSA) **6**(4), 16 (2017)
129. Simmons, G.F.: Introduction to Topology and Modern Analysis. McGraw Hill International Edition (1963)
130. Som,T.: On the theory of soft sets, soft relation and fuzzy soft relation. In: Proceedings of the National Conference on Uncertainty: A Mathematical Approach, UAMA-06, Burdwan, pp. 1–9 (2006)
131. Sreedevi, A., Shankar, N.R.: Key properties of hesitant fuzzy soft topological spaces. Int. J. Sci. Eng. Res. **7**(2), 149–156 (2016)
132. Sreedevi, A., Shankar, N.R.: Results on hesitant fuzzy soft topological spaces. Int. J. Adv. Res. **4**(3), 1–9 (2016)
133. Sun, Q.M., Zhang, Z.L., Liu, J.: soft sets and soft modules. In: Wang, G., et al. (eds.) RSKT, LNAI 5009, pp. 403–409. Springer, Berlin, Heidelberg (2008)
134. Tanay, B., Kandemir, M.B.: Topological structures of fuzzy soft sets. Comput. Math. Appl. **61**, 412–418 (2011)
135. Teruo Sunaga.: Theory of an interval algebra and its application to numerical analysis. RAAG Memoirs **2**, 29–46 (1958)
136. Thomas, J.: On Generalized and Soft Generalized Topological Spaces. Ph.D. Thesis, National Institute of Technology Calicut (2015)
137. Thomas, J., Sunil, J.J.: A note on soft topology. J. New Results Sci. **5**(11), 24–29 (2016)
138. Thomas, J., Sunil, J.J.: Soft π-open sets in soft generalized topological spaces. J. New Theory 53–66 (2015)
139. Thomas, J., Sunil, J.J.: On soft generalized topological spaces. J. New Results Sci. **4**, 1–15 (2014)
140. Thomas, J., Sunil, J.J.: Soft generalized separation axioms in soft generalized topological spaces. Int. J. Sci. Eng. Res. **6**(3), 969–974 (2015)
141. Thomas, J., Sunil, J.J.: On soft μ-compact soft generalized topological spaces. J. Uncertain. Math. Sci. **2014**(6), 1–9 (2014)
142. Thumbakara, R.K., George, B.: Soft graphs. Gen. Math. Notes **21**(2), 75–86 (2014)

143. Torra, V.: Hesitant fuzzy sets. Int. J. Intell. Syst. **25**(6), 529–539 (2010)
144. Torra, V., Narukawa, Y.: On hesitant fuzzy sets and decision. In: Proceedings of the IEEE International Conference on Fuzzy Systems, Jeju-do, Republic of Korea, pp. 1378–1382 (2009)
145. Tripathy, B.K., Arun, K.R.: Soft sets and its applications. In: Sunil J.J. (ed.) Handbook of Research on Generalized and Hybrid Set Structures and Applications for Soft Computing, IGI Global USA, pp. 65–85 (2016)
146. Varol, B.P., Aygun, H.: Fuzzy soft topology. Hacet. J. Math. Stat. **41**(3), 407–419 (2012)
147. Wang, F., Li, X., Chen, X.: Hesitant fuzzy soft set and its applications in multicriteria decision making. J. Appl. Math. **643785**, 10 (2014)
148. Webb, J.N.: Game Theory Decisions. Interaction and Evolution, Springer Undergraduate Mathematics Ser (2006)
149. Willard, S.: General topology. Addison-Wesley, Boston (1970)
150. Williams, J., Steele, N.: Difference, distance and similarity as a basis for fuzzy decision support based on prototypical decision classes. Fuzzy Sets Syst. **131**, 35–46 (2002)
151. Xia, M., Xu, Z.: Hesitant fuzzy information aggregation in decision making. Int. J. Approx. Reason. **52**(3), 395–5407 (2011)
152. Xiao, Z., Gong, K., Xia, S., Zou, Y.: Exclusive disjunctive soft sets. Comput. Math. Appl. **59**(6), 2128–2137 (2010)
153. Xie, N.X.: An Algorithm on the parameter reduction of soft sets. Fuzzy Inf. Eng. **8**(2), 127–145 (2016)
154. Xu, Y., Sun, Y., Li, D.: Intuitionistic fuzzy soft set. In: 2nd International Workshop on Intelligent Systems and Applications, Wuhan, pp. 1–4 (2010)
155. Yang, H.L., Guo, Z.L.: Kernels and closures of soft set relations, and soft set relation mappings. Comput. Math. Appl. **61**(3), 651–662 (2011)
156. Yao, B., Liu, J., Yan, R.: Fuzzy soft set and soft fuzzy set. In: Proceedings of Fourth International Conference on Natural Computation (2008)
157. Yager, R.R.: On the theory of bags. Int. J. General Syst. **13**, 23–37 (1986)
158. Yager, R.R.: Pythagorean fuzzy subsets. In: Proceedings of the 9th Joint World Congress on Fuzzy Systems and NAFIPS Annual Meeting, IFSA/NAFIPS, Edmonton, Canada, pp. 57–61 (2013)
159. Yao, Y.Y.: Constructive and algebraic methods of the theory of rough sets. Inf. Sci. **109**, 21–47 (1998)
160. Zadeh, L.A.: Fuzzy sets. Inf. Control **8**, 338–353 (1965)
161. Zadeh, L.A.: Fuzzy sets and systems. In: Proceedings of the Symposium on Systems Theory, Polytechnic Institute of Brooklyn, NY, pp. 29–37 (1965)
162. Zadeh, L.A.: The concept of a linguistic variable and its application to approximate reasoning - 1. Inf. Sci. **8**, 199–249 (1975)
163. Zahiri, O.: Category of soft sets. Ann. Univ. Craiova-Math. Comput. Sci. Ser. **40**(2), 154–166 (2013)
164. Zhang, Y.H., Yuan, X.H.: Soft relation and fuzzy soft relation. In: Fuzzy Information & Engineering and Operations Research & Management. pp. 205–213. Springer, Berlin (2014)
165. Zhang, W.R.: Bipolar fuzzy sets and relations: a computational framework for cognitive modeling and multi agent decision analysis. In: Proceedings of the Industrial Fuzzy Control and Intelligent Systems Conference, and the NASA Joint Technology Workshop on Neural Networks and Fuzzy Logic, Fuzzy Information Processing Society Biannual Conference, San Antonio, Tex, USA, pp. 305–309 (2014)
166. Zhan, J., Davvaz, B.: A kind of new rough set: Rough soft sets and rough soft rings. J. Intell. Fuzzy Syst. **30**(1), 475–483 (2016)
167. Zhou, M., Li, S., Akram, M.: Categorical properties of soft sets. Sci. World J. **783056**, 10 (2014)
168. Zorlutuna, I., Akdag, M., Min, W.K., Atmaca, S.: Remarks on soft topological spaces. Ann. Fuzzy Math. Inform. **3**(2), 171–185 (2012)

169. Zhu, P., Wen, Q.: Probabilistic soft sets. In: Proceedings of the 2010 IEEE International Conference on Granular Computing, San Jose, CA, pp. 635–638 (2010)
170. Zou, Y., Xiao, Z.: Data analysis approaches of soft sets under incomplete information. Knowl. Based Syst. **21**, 941–945 (2008)

Index

S. J. John, *Soft Sets*, Studies in Fuzziness and Soft Computing 400,
https://doi.org/10.1007/978-3-030-57654-7

T

U

Printed in the United States
by Baker & Taylor Publisher Services

Printed in the United States
by Baker & Taylor Publisher Services